Traditional Organic Farming Practices

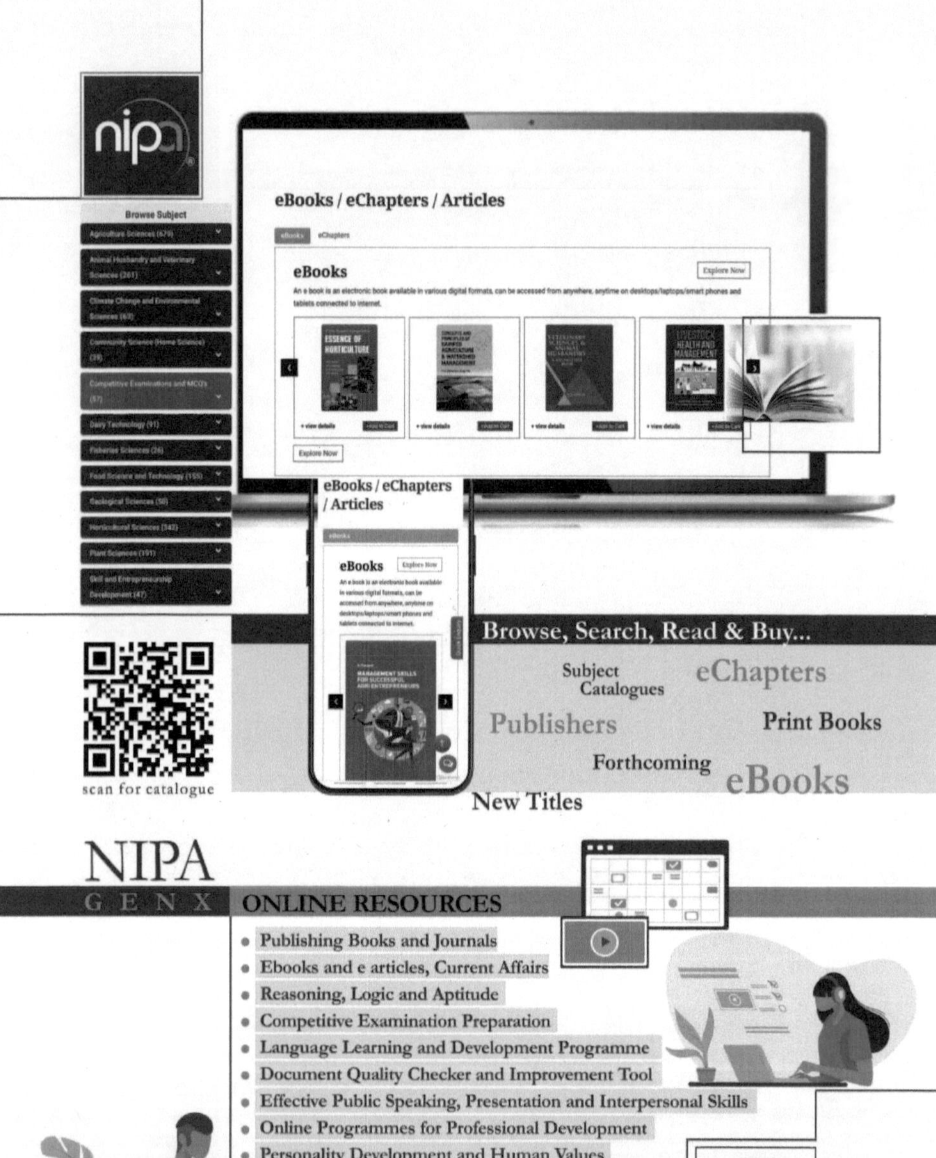
nipa
Browse Subject
Agriculture Sciences (679)
Animal Husbandry and Veterinary Sciences (261)
Climate Change and Environmental Sciences (63)
Community Science (Home Science) (28)
Competitive Examinations and MCQ's (57)
Dairy Technology (91)
Fisheries Sciences (26)
Food Science and Technology (155)
Geological Sciences (58)
Horticultural Sciences (343)
Plant Sciences (191)
Skill and Entrepreneurship Development (47)
eBooks / eChapters / Articles
eBooks
Explore Now
An e book is an electronic book available in various digital formats, can be accessed from anywhere, anytime on desktops/laptops/smart phones and tablets connected to internet.
Explore Now
scan for catalogue
Browse, Search, Read & Buy...
Subject Catalogues
eChapters
Publishers
Print Books
Forthcoming
eBooks
New Titles
NIPA
GENX
ONLINE RESOURCES
Publishing Books and Journals
Ebooks and e articles, Current Affairs
Reasoning, Logic and Aptitude
Competitive Examination Preparation
Language Learning and Development Programme
Document Quality Checker and Improvement Tool
Effective Public Speaking, Presentation and Interpersonal Skills
Online Programmes for Professional Development
Personality Development and Human Values
UPI
PAY USING
PayPal

Traditional Organic Farming Practices

E. Somasundaram
D. Udhaya Nandhini

NEW INDIA PUBLISHING AGENCY
New Delhi – 110 034

NEW INDIA PUBLISHING AGENCY
101, Vikas Surya Plaza, CU Block, LSC Market
Pitam Pura, New Delhi 110 034, India
Phone: + 91 (11) 27 34 17 17 Fax: + 91 (11) 27 34 16 16
Email: info@nipabooks.com
Web: www.nipabooks.com

Feedback at feedbacks@nipabooks.com

ISBN: 978-93-86546-17-3

Composed and Designed by NIPA

Preface

The quantitative sufficiency of food in our country has led into think the maintenance of soil health and crop husbandry techniques, which maintain the nature's balance. Understanding the basic principles of organic agriculture is as much important as that of knowing the latest developments scenario in the field of agriculture. It is strenuous strive to keep pace with the progress of such a vast area like organic farming which is in practice throughout the globe. The budding farmer / scientist have to brace him with the fundamentals of organic agriculture. Under this circumstance, important and relevant information about the various principles and practices of organic agriculture are compiled in a book form entitled "**Traditional Organic Farming Practices**". The book is divided in to 20 chapters and which covers comprehensively the syllabus of organic farming offered for B.Sc (Agri) as stepping stone to other courses in agriculture. This book provides attention of one and all concerned to promote organic farming as a measure to provide the elutes to posterity and to save our farm land that we inherited from our forefathers from being degraded and made in to wastelands through our excessive interventions. Opportunity for employment generation in rural India and making the rural roles empowered to produce their farm inputs and a message to live healthy by eating organic food. In addition this publication guides the farmer interested in organic farming.

The authors acknowledge their indebtedness to the authors/ publishers of various books from which they have drawn the matter for compiling this book. The list of which is attached at the end.

The authors thank Dr. K. Ramasamy, Vice-Chancellor, Tamil Nadu Agricultural University for providing necessary guidelines for bringing out this book.

The authors are much indebted to the Tamil Nadu Agricultural University, Coimbatore for providing an excellent opportunity to bring out this piece of work.

The Authors thank their family members for their help, physical and moral support and M/s New India Publishing Agency for their sincere efforts in bringing out this book in time.

E. Somasundaram
D. Udhaya Nandhini

Contents

List of Tables

1

Towards Sustainable Organic Agriculture

Presently the farming situation urges need to develop farming techniques, which are sustainable from environmental, production, and socio-economic points of view. Modern agricultural production throughout the world does not appear to be sustainable in the long run. Sustainable agricultural development "is the management and conservation of the natural resource base and the orientation of technological and institutional change in such a manner to assure the attainment and continued satisfaction of human needs for the present and future generations". Such sustainable development in the agriculture, forestry and fishery sectors, conserves land, water, plant and animal genetic resources, is environmentally non-degrading, technically appropriate, economically viable and socially acceptable. Such concerns imparted a way to organic farming. It is the need of the day to understand the prospects and problems of organic farming to launch a successful and flawless organic production programme in the farm of environment.

Organic farming

It is the most widely recognized alternative farming system to chemical agriculture evolved during 1940's largely in response to the publication of J.I Rodale in the U.S., Lady Eve Balfour in England and Sir Albert Howard in India. In 1980, USDA released a landmark report on organic farming. The report defined organic farming as follows;

"Organic farming is a production system, which avoids or largely excludes the use of synthetically compounded fertilizers, pesticides, growth regulators, and livestock feed additives. To the maximum extent feasible, organic farming systems rely upon crop rotations, crop residues, animal manures, legumes, green manures, off-farm organic wastes, mechanical cultivation, mineral bearing rocks, and aspects of biological pest control to maintain soil productivity and tilth to supply plant nutrients, and to control insects, weeds, and other pests".

Organic agricultur esystem primarily aims at managing the agro- ecosystem as an autonomous system, based on the primary production capacity of the soil under the local conditions. It implies treating the system on any scale, as a living organism supporting its own vital potential for biomass and animal production, along with biological mechanisms for mineral balancing, soil improvement and pest and disease control.

Organic agriculture is a unique production management system which promotes and enhances agro-ecosystem health, including biodiversity, biological cycles and soil biological activity, and this is accomplished by using on-farm agronomic, biological and mechanical methods in exclusion of all synthetic off-farm inputs (as reported by FAO).

1.1. Concepts of Organic Farming

1. Use of on- farm resources and avoiding / minimizing the use of off – farm resources.
2. Integrating farm resources for reducing the use of external inputs.
3. Soil health management using farm waste (crop, weed and animal wastes)
4. Generating feed resources in the farm: by growing forage crops and fodders (shrub and trees).
5. Green biomass generation for soil fertility management by growing high foliage producing plants and trees.
6. Composting of crop residues and livestock manures: with the help of earth worms converting them into value added vermi composts.
7. Preparing plant growth promoting and disease tolerant/ controlling substances (like Panchagavya, Coconut milk slurry, Amudhakaraisal etc.).
8. Use of eco friendly pest management / crop protection measures (like use of predators and parasites, herbal extracts, crop diversification etc.).

1.2. Other Possible Aims are

1. To work as much as possible within a closed system and draw upon local resources.
2. To maintain the long term fertility of the soil.
3. To avoid all forms of pollution caused by agricultural techniques.
4. To provide a food stuff of high nutritional quality in sufficient quantity.
5. To reduce the use of fossil energy in agricultural practices to be minimum tending to zero.
6. To give to all livestock the conditions of life that conform to their physiological needs.

7. To make it possible for agricultural families to earn a living through their work and develop their potentialities as human beings.
8. To maintain the rural environment and also preserve non- agricultural ecological habitats.

1.3. Key Principles of the Organic Farming System

It is the automonous ecosystem management which is based on three strongly interrelated principles, which take place under:

- Mixed farming
- Crop rotation and
- Organic cycle optimization

1.4. Basic Steps

Organic farming approach involves following five steps

- Conversion of land from conventional management to organic management.
- Management of the entire surrounding system to ensure biodiversity and sustainability of the system.
- Crop production with the use of alternative sources of nutrients such as crop rotation, residue management, organic manures and biological inputs.
- Management of weeds and pests by better management practices (physical and cultural means and by biological control system).
- Maintenance of live stock in tandem with organic concept and make them an integral part of the entire system.

1.5. Prospects of Organic Farming

Deficiencies of at least five out of the critical soil nutrients are widespread in Asia due to imbalance in application of fertilizers and very limited use of organic manures. Organic techniques alone can help to regenerate the degraded soils and ensure sustainability in crop production.

Soil organic matter is the life source of dynamic soil. The decline in soil organic matter in the recent times in Indian soils often associated with crop yield loss of about 30 per cent and organic manures alone can sustain the productivity of the soil. There is an increase in crop yield by 12 % for every 1% increase in organic matter.

As organic farming is attracting worldwide attention, and there is a potential for export of organic agricultural produce, this opportunity has to be tapped with adequate safeguards so that the interest of small and marginal farmers is not harmed.

Organic farming may be practiced in crops, commodities and regions where the country has comparative advantage. To begin with, the practice of organic farming should be for low volume high value crops like spices, medicinal plants, fruits and vegetables and also in rainfed areas.

Besides the identification of regions suitable for the adoption of organic farming, the crops and their products should also be identified which are amenable for production through organic ways and have the potential to fetch a premium price in the international organic market.

Organic farming should not be confined to the age old practice of using cattle dung, and other inputs of organic/biological origin, but an emphasis needs to be laid on the soil and crop management practices that enhance the population and efficiency of belowground soil biodiversity to improve nutrient availability. Performance of cultural techniques for weed control and that of bio-pesticides for pest management need to be evaluated under field conditions, preferably under cultivators.

Indian agricultural activity results in abundant crop residues. The residue turnover is 273.63 mt and the nutrient potential is 5.67 mt of NPK. Proper residue recycling can serve as effective substitute for inorganic fertilizers. Large potential of organic resources remain untapped in India. Nearly 750 mt cow dung and 250 mt of buffalo manures are available.

Crop rotation including pulses / green manures, which fixes atmospheric nitrogen and leave root nodules in the soil and help in improving residual nitrogen content thereby economizing nitrogen use.

Use of forest leaf litters, in places of availability (For example: Andaman and Nicobar Islands) greatly improvises the soil organic matter and in turn soil organic carbon status.

Integration of traditional knowledge with scientific non chemical inputs and methods brings sustainability in farming.

Solvable problems of organic farming

- It is true that sudden conversion of lands from conventional to organic farming results in decline in yields in irrigated lands. But in the long run, organic farming has resulted in spectacular increase in the productivity of several farmlands. In traditional rainfed agriculture with low external inputs, organic agriculture has shown greater potentials to increase the yield whereas in intensive modern agriculture yield decline is witnessed in the initial years of conversion but with a steady and sustainable increase on continuous organic farming.

- Organic farming had been an integral component of crop cultivation in the past. Application of organic manures is now limited owing to the non-availability of organic manures in sufficient quantities, higher cost, flimsiness in application and transportation expenses.
- The availability of organic manures in adequate amounts and at costs affordable by the farmers is a major problem. The increased mechanization has further reduced the availability of manures with the farmers and this problem will become more acute in future. In such circumstances, postharvest residues should be exploited to partly supplement plant nutrient needs of the organic farming systems.
- Changing cropping patterns with area under legumes is going down, shrinking area under green manures due to economic considerations and water availability and reduced availability of lopping's from forests seriously restrict wide scale use of green manures. Inclusion of legumes in intensive cereal-cereal production systems as short duration grain or forage crops, as substitute to one of the cereals or as break crops needs to be promoted which can cater to the nutrient demands of crops under organic farming system.
- Organic manures are bulky and there is great difficulty in transporting and handling organic manures. However, composting of organic manures can reduce their bulky nature.
- Due to differential availability of nutrients in manures, there is difficulty in standardization.
- In India, the relative lack of national rules, regulations and specific standards relating to organic input and organic food production, inadequate certifying agencies, unrecognized 'green' marketing and retailing channels are preventing the farmers to exploit the export advantages of organic production.

2

Organic Inputs for Nutrient Management

Manure is the organic material derived from animal, human and plant residues which contain plant nutrients in complex forms. Manures are used as source of plant nutrients. They release nutrient after their decomposition. They are applied in large quantities

Advantages of Organic Manures

- Organic manure provides all the nutrients that are required by plants but in limited quantities.
- It helps in maintaining C:N ratio in the soil and also increases the fertility and productivity of the soil.
- It improves the physical, chemical and biological properties of the soil.
- It improves both the structure and texture of the soils.
- It increases the water holding capacity of the soil.
- Due to increase in the biological activity, the nutrients that are in the lower depths are also made available to the plants through manures.
- It acts as mulch, thereby minimizing the evaporation losses of moisture from the soil.

2.1. Composting - An Overview

Composting is the natural process of 'rotting' or decomposition of organic matter by microorganisms under controlled conditions. Raw organic materials such as crop residues, animal wastes, food garbage, some municipal wastes and suitable industrial wastes, enhance their suitability for application to the soil as a fertilizing resource after having undergone composting.

A mass of rotted organic matter made from waste is called compost. The compost made from farm waste like sugarcane trash, paddy straw, weeds and other plants and other waste is called farm compost. The compost made from town refuses like night soil, street sweepings and dustbin refuse is called town

Fig. 1: Final product of Vermicompost for field use

compost. Composting is essentially a microbiological decomposition of organic residues collected from rural area (rural compost) or urban area (urban compost).

2.1.1. Vermicomposting

Vermicomposting is the process of turning organic debris into worm castings. The worm castings are very important to the fertility of the soil. The castings contain high amounts of nitrogen, potassium, phosphorus, calcium, and magnesium. Several researchers have demonstrated that earthworm castings have excellent aeration, porosity, structure, drainage, and moisture-holding capacity. The content of the earthworm castings, along with the natural tillage by the worms burrowing action, enhances the permeability of water in the soil. Worm castings can hold close to nine times their weight in water.

In the 1996, Summer Olympics in Sydney, Australia, the Australians used worms to take care of their tons and tons of waste. They then found that waste produced by the worms was could be very beneficial to their plants and soil. People in the U.S. have commercial vermicomposting facilities, where they raise worms and sell the castings that the worms produce.

Earthworms have been on the Earth for over 20 million years. In this time they have faithfully done their part to keep the cycle of life continuously moving. Their purpose is simple but very important. They are nature's way of recycling organic nutrients from dead tissues back to living organisms. Many have recognized the value of these worms. Ancient civilizations, including Greece and Egypt valued the role earthworms played in soil. The Egyptian Pharaoh, Cleopatra said, "Earthworms are sacred." She recognized the important role the worms played in fertilizing the Nile Valley croplands after annual floods. Charles Darwin was intrigued by the worms and studied them for 39 years. Referring to an earthworm, Darwin said, "It may be doubted whether there are many other animals in the world which have played so important a part in the history of the world. The earthworm is a natural resource of fertility and life."

2.1.1.1. Materials for preparation of vermicompost

Any types of biodegradable wastes viz.,

1. Crop residues
2. Weed biomass
3. Vegetable wastes
4. Leaf litter
5. Biodegradable hostel refuses
6. Waste from agro-industries
7. Biodegradable portion of urban and rural wastes
8. Earthworms

2.1.1.2. Selection of suitable earthworm

For vermicompost production, the surface dwelling earthworm alone should be used. The earthworm, which lives below the soil, is not suitable for vermicompost production. The African earthworm (*Eudrillus eugeniae),* Red worms (*Eisenia foetida)* and composting worm (*Perionyx excavatus)* are promising worms used for vermicompost production. All the three worms can be mixed together for vermicompost production. The African worm (*Eudrillus eugeniae)* is preferred over other two types, because it produces higher production of vermicompost in short period of time and younger ones in the composting period.

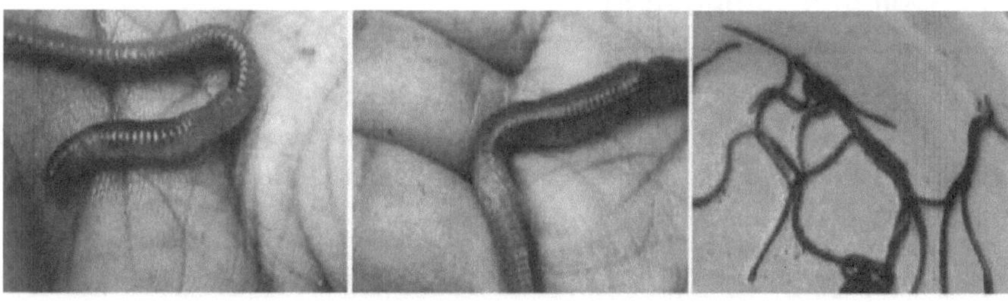

African earthworm (*Eudrillus euginiae*) | Tiger worm or Red wrinkle (*Eisenia foetida*) | Asian worms (*Perinonyx excavatus*)

Fig. 2: Suitable earth worms for vermicompost preparation

Steps for vermicompost preparation

Select a thatched roof or place with shade, high humidity and cool

⇩

A cement tub may be constructed to a height of 2½ feet and a breadth of 3 feet or over the hand floor, hollow blocks / bricks may be arranged in compartment to a height of one feet, breadth of 3 feet and length to a desired level.

⇩

Worm bed (3 cm) - by placing after saw dust or husk or coir waste or sugarcane trash in the bottom of tub / container.

⇩ Moistened with water

A layer of fine sand (3 cm) should be spread over the culture bed followed by a layer of garden soil (3 cm).

⇩ Moistened with water

Keep it all for predigestion (frequent turnings) for 20 days

⇩

The predigested waste material should be mud with 30% cattle dung either by weight or volume.

⇩ Maintained at 60% moisture level

1 kg of worm (1000 Nos.) is released. Water should be sprinkled over the bed rather than pouring the water.

⇩

The castings formed on the top layer are collected periodically

2.1.2. Coir pith composting

The largest by products of coconut is coconut husk from which coir fibre is extracted. This extraction process generates a large quantity of dusty material called coir dust or coir pith. Large quantity of coir waste of about 7.5 million tonnes is available annually form coir industries in India. In Tamil Nadu, 5.5 lakh tonnes of coir dust is available at present.

Coir pith is collected from the coir industry without any fibre. If fibrous materials are present, it is removed by sieving at the source itself.

It is better to have an elevated shady place for composting. If unavailable, inbetween coconut trees or shade under any tree is good for composting.

Coir pith composting is an aerobic process. So it should be heaped above the soil. Coir pith should be spread to the length of 4 feet and breadth of 3 feet. Initially coir pith should be put up for 3 inch height and thoroughly moistened.

After moistening, nitrogenous source material should be added. The nitrogenous source may be in the form of urea (5 kg/ton) or fresh poultry litter (200 kg/ton). This 5 kg equally divided into five portions and in alternative layer of coir pith one kg of urea should be applied.

One has to proportionally divide and put the required amount of poultry litter over the coir pith. For example if one ton coir pith is divided into 10 portions, in the first layer, 100 kg poultry litter is added.

After adding, the nitrogen source, the microbial inoculums *Pleurotus* and TNAU biomineralizer (2%) are added over the material @ 2 kg per ton of coir pits. Over this one portion of coir pith is added and the same input mentioned above should be added. It is advisable to make a heap up to minimum of 3 feet height. But beyond 4 feet, it requires machinery to handle the materials.

The compost heap should be turned once in 10 days to allow the stale air trapped inside the compost material to go out and fresh air will get in. The other way of giving aeration is inserting perforated unused PVC or iron pipe in the composting material both vertically and horizontally.

Maintaining optimum moisture (60 % moisture) is the pre-requisite for uniform composting of waste material.

After 60 days the composted material which is obtained from sieving is ready for use. It is recommended that 5 tons of composted coir pith per hectare;

2.1.3. Sugarcane trash composting

Sugarcane produces about 10 to 12 tonnes of dry leaves per hectare per crop. The detrashing is done on 5th and 7th month during its growth period. The sugarcane trash incorporation in the soil reduces soil EC, improves the water holding capacity, better soil aggregation and thereby improves porosity in the soil. Sugarcane trash incorporation reduces the bulk density of the soil and there is an increase in infiltration rate and decrease in penetration resistance. The direct incorporation of chopped trash increases the availability of nutrients leading to soil fertility.

2.1.3.1. Collection of trash

The detrashed material has to be pooled together and transported to the compost yard. If no compost yard is available to farmer, anyone of the corner area in the sugarcane field itself can be used for making composting. Sugarcane trash is lengthy one. It is recommended to shred the waste into small particles (Mechanical Shredders are available in the market). This process reduces the volume of material, increases the surface area of the waste. If the waste material contains more surface area, more microorganisms work effectively on the surface and degradation will be faster. Shredder is the ideal instrument to shred all the sugarcane trash. Chop cutter machine can also be used. If no machinery is available manual shredding is recommended.

2.1.3.2. TNAU biomineralizer

TNAU biomineralizer is the consortium of microorganism recommended for composting all the agro wastes. For one ton of trash, two kg inoculums are recommended. Without the inoculation of microbial consortium, the composting process will take its own time. Animal dung or fresh poultry litter can be used as a source of nitrogen to reduce the C: N ratio. For one ton of sugarcane trash 50 kg fresh dung is recommended. The dung can be mixed with 100 litres of water and thoroughly mixed with sugarcane trash. Rock phosphate at @ 5 kg/ ton waste can be added to increase the phosphorus content of the compost.

After mixing all the inputs with sugarcane trash, heap should be formed with a minimum height of 4 feet. This height is required to generate more heat in the composting process, and the generated heat will be retained long time inside the material.

The compost material should be turned periodically once in 15 days to allow more aeration inside the material. In the turning process, bottom layer comes to top and top layer comes to bottom. So that uniform composting will occur.

Throughout the composting period about 60% moisture should be maintained. If composting material is allowed to dry, all the established microorganisms get killed and composting process will be terminated.

Volume reduction, earthy odour, brownish black colour and reduction in particle size are important parameters to be observed for assessing compost maturity. Once the compost attained the maturity, the compost heap should be disturbed and spread the material for curing. After 24 hours the composted material can be sieved through 4 mm sieve to get uniform compost material. The residues available after sieving will be recycled to the next composting batch for further composting.

The enriched compost can be applied at the rate of 5 tons per hectare as basal application to the field. Whatever the compost derived from sugarcane trash, it can be ploughed back into sugarcane field to enrich the soil.

2.1.4. Crop residue composting

Crop residues are the non-economic plant parts that are left in the field after harvest. The harvest refuses include straws, stubble, stover and haulms of different crops. Crop remains are also from thrashing sheds or that are discarded during crop processing. This includes process wastes like groundnut shell, oil cakes, rice husks and cobs of maize, Jowar (Sorghum) and Pearl millet . The greatest potential as a biomass resource appears to be from the field residues of sorghum, maize, soybean, cotton, sugarcane etc. In Tamil Nadu, at present 190 lakh tones of crop residues are available for use.

Shred the waste into small particles. This process reduces the volume of material, increases the surface area of the waste. Carbon and nitrogen ratio decides the initiation of composting process.

Green coloured waste materials like *Glyricidia* leaves, *Parthenium*, freshly harvested weeds; *Sesbania* leaves are rich in nitrogen, whereas brown coloured waste material like straw, coir dust, dried leaves and dried grasses are rich in carbon. Animal dung is also a good source of nitrogen. While making heap formation, alternative layers of carbon rich material, animal dung and nitrogen rich material are to be heaped to get a quicker result in composting.

Minimum 4 feet height at an elevated place to have a sufficient shade should be maintained for composting. While heap formation, all the crop residues should be mixed in an alternate layers of carbon and nitrogen rich material with intermittent layers of animal dung are essential. After heap formation the material should be thoroughly moistened.

For one ton of crop wastes 2 kg of TNAU biomineralizer is recommended. This two kg biomineralizer should be mixed with 20 litres of water and made slurry. When the compost heap is formed in between layers the slurry should be inoculated, so that it mixes with the waste material thoroughly for uniform coating of microorganism on the waste material.

Sufficient quantity of oxygen should be available inside the compost heap. Normally to allow the fresh air to get inside, the compost heap should be turned upside down, once in fifteen days. Throughout the composting period about 60% moisture should be maintained.

Volume reductions, black colour, earthy odour, reduction in particle size are all the physical factors to be observed for compost maturity. After satisfying with the compost maturity index, the compost heap can be disturbed and spread on the floor for curing. After curing for one day, the composted material is sieved through 4 mm sieve to get uniform composted material. The residues collected after composting has to be again composted to finish the composting process.

Compost enrichment

The harvested compost should be heaped in a shade, preferably on a hard floor. The beneficial microorganisms like *Azotobacter* or *Azospirillum*,

Pseudomonas, Phosphobacteria (0.2%) and rock phosphate (2%) have to be inoculated for one ton of compost. About 40 % moisture should be maintained for the maximum growth of inoculated microorganism. This incubation should be allowed for 20 days for the organism to reach the maximum population. Now the compost is called as enriched compost. The advantage of enriched compost over normal compost is the quality manure with higher nutrient status with high number of beneficial microorganisms and plant growth promoting substances.

For one hectare of land 5 tons of enriched biocompost is recommended. It can be used as basal application in the field before sowing/planting.

2.1.5. Tricho-composting

A composting method developed by Central Cotton Research Institute, Nagpur. Composting of Cotton Stalks of five ha in a pit of 6 x 3 x 1m using the cellulytic fungus *Trichoderma viride* is worth mentioning.

Conversion of solid waste with the help of microorganisms like *Trichoderma viride* needs no heavy equipment or investment. All work is proposed to be done by hands, which are available in plenty in all villages provided they are paid a reasonable wage.

All biodegradable materials, crop dry wastes are not suitable as feed for the livestock such as crop residues of cotton, pigeonpea, sorghum, sugarcane all weeds including (*Parthenium hysterophorus*) is chopped into small pieces of say 3-5 cms length by hand or chaff cutter or a threshing machine.

These small pieces of material can then be soaked in water, for a few minutes and then laid on the site in a raised bed of 2 x 3 m size with a height of only 30 cm.

The mixture of compost making micro- organisms like *Trichoderma viride* is prepared with 200 litres of water, 30 kg cow dung and 30 kg soil. About 40 litres of sprinkled on the raised heap.

Another layer of the raw material is then laid on the first layer and again the same mixture is sprinkled. The process is repeated for the next 3 layers.

The total height of the heap will be about 2 m on the first day. The heap is then covered by mud or grass.

The micro organisms become active on the substrate immediately and in the process they themselves also multiply.

If culture is not available about 100 kg of the prepared compost could be used as a culture.

In about 6 week's time, the compost will be ready after turned upside down. If needed compost can be sieved through a mesh of 2.5 x 2.5 cm size and transported to the field loose or in bags.

Approximately the total cost of production will be Rs1500/ ton of compost.

With the adoption of organic farming by a large majority of farmers in the years to come demand for all organic manures is bound to increase phenomenally. The farmers are not willing to exert themselves and they need something, ready to use like bags of urea or bottles of pesticide: Therefore it can be safely assumed that such production activity will flourish in any farm / village.

2.1.6. Preparation of leaf mould

- Dig a pit of 3× 3 feet in the corner of the garden
- Bottom should be inclined by an inch for rainwater to drain away
- Fill the pit with vegetable matters like leaves weeds etc.
- After one feet of filling this spread 1 inch of well rotten cow manure and half inch of lime powder over the waste.
- Then repeat this again to get one more layer. Cover the top layer with soil.
- Stir it with bamboo or a long stick once in 2 weeks.
- This will be ready in 3 months.

2.1.7. Preparation of nutri rich compost

Cowdung + urine (Slurry)	- 1000 kg
Daincha (Green manure)	- 500 kg
Sunhemp (Green manure)	- 500 kg
Calotropis (Erukku)	- 200 kg
Poultry waste	- 500 kg
Tank silt/ Kulathu vandal	- 250 kg
Parthenium (Before seed set)	- 200 kg
Panchagavya	- 1 litre

Compost these all under usual method. After composting it is possible to get approximately 900 kg of nutria rich compost

2.1.8. Preparation of herbal compost

Cowdung + urine (slurry)	- 1000 kg
Daincha (Green manure)	- 500 kg
Sunnhemp (Green manure)	- 500 kg
Neem leaves (*Azadirachta*)	- 500 kg
Black notchi leaves	- 100 kg
Tank silt	- 200 kg
Panchagavya	- 1 litre

Compost these all under usual method. After composting it is possible to get approximately 520 kg of nutria rich compost

2.1.9. Cowdung coconut leaf sheath compost

Cowdung + urine (slurry)	- 1000 kg
Coconut leaf sheath powder (base fronds)	- 500 kg
Calotropis spp.	- 200 kg
Daincha (*Sesbania aculeata*)	- 500 kg
Maize crop waste (Left from cattle shed)	- 200 kg
Tank silt	- 200 kg

Compost these all under usual method. After composting it is possible to get approximately 815 kg of nutrients rich compost.

2.1.10. *In situ* trash composting

Shri Subhash, a member of Innovative Farmers' Club, established by KVK, Ahmednagar, (Shri Subhash Divanchand Karir, Maharashtra, Phones: 02422-273511, 09271940320) during 1996 has developed a method of *in situ* trash composting. In ratoon sugarcane, he placed trash in between cane rows and then applied on spot chemical fertilizers in the form of briquette, 10-15 cm near the cane stumps followed by 10-12 tonnes of press-mud on the trash. After 15 days, 10 kg decomposed microbial culture mixed with cow-dung was applied on trash and irrigated with sprinkler irrigation for 2 to 3 months. With the help of special type of tool called Pahar the trash was placed. This tool is made from the porous MS pipe having 2 inch diameter and 5 feet length. The lower side of the pipe is pointed. The round MS plate was fitted 15 cm above the bottom side of the tool so that the placement is at 10 to 15 cm in the soil. Excluding the chemical fertilizers, these method of supercane trash composting can be attempted of the small scale study to verify its suitability.

2.1.11. Composting of dropped guava leaves

- Make trenches of (1 feet – depth, breadth – 2 feet) and length desirable (10m) in between the trees.
- Then fill the trenches once in a week with the fallen leaves of guava.
- Then surface of the trench to be covered with a thin layer of soil.
- The compost will be ready in about 40 – 60 days with little moisture available in the soil

2.1.12. Rice husk compost (Nel makku uram)

- Dig a pit size of 5x3x3 feet length in the backyard of house or corner of the field.
- Along with the soil, apply husk, ill filled grains, and chaffs inside the pits in layers.
- After 3-4 months, the composting process would be completed and ready for the application in any cropped field.
- Within this period cow's urine and cowdung can also be applied inside the pit for easy decomposition.

2.1.13. Wood ash and tea dust compost

- Make a pit with a size of 10" length, 5" width and 5" depth.
- Collect the used tea dust from local tea stalls where as the wood ash from brick kilns or from their own houses.
- Dump all the collected materials in pit.
- The materials in the pit get decomposed within 30-45 days
- During vegetative phase, each tree would be supplied with 5 kgs of the manure whereas during reproductive stage about10 kg of the manure/tree is required.
- These manures are applied to each tree at a distance of 3-5 feet away from the stem, at a depth of ½ a foot and covered with soil.

2.1.14. Indigenous organic manure pit

- Dig a shallow pit of 95-105 cm approximately (3-3.5 feet) depth and 1.5 m (5 feet) width
- Collect the animal wastes and fill the pit till it reashces to height of 2 feet and then apply black soil over it up to 2 feet.
- As such, these are to be applied alternatively.

- Spray water on the pit once in a week and give turnup once in a month.
- The compost will be ready in 4-6 months for application.

2.1.15. Hot composting

- It includes creating heaps of organic matter up to a height of 3-4 ft. with base of 5ft x 8ft. Organic matter includes any natural substance like husk, dry foliage, dry grass, left-over food *etc*.
- Organic matter is covered with cow dung slurry and layers are stacked and left for natural decomposition. The heap is turned in regular intervals, so that the aerobic digestion takes place in 2-3 months, which would otherwise take 6 months.

2.1.16. Heap Composting

Materials required

Green biomass – 1 ton

Dry Biomass – 1 ton

Cow dung – 200 kg

Water

Method

1. Demark an area 2 metres width and 3 metres length. Depending on the availability of the materials the length of the heap can be extended.
2. A layer of stones or wooden logs are placed within the demarked area as a basal layer which helps in providing aeration.
3. Dung slurry is prepared by mixing 25 kg of dung in 100 liters of water.
4. A layer of dry biomass is uniformly spread to the height of 30 cm and moistened with dung slurry.
5. On the top of the dry biomass layer, green biomass is spread and moistened with dung slurry.
6. Similarly, alternate layers of dry and green biomass are laid to the height of about 1-1.5 metres height. Each layer of biomass is thoroughly moistened with dung slurry.
7. Similarly, alternate layers of dry and green biomass are laid to the height of about 1-1.5 metres height. Each layer of biomass is thoroughly moistened with dung slurry.

8. When the heap has reached a desired height it is covered with soil or straw.
9. To prevent the loss of nutrients by volatilization, the heap is made under shade. In certain places wherein there is no natural shade, thatched roofs or shade nets are to be made available.
10. To hasten the process of composting the heap can be turned after a month.
11. Depending on the weather conditions the compost will be ready in 2-3 months. In warm weather the composting process is faster than in winter months.

2.1.17. Water hyacinth compost

- Make a pit size of 2 x 1.5 x 0.5 m.
- Water hyacinth plant materials collected from water bodies and are spread to the ground for a day or two to reduce the water content of it.
- First, a layer of water hyacinth is spread at the bottom of the trench and over this cow dung mixed water is sprinkled to hasten decomposition.
- The process is repeated till the pit is filled to 50-60 cm high above ground level. The top of the heap should be made dome shaped and plastered with 2.5-50 cm thick layer of soil mixed with cow dung.
- Compost will be ready in about 2 months.

2.1.18. Phospho compost

The phospho-compost is a valuable amendment which could be prepared by mixing farm wastes, cattle dung, soil, compost, chopped grasses, crop residues and plant leaves with mussoorie rock phosphate at the rate of 30 per cent of the compostable materials. This mixture is made into a slurry so as to provide adequate moisture and after uniform mixing with decomposer fungus/bacteria. The slurry is allowed to decompose in a compost pit for about 60 to 90 days. The moisture is maintained at 60 per cent throughout the period of composting and the compost will be ready for use in 60 - 90 days. Once the compost is ready, 4kg/tonne of solid, liquid 400 ml/tonne of Phosphate solubizling bacteria is added and allowed for 21 days for release of fixed phosphate. Phospho-compost has better performance than other sources of phosphorus and it can be also used as an effective organic amendment in problematic soils.

2.1.19. Potassic compost

- Potassic compost is a valuable amendment to the soil which could be prepared by mixing rich farm wastes, cattle dung, sugar cane trash, soil, potassium rich ores, banana wastes etc. The waste materials first decomposed with the help of fungus/bacteria (Silicate solubilizing bacteria) using 400ml liquid cultures/tonne of waste. The use of decomposer culture decompose the materials within 45 days. About 60% moisture is maintained throughout decomposition process. Once the material is decomposed, 200ml/tonne of waste material potash mobilizing bacteria is to be added and allowed for 20 days. The compost will be rich in potassium, and hormones required for plant growth. Potassic-Compost has better performance than other sources of potassium, as it is in organic form.

2.2. Organic Manures

As mentioned earlier manures are plant and animal wastes that are used as sources of plant nutrients. The art of collecting and using wastes from animal, human and vegetable sources for improving crop productivity is as old as agriculture. Manures are the organic materials derived from animal, human and plant residues which contain plant nutrients in complex organic forms. Manures with low nutrient, content per unit quantity have longer residual effect besides improving soil physical properties compared to fertilizer with high nutrient content. Baed on nutrient comcentration manures are grouped into two types viz., bulky and concentrated.

2.2.1. Bulky organic manures

Bulky organic manures contain small percentage of nutrients and they are applied in large quantities. Farmyard manure (FYM), compost and green-manure are the most important and widely used bulky organic manures.

2.2.1.1. Farmyard manure

Farmyard manure refers to the decomposed mixture of dung and urine of farm animals along with litter and left over material from roughages or fodder fed to the cattle.

The present method of preparing farmyard manure by the farmers is defective. Urine, which is wasted, contains 1.0 per cent nitrogen and 1.35 per cent potassium. Trenches of size 6 m to 7.5 m length, 1.5 m to 2.0 m width and 1.0 m deep are dug. All available litter and refuse is mixed with soil and spread in the shed so as to absorb urine. The next day morning, urine soaked refuse bedding material along with dung is collected and placed in the trench.

A section of the trench from one end should be taken up for filling with daily collection. When the section is filled up to a height of 45 cm to 60 cm above the ground level, the top of the heap is made into a dome and plastered with cow dung earth slurry. The process is continued and when the first trench is completely filled, second trench is prepared. The manure becomes ready for use in about four to five months after plastering.

2.2.1.2. Sheep and goat manure

The droppings of sheep and goat contains higher nutrients than farmyard manure and compost. The sweeping of sheep or goat sheds are placed in pits for decomposition and it is applied later to the field. The nutrients present in the urine are wasted in this method.

The second method is sheep penning, wherein sheep and goats are kept overnight in the field and urine and fecal matter added to the soil is incorporated to a shallow depth by working blade harrow or cultivator or cultivator. Sheep penning is an age old practice to revie fertility of the land.

2.2.1.3. Poultry manures

The excreta of birds ferment very quickly. If left exposed, 50 percent of its nitrogen is lost within 30 days. Poultry manure contains higher nitrogen and phosphorus compared to other bulky organic manures. Compost, vermicompost pressmud, sewage sludge, nightsoil *etc.* are the other forms of bulky organic manures of time.

2.2.2. Concentrated organic manures

Concentrated organic manures have higher nutrient content than bulky organic manures. The important concentrated organic manures are oilcakes, blood meal, meat/bone meal, fish manure *etc.* These are also known as organic nitrogen fertilizers. Before their organic nitrogen is used by the crops, it is converted through bacterial action into readily usable ammoniacal nitrogen and nitrate nitrogen. These organic fertilizers are, therefore, relatively slow acting, but they supply available nitrogen for a longer period of time.

2.2.2.1. Oil cakes

After oil is extracted from oilseeds, the remaining solid portion is dried as cake which can, be used as manure. The oil cakes are of two types:

- Edible oil cakes which can be safely fed to livestock; *e.g.*: Groundnut cake, Coconut cake *etc.* and
- Non edible oil cakes which are not fit for feeding livestock; *e.g.*: Castor cake, Neem cake, Mahua cake *etc.*,

Both edible and non-edible oil cakes can be used as manures. However, edible oil cakes are fed to cattle and non-edible oil cakes are used as manures especially for horticultural crops. Nutrients present in oil cakes, after mineralization, are made available to crops 7 to 10 days after application. Oilcakes need to be well powdered before application for even distribution and quicker decomposition.

2.2.2.1.1. Groundnut cake solution spray

- To prepare this extract about 25 kg of groundnut cake is powdered and mixed with 150 litres of water and kept overnight in a closed vessel.
- Next day, only the cleared filtrate was collected and used as a foliar spray solution (10%).
- About 1 lit from this solution was taken and mixed with 9 litres of water to fill a tank. This 10 litre was enough for spraying l acre of tomato field.
- Spray this groundnut cake solution in the fields 15 days after tomato transplantation using knapsack sprayer.

2.2.2.2. Other concentrated organic manures

Blood meal when dried and powdered can be used as manure. The meat of dead animals is dried and converted into meat meal, which is a good source of nitrogen.

2.2.2.3. Concentrated organic manure

- Rice bran 10 parts
- Fish meal 1 Part
- Oil cake 1 part
- Egg shell 1 percent of total weight
- Rock phosphate 1-5%
- Molasses and Undisturbed forest soil for microorganisms on mixed microbial inoculants

In place of fishmeal, bone meal, blood meal, slaughterhouse refuse also can be used. In place of forest soil wide range of microorganisms from different sources can also be used. Mixture of at least 10-12 different types of decomposing bacteria, Fungi and actinomycetes, which are available commercially, can also be used. Partly decomposed material from a compost pit mixed with sour milk or curd, fermented coconut milk etc can also be used as microorganism's source.

Process

- Mix all the contents except molasses in the appropriate proportion as shown above. Dilute molasses with water (1:500). Add molasses water in the mixture to obtain 50-55% moisture.
- Keep the contents in a container or make a heap on cement floor and cover with polythene sheet.
- Give first turning after 24 hours

- Thereafter turn the mixture twice a day.
- Maintain the temperature of the mixture below 40-45°C
- Compost will be ready within 4-5 days.

At the end of 4-5 days, there should not be any foul odour. If there is foul odour, it means that something has gone wrong. In that case thoroughly turn the mixture three times a day. The mixture will become healthy and foul odour will disappear. If foul odour still persists then inoculate the mixture with fresh formulation of microorganisms and turn three times a day. Compost will be ready in next 2-3 days.

2.2.3. Green manures

Green undecomposed material used as manure is called green manure. It is obtained in two ways: by growing green manure crops or by collecting green leaf (along with twigs) plants grown in wastelands, field bunds and forest. Green manuring is growing in the field plants usually belonging to leguminous family and incorporating into the soil after sufficient growth. The plants that are grown for green manure are known as green manure crops. The most important green manure crops are sunnhemp, daincha (*Sesbania aculeta*), pillipesara, cluster beans and *Sesbania rostrata* (Manila agathi).

Application of green leaves and twigs of trees, shrubs and herbs collected from elsewhere is known as green leaf manuring. Forest tree leaves are the main sources for green leaf manure. Plants growing in wastelands, field bunds etc., are another source of green leaf manure. The important plant species useful for green leaf manure are neem, mahua, wild indigo (*Tephrosia purpurea*), Glyricidia, Karanji (*Pongamia glabra*) calotropis, indigo (*Indigofera tinctoria)*, agathi (*Sesbania grandiflora*), subabul and other shrubs.

Advantages of green manures

Usage of green leaf manure is advantageous both for crops and soil. The advantages are:

- As they decompose rapidly, it is easy to retain the organic matter in the soil.
- Green manures improve both physical and chemical properties of the soil.
- They provide energy to (Source of quality available carbon and nitrogen).
- They provide nutrients to the standing crop and also to the next crop.

Fig. 3: Common green manure crops

Fig. 4: Green leaf manures (*Courtsey*: TNAU Portal)

- Addition of green manure crops to the soil, acts as much and prevent soil erosion.
- Leaching of nutrients in light soils can be prevented by addition of green manure.
- Cultivating green manure crops can control weeds.
- Majority of green manure crops being legumes, use of nitrogenous fertilizers can be minimized/avoided.

Criteria for Selection of Green Manure

1. Possibility to grow after main cropping season on residual soil moisture or with less rainfall
2. Non-host for crop related pests and diseases
3. Decrease in pest and disease populations
4. No rhizomes
5. Controllable growth
6. Easy and abundant seed formation
7. Useful by products(*e.g.* fodder, wood) Integration of animal husbandry and forestry

2.2.3.1. Basal manure application in coconut (Adi uram)

- Make a ring like channel around the trees in 2 feet depth.
- Apply dried fish (½ kg) of any type cheaply available in their locality, kolinji leaves (1 kg) and salt (2 kg) per tree as basal dose around the trees (ring application method)

2.2.3.2. Cactus as organic manure in coconut

- Cut the cactus plant and chop into pieces and burry in the soil under the coconut trees up to 2 feet depth by digging the soil.
- About 2-3 plant have to be buried per trees and closed with soil again

2.2.3.3. Rice husk burnt ash manure

- After processing the paddy collected rice husk and burnt it to ash. Then this burnt ash (25 kg/acre) was applied in the field as basal manure for wide range of crops like cereals, pulses and millets.

2.2.3.4. Cotton boll residue

- After 2-3 pickings, the plants are to be later uprooted leaving the dried bracts and leaves sticking with cotton bolls in the field itself.
- Then the field is ploughed after the receipt of monsoon.

2.2.4. Use of mixture of ash and manure

Ash dust has to mix with farmyard manure and applied into the field for vegetable cultivation.

2.2.5. Burying dead animals in lemon garden

When any of the domestic animals die, burry the dead bodies of the animal under the trees (at a depth of 4 feet and distance of 15 feet from base) which is believed to enhance the calcium content in the soil.

2.2.6. Cow-pat Pit (CPP)

- Prepare a brick lined pit measuring 90 x 60 cm and 30 cm deep without any lining in the bottom. Mix 60 kg fresh cow dung with 200gm crushed and powdered egg shells and 300 gm basalt dust (or blue granite dust or bore well soil). Mix thoroughly to obtain smooth paste. Fill the mixture in to pit up to 12 cm height. Dog 5 holes in the paste and put one teaspoon full (3 gm each) of preparation 502 to 506 in each hole.
- Preparation 507 is mixed with water and half is poured in one hole and half sprinkled over the entire surface. Cover the surface with wet gunny bag.
- After four weeks, aerate the dung by turning it with the help of a fork. Smooth out again and cover. Thereafter turn every week. CPP compost will be ready in 12 weeks time.
- CPP can be used in various ways depending upon the requirement and crop/plants.
- Use 100 gm CPP/acre, mix with BD 500 or 501 and use as spray. CPP can be used as soil inoculants (@ 2 kg/acre) mixed with composts. CPP can also be used as foliar spray (@ 5kg/acre) right from the beginning of crop to up to fruit/pod formation stage with an interval of 7 to 15 days. CPP can also be used as paste on stem of fruit trees

2.3. Biofertilisers

Biofertilizers are defined as preparations containing living cells or latent cells of efficient strains of microorganisms that help crop plants for uptake of nutrients by their interactions in the rhizosphere when applied through seed or soil. They accelerate certain microbial processes in the soil, which augment the extent of availability of nutrients in a form easily assimilated by plants.

Use of biofertilizers is one of the important components of integrated nutrient management, as they are cost effective and renewable source of plant nutrients to supplement the chemical fertilizers for sustainable agriculture. Several microorganisms and their association with crop plants are being exploited in the production of biofertilizers. They can be grouped in different ways based on their nature and function.

Application of Biofertilizers

1. Seed treatment or seed inoculation
2. Seedling root dip
3. Main field application

Seed treatment

One packet of the inoculant is mixed with 200 ml of rice kanji to make slurry. The seeds required for an acre are mixed in the slurry so as to have a uniform coating of the inoculant over the seeds and then shade dried for 30 minutes. The shade dried seeds should be sown within 24 hours. One packet of the inoculant (200 g) is sufficient to treat 10 kg of seeds.

Seedling root dip

This method is used for transplanted crops. Two packets of the inoculant (400g) are mixed in 40 litres of water. The root portion of the seedlings required for an acre is dipped in the mixture for 5 to 10 minutes and then transplanted.

Main field application

Four packets of the inoculant is mixed with 20 kgs of dried and powdered farm yard manure and then broadcasted in one acre of main field just before transplanting.

2.3.1. Rhizobium

For all legumes *Rhizobium* is applied as seed inoculant.

Rhizobial strains for various crops

Groundnut - TNAU 14, SK-1

Soyabean - Co S 1

Balck gram - PMBS 47, CRU-7

Green gram - GMBS1, Co C 10

Bengal gram - Co B 13

Red gran - CC 1

Methods to use

- For 10 kg of seed 1 pocket of rhizobium (200 g) is sufficient.
- Mix this rhizobium in well fertile soil along with 200 ml rice kanchi and mix well
- Shade dry for 30 minutes and take up sowing immediately

2.3.2. *Azospirillum/Azotobacter*

In the transplanted crops, *Azospirillum* is inoculated through seed, seedling root dip and soil application methods. For direct sown crops, *Azospirillum* is applied through seed treatment and soil application. For every hectare of crop, two kg of *Azospirillum* should be mixed with 25 kg of well decayed FYM manure or wet sand and applied before transplanting.

Seed treatment

For 1 acre seeds, 2 pockets. Mix well with rice kanji and shade dry for 30 minutes.

For transplanted crop

4 pockets of *Azospirillum* should be mixed with well decomposed FYM and broadcast it over 1 acre

Seedling root tip

Two packets of the inoculant are mixed in 40 litres of water. The root portion of the seedlings required for an acre is dipped in the mixture for 20 minutes and then transplanted.

For trees

For a grown tree 20 to 50 of *Azospirillum* should be mixed with well decomposed farm yard manure and apply over the root zone.

2.3.3. Phosphobacteria

Inoculated through seed, seedling root dip and soil application methods as in the case of *Azospirillum*. For every hectare of crop, two kg of *Azospirillum* or *Azotobacter* and two kg of phosphobacteria should be mixed with 25 kg of well decayed manure or wet sand and applied before transplanting. This helps to increase the population of these microorganisms in the soil.

Points to remember

- Bacterial inoculants should not be mixed with insecticide, fungicide, herbicide and fertilizers.
- Seed treatment with bacterial inoculant is to be done at last when seeds are treated with fungicides.

Table 1: Biofertilizers recommendation (one packet - 200 g)

Crop	Seed	Nursery	Seedling dip	Main field	Total requirement of packets per ha
Rice	5	10	5	10	30
Sorghum	3	-	-	10	13
Pearl millet	3	-	-	10	13
Ragi	3	-	5	10	18
Maize	3	-	-	10	13
Cotton	3	-	-	10	13
Sunflower	3	-	-	10	13
Castor	3	-	-	10	13
Sugarcane	10	-	-	36 (3 splits)	46
Turmeric	-	-	-	24 (2 splits)	24
Tobacco	1	3	-	10 g/pit	14
Papaya	2	-	-	10	-
Mandarin Orange	2	-	-	10 g/pit	-
Tomato	1	-	-	10	14
Banana	-	-	5	10 g/pit	-

Table 2: A key to biofertilizer use

For Crops	Biofertilizers recommended & Doses
Pulse crops like moong, urd, arhar, cowpea, lentil, pea, bengal gram, all beans, ground nut, soybean, leucern, berseem and other legume crops.	*Rhizobium* 200 gms + PSB 200 gm for every 10 kg of seed as seed treatment.
All nonlegume crops like wheat, seedsown upland paddy, barley, maize, cotton,sorghum, bhindi, mustard, sunflower, niger etc. and other non legume crops taken by direct seed sowing.	*Azotobacter* 200 gms + PSB 200 gms for every 10 kg of seed as seed treatment.
Jute	*Azospirillum* 200 gms + PSB 200 gms for every 10 kg of seed as seed treatment.
Vegetables like tomato, brinjal, chilli, cauliflower, cabbage etc. and othertransplanted crops.	*Azotobacter* 1kg + PSB 1 kg for one acre as seeding root dip method.

Lowland transplanted paddy.	*Azospirillum* 2 kg + PSB 2 kg for one acre as seedling root dip for 8 -12 hours.
Potato, ginger, colocassia, turmericand jhum paddy.	*Azotobacter* or *Azospirillum* 4 kg + PSB 4 kg/ acre mixed with 100-200 kg compost and applied in soil.
Standing plantation crops like tea, coffee, rubber, mulberry and fruit trees.	2-3 kg *Azotobacter/Azospirillum* + 2-3 kg PSB mixed with 200 kg Compost for one acre and applied as soil treatment. This treatment is to be done 2 to 3 times a year with a gap of 4-6 months
Sugarcane	5 kg *Acetobacter* mixed in sufficient water for setts dipping treatment

2.3.4. Arbuscular mycorrhiza (AM)

Arbuscular mycorrhiza colonizes all most all crops in all stages and creates resistance to entering pathogens and nematodes in plants. On the other hand this fungus helps to plants to absorb nutrients altering root anatomy, modifying root exudations and root system morphology by which plants gets good health as well as less disease incidence.

2.3.4.1. AM Inoculation

Optimal spore count 60-100 spores/ 100g soil

Rate of Inoculation

Vegetables	-	100 g/m^2 nursery
Fruit trees & Coconut	-	100-200 g/tree
Other crops	-	10% of the seed rate
Established plants	-	10 g/plant
Nursery	-	750 – 1000 g /m^2
Nursery poly bag	-	10 g/ poly bag

2.3.5. Alage

Growing algae along with paddy supplies the crop with the required nitrogen and phosphorus. Azolla or blue green algae should be strewn in the field 5–10 days after transplantation of paddy. The field should be drained twenty-five days and 45–50 days after strewing and the algae should be stamped into the soil.

Five to seven kilograms of azolla are required per hectare of land. First, the water should be allowed to stagnate in the land that is to be cultivated. Then azolla should be sown. After one week, these plants are stamped into the soil before transplanting is done. It degrades in about 7–10 days and thereby provides nitrogen to the rice crop.

Fig. 5: Azolla in rice field

2.3.5.1. Blue green algae

BGA are a type of photosynthetic cyanobacteria that belong to the plant kingdom. They are found in paddy fields where good sunlight, water, high temperature and high nutrients are found. They fix atmospheric nitrogen and can be seen floating as dense mats in a filamentous form on the water surface in paddy fields. BGA grows well in clayey and alluvial soil.

Method of application

BGA are added to the soil within ten days of transplanting at the rate of 10 -12 kg/ha. They are available as small bits in plastic packets. This should be powdered and directly added to the soil. Water should be allowed to stagnate to a depth of 3–5 cms in fields where algae are grown. Blue green algae should be added to the field continuously for four cropping seasons. Thereafter, it grows naturally in the soil and produces the desired results.

Method of preparation

The algal inoculums can be prepared by the farmers at their fields by using rural oriented low cost production technology without any appreciable investment.

Trough method

i) Prepare a shallow tray (2m x 1m x 20 cm) of galvanized iron sheet or cement or polythene lined pit on ground. The size can be increased if more material is to be produced.

ii) Spread 5-6 kg of sieved field soil which is free from any organic material. Fill the trays with water up to 5-10 cm and mix well with the help of long handle brush so that a uniform layer of soil is formed at the bottom. Spread 300-400 gm of rock phosphate.

iii) Sprinkle the starter culture of algae on the surface of the standing water, keeping trays in the open air and completely exposed to sun.

iv) In hot summer months, the growth of the algae will be rapid and in about 7-10 days a thick algal mat will be formed on the surface of water. The water level is maintained at 5-7 cm by adding water intermittently, when the growth of algae becomes very thick watering has to be stopped.

v) Allow the water to evaporate completely in the sun and dry algae cracks into flakes. The dry algal flakes can be stored in the polythene bag and used in the fields when required.

vi) Fill the trough with water and add very small amount of the dry algal flakes as inoculum without adding fresh soil. After 10-15 days, harvest the algal mat and repeat the process by adding fresh soil and other required material as mentioned above.

Growing azolla along with paddy

After ten days of planting, azolla should be strewn in the rice fields. It grows along with rice as an intercrop. It grows well in about 25 days and spreads uniformly over the field. Then the water in the field is allowed to drain and azolla is stamped beneath the soil. Once again, azolla resumes profuse growth. This is again stamped at the time of the second weeding. Burying azolla beneath the soil once helps in fixing 15–20 kg of nitrogen.

Growing Azolla in nursery

- Azolla nursery should be prepared simultaneously with the paddy nursery. The area of Azolla nursery should be half of that of paddy nursery.
- Select the proper nursery field and divide in plots (20 m x 2 m) with bunds and irrigation channels. The nursery should be well prepared and levelled.
- Apply 250 g rock phosphate/plot (acidic soil) and irrigate the field for dissolution of phosphates.
- After 10 days of Rock Phosphate application, each plot should be flooded with irrigation water up to a depth of 8-10 cm.
- Prepare a cattle dung by mixing 10 kg fresh dung in 10 litres of water and spread it on each plot.

- Apply 8-10 kg fresh Azolla/plot.
- When temperature is favourable, depth of water should be maintained. Azolla grow to double by weight in 3-4 days.
- Azolla can be harvested after 15 days of application and introduced in the rice field as source of nitrogen.
- About 40-50 kg fresh Azolla can be harvested from each plot. The same Azolla can be used for further production in the nursery field

2.3.6. Root fungus

- Seedlings – 100 g of root fungus is sufficient. Before sowing it must be incorporated at a depth of 2-3 cm/m^2.
- For poly bag nursery seedlings – 10 g/ bag is sufficient
- For a well grown tree – 200 g / tree

2.4. Nutrient Management in Cropping Systems

2.4.1. Legumes in cropping systems

Soil nitrogen is primarily in the organic fraction of the soil. The legumes which have symbiotic nitrogen fixing micro-organisms such as rhizobium in the nodules of their roots, at thus considered to improve the soil fertility and productivity and their inclusion in the cropping sequences is thereafter generally recommended. In sequential cropping system involving pulses 18 – 70 kg N /ha as preceding crop, was added to the soil.

Fig. 6: Intercropping of pigeonpea in pearl millet (Bajra)

Table 3: Legumes in cropping system

Proceeding legume	Following cereal	Fertilizer N – equivalents (kg N/ha)
Chickpea	Maize Pearl millet	60-70-40
Pigeon pea	Wheat Maize Pearl millet	40,20-49,30
Lentil	Pearl millet Maize	40,20-32
Green gram	Pearl millet	30
Lathyrus	Maize	36-48
Cowpea	Pearl millet	60

2.4.2. Nutrient management through mixed cropping, intercropping and crop rotation

i). Mixed cropping

It has been found feasible to cultivate green manure mixed with rice fallow pulses. Green manure crops like indigo and pulses (Black gram) can be mixed with the ratio of 1:3. By adopting such practice in rice fallow pulses, no yield reduction has been noticed. Growth of green manure crop is also not affected. By the time pulse is harvested, green manure put forth growth only to a few cm. after the harvest of pulses, the green manure put on growth.

ii). Intercropping

Seedlings of *S. rostrata* were planted at interval of 1.5 m in rice. On 30 days after planting, the green manure is cut at a height of 15 cm from the ground

Fig. 7: *Sesbania* intercropped in rice

level and incorporated as manure. The ratoon is allowed and the reflesh incorporated while planting second crop of rice.

iii). Crop rotation

It invariably includes pulses or green manure crops which leave root nodules in the soil and help in improving residual nitrogen content facilitating to economise nitrogen use. Pulses at plants fixing atmospheric nitrogen find its importance as sequential crop, mixed crop, inter crop, alley crop in cropping system depending upon the need and resource availability of the locality.

2.4.3. Nutrient management in rice based cropping system

Rice – Rice – grain legume

A fast growing grain legume such as mung bean is planted following the second rice crop. The legume supplies food and crop stover serves as green manure for the next rice crop.

Rice – Rice – Sesbania rostrata

Sesbania rostrata which can stand water logging, and after 45 days of its growth, it can be ploughed under *in-situ* to increase soil fertility.

Rice – Maize – Grain legume

In this system after the maize harvest, mung bean or cowpea were raised. After gravest the stover used as green manure will supply 50 – 60 kg nitrogen per hectare. Plough in rice and maize crop residues for additional nutrients.

2.5. ITK Based Manures

2.5.1. Basal application in coconut (Paramakudi)

- Apply dried fish (½ kg) of any type cheaply available, kolinji leaves (1 kg) and salt (2 kg) per tree as basal dose around the trees (ring application method) in 2 feet depth (once in a year).
- Make a ring like channel around the trees so that applied manure would reach the root zone area and watering become easier.
- The flowering and fruit set will be increased in the subsequent years by this low cost practice.

2.5.2. Cactus as organic manure in Coconut

- Cactus plant was chopped to pieces and buried in the soil under the coconut trees up to 2 feet depth by digging the soil.

- About 2-3 plant was buried per trees and closed with soil again.
- Repeat the practice twice in a year

2.5.3. Rice husk compost

- Dig manure pits of size 5 x 3 x 3 feet length, width respectively in the backyards. Along with soil, apply these materials (Husk, ill filled grains, and chaffs) inside the pits in layers.
- After 3-4 months, the composting process would be completed and ready for the application in any cropped field.
- During these three months period, farmers owning cattle had applied cow's urine and cowdung inside the pit for easy and earlier decomposition.

2.5.4. Rice husk ash

- Collect rice husk and burnt it to ash.
- Then this burnt ash (25 Kg/acre) can be applied in the field as basal manure for wide range of crops like cereals, pulses and millets.

2.5.5. Manuring for fruit crop (Tamil Nadu)

- Collect the used tea dust from local tea stalls and wood ash from brick kilns or from their own houses.
- The collected materials are subjected to decomposition in manure pits (10' × 5' × 3" depth) minimum for duration of 100 days at the corner of a farm.
- The application of the manure varied according to the stage of the crop. During vegetative phase, each tree would be supplied with 5 kgs of the manure whereas during reproductive stage about 10 kg of the manure/ tree were applied.
- These manures are applied to each tree at a distance of 3-5 feet away from the stem, at a depth of ½ a foot and covered with soil.
- Normally this type of manuring is done during October-November so that the crops can utilize north-east monsoon or otherwise the crop has to be irrigated.

2.5.6. Sheep penning (Patti Poduthal)

- Construct a 'Patti' or house in the field using bamboo sticks with length of 450 cm, Breath of 900 cm and weight of 120 cm without any roof.

- In the off-season 'patti' is constructed in the field and sheeps numbering 80-100 can be housed in it usually during the night time. This practice would allow animal wastes to pass into the land.
- After one-month interval 'patti' is moved to other stretch of land. This sheep penning was continued for about 2-3 months so that the land was totally covered and all the organic materials got decomposed completely.
- Afterwards the manure is incorporated in the soil by working 2-3 times with country plough. This practice also helped to keep the environment clean.
- Farmers twice in a year followed this practice to build soil fertility.

2.5.7. Asafoetida extracts (Perungkaya Thannir Thelithal)

- Asafoetida extract solution is prepared by mixing the Asafoetida powder (50g) with 1 litre of water.
- This solution has to be prepared freshly and then pour to individual moringa tree at the root zone region.
- Spray the asafoetida extract solution in early morning or in the late evening time.

2.6. Biogas Slurry Application

The following are the different methods of applying biodigested slurry as manure:

a) Air dried biogas slurry can be applied by spreading on the agricultural land at least one week before sowing the seeds or transplanting the seedlings.

b) The liquid slurry can be mixed directly with the running water in irrigation canal which will enable spreading of the slurry uniformly in the cropped area or in cultivation land

c) Biogas slurry can also be coated on the seeds prior to sowing. This acts as insecticide and prevents seeds or plants from insect attack. This helps in early germination and healthy growth of seedlings.

d) The digested slurry is fed through the channel, flowing over a layer of green or dry leaves and filtered in the bed. The water from the slurry filters down which can be reused for preparing another fresh dung slurry. The semi-solid slurry can be transported easily as it was in the consistency of fresh dung and used for top dressing of crops like sugarcane and potato.

e) Biodigested slurry is also being used to fish culture, which acts as a supplementary feed. On an average, 15-25 litre of wet slurry can be applied per day in a 1200 sq pond. Slurry mixed with oil cake or rice-bran in the 2-1 ratio increases the fish production remarkably. In general, organic manures about 10t/ha in the form of FYM or compost or biodigested slurry is recommended to be applied once in three years to maintain the organic content of soil, besides providing nitrogen, phosphorus and potassium in the form of organic fertilizers to the crop.

f) The digested slurry, if mixed with Azospirillum, KMB and PSM @ 200ml each/acre ensures increase of yield minimum 30% over slurry alone.

g) Less weed occurence is noticed in places BGS application as it earry no weed seeds like FYM.

2.7. Organic Urea (Sand urea preparation)

- Construct a pit of convenient size 3 feet length X 1.5 feet width X 1 feet depth lined with brick and cement.
- Fill ¾ of the brick tank with sand. Pour 5-10 litres of lo-cally bred cow's urine over this sand. Keep pouring the locally bred cow's urine for 20 consecutive days in the tank filled with sand.
- Close the tank with wooden / metal lid to avoid evaporation of urea fumes. Allow the sand to react with cow's urine in the tank. The sand turns black in color within 25-30 days.
- Remove this sand urea from the tank and dry it under the shade for 3-4 days. Now the organic sand urea is ready for use. Compost should be applied as a basal dose to the crops.
- As a top dress, sand urea is applied during vegetative and flowering stage. In case of ragi, 100 kg/acre (twice or thrice the quantity of chemical urea) is applied two times – once after 25 days of sowing and another after 45 days of sowing.
- In case of paddy, 250 kg/acre (twice or thrice the quantity of chemical urea) is applied three to four times. two times during growing stage and once or twice during flowering stage of the crop

2.8. Multivarietal Seed Technique (Navathaniyam)

Soil fertility can be enhanced non- chemically using multivarietal seed technique with in $6\frac{1}{2}$ month (200 days) even under highly chemical fertilizer applied and deteriorated soil conditions.

Mixed sowing of the following five crops is called as multivarietal seed technique

1. **Cereals:** Any four crops under this category

    Eg: Sorghum- 1kg

    Pearl millet (cumbu) - $\frac{1}{2}$ kg

    Foxtail millet (Thenai) - $\frac{1}{4}$ kg

    Little millet (samai)- $\frac{1}{4}$ kg

2. **Pulses:** Any four crops under this category

    Eg: Blackgram -1kg

    Green gram - 1kg,

    Cowpea -1kg,

    Bengal gram -1kg

3. **Oilseeds:** Any four crops under this category

    Eg: Gingelly - kg

    Groundnut -2kg

    Sunflower -2kg,

    Castor -2kg

4. **Green manures:** Any four crops under this category

    Eg: Daincha (Sesbania aculeate) -2kg

    Sunnhemp (Crotalaria juncea) -2kg

    Naripayaru (Wild indigo) - $\frac{1}{2}$ kg

    Horse gram (Macrotyloma uniflorum) -1kg

5. **Spices and condiments:** Any four crops under this group

    Eg: Mustard - $\frac{1}{2}$ kg

    Fenugreek - $\frac{1}{4}$ kg

    Cumin - $\frac{1}{4}$ kg

    Coriander -1kg

Grow the above crop combinations as sole pure crop upto 60 days and incorporate into the soil through substantial ploughing which provides all major and micro nutrients to the soil and support soil life and this in turn better crop growth (as this is being practiced in Maharashtra on seeing the blooming effect of this technique developed by Mr. Dabolkar). This technique rejunuvate the soil through improvement its physical, chemical and biological properties, enhances biomass addition to the soil and helps to restore its fertility.

2.9. Microorganisms Enriched Mixture (MEM)

Ingredients

- Group 1: 70 kg fully digested compost or vermi compost, 10 kg ash or rice bran ash and 20 kg saw dust.
- Group 2: (a) Five liters panchakavya, (b) five liters concentrated amudham solution, (c) five liters coconut-buttermilk or arappu-buttermilk or soapnut-buttermilk solution, (d) ten liters ETFPE, (e) five liters archae bacterial solution.
- Group 3: Bio-fertilizers – *Azospirillum*, *Rhizobium*, *Phospobacteria*, Potash Bacteria: 500 gms to 1 kg each and VAM 5 to 10 kgs.
- Group 4: (To control root rot, rhizome rot and fuzarium wilt): 500 gms-1kg each of Pseudomonas Fluorescence, Trichoderma Viridi, *Trichoderma harzianum* and *Bassillus subtillus.*
- Group 5: (To control nematode) 1-2 kg of Paicyllomycis.
- Group 6: (To control root grub, white grub, rhinoceros beetle and other soil-dwelling beetles and grubs): 500 gms – 1kg each of *Beauveria* and *Metarrhizium*.

Preparation

- Mix well the items in group 1
- Mix the solutions mentioned in group 2
- Mix the powders in groups 3,4,5and6 well together; choose the powders based on yours crop'scondition.
- Add the mixtures from steps (a) and (c). On this combination sprinkle the solutions mixture from (b) and mix thoroughly so the combined mixture is uniformly moist.

Usage

Use within 30 days. When you need to store it for longer periods, store it in a heap that is about 2 feet broad and nine inches tall; the length could be chosen based on convenience. Cover it with wet gunny bags, coconut leaves, or sugarcane leaves. Take care to maintain uniform moisture by sprinkling water as often as necessary. This heap should be in a shed or in the shade of a tree. This can be used as basal application or as a top dressing, depending on the need. Use it as a precautionary measure according to the condition of the crop.

Note

MEM can be used at the rate of 100-500 kg per acre. The ingredients given above are for preparing 100 kg MEM. To prepare larger quantities increase the quantities in Group 1 accordingly maintaining the same ratio of the ingredients in that group. This increases the quantity of the mixture. To make this mixture uniformly moist add sufficient quantity of archae solution. Do not change the quantity of other items in Groups 2 to 5.

If the crop growth is not healthy and you cannot irrigate the crop because of rain, use the mixture at least twice in fifteen days interval. If the crop is healthy, use it once in 1-2 months during the growth period.

For bed crops like vanilla, pepper and cardamom use MEM over the bed and cover it with leaves.

In rainy season, move the mulch away from the stem for effective drainage and spread MEM over the feeder roots to protect these roots.

2.10. Micronutrients

2.10.1. Thirami-fermented plant extract (TFPE) for rectifying micronutrient deficiency

Ingredients

Collect tender leaves of the following: (a) tamarind or vadhanarayanan (source of zinc), (b) (*Cassia* spp.) Avarai, *Hibiscus* spp., (copper), (c) curry leaf, drumstick leaf, or any other leafy greens (iron), (d) Calotropis (boron), (e) all types of flowers (molybdenum), (f) *Abutilon indicum* (calcium), (g) gingelly or mustard plants (sulphur), (h) ladies finger plant (iodine), (i) *Lantana camera*, Casurina, or bamboo (silica), (j) neyveli kattamaaakku (Ipomea) (mercury), (k) glyricidia (nitrogen), (i) thulasi, nochi, neem, aloe vera (to build resistance to fungal, bacterial, and powdery mildew diseases).

Preparation

- Collect 5 kg leaves and plants from the above list. Choose any combination depending on micronutrient deficiency of the crops.
- Chop into small pieces and crush.
- Add 250 gms jaggery in ten litres of water.
- Add 250-300 ml ET.
- Set aside the mixture for 7-10 days for fermentation. This provides ten litres solution.

Usage

Use within 90 days. Spray: 5-10% solution. Irrigation: 10-20 litres per acre.

Benefits

Using EMFPE with other growth promoters as prophylactic measures solves all problems for any crop: (a) rectifies micronutrient deficiency, (b) acts as pest repellent, (c) prevents pests from feeding and (d) induces disease resistance.

Table 4: Average nutrient content of Organic Manures

Manure	Percentage content		
	Nitrogen (N)	Phosphoric acid (P_2O_5)	Potash (K_2O)
Non edible oil cakes			
Castor cake	4.3	1.8	1.3
Cotton cake	3.9	1.8	1.6
Filter-press cake	1.0-1.5	4.0-5.0	2.0-7.0
Karanj cake	3.9	0.9	1.2
Mahua cake	2.5	0.8	1.8
Neem cake	5.2	1.0	1.4
Safflower cake	4.9	1.4	1.2
Edible oil cakes			
Coconut cake	3.0	1.9	1.8
Groundnut cake	7.3	1.5	1.3
Niger cake	4.7	1.8	1.3
Rape seed cake	5.2	1.8	1.2
Sesame cake	6.2	2.0	1.2
Crop Residues			
Banana, dry	0.61	0.12	1.00
Cotton	0.44	0.10	0.66
Groundnut husks	1.6-1.8	0.3-0.5	1.1-1.7
Maize	0.42	1.57	1.65
Paddy	0.36	0.08	0.71

Contd.

Manure	Percentage content		
	Nitrogen (N)	Phosphoric acid (P_2O_5)	Potash (K_2O)
Pigeon pea	1.10	0.58	1.28
Rice hulls	0.3-0.5	0.2-0.5	0.3-0.5
Sugarcane trash	0.35	0.10	0.60
Tobacco	1.12	0.84	0.80
Tobacco dust	1.10	0.31	0.93
Wheat	0.53	0.10	1.10
Tree leaves dry			
Erukku leaves (*Calotropis gigantean*)	0.35	0.12	0.36
Wild Guava (*Careya arborea*)	1.67	0.40	2.20
Tanners cassia (*Cassia auriculata*)	0.98	0.12	0.67
Dog Teak (*Dillenia pentagyna*)	1.34	0.50	3.20
Butter Tree (*Madhuca indica*)	1.66	0.50	2.00
Pungam (*Pongamia pinnata*)	3.69	2.41	2.42
Indian Kino Tree (*Pterocarpus marsupium*)	1.97	0.40	2.90
Black- or *chebulic myrobalan* (*Terminalia chebula*)	1.46	0.35	1.35
Kindal Tree (*Terminalia paniculata*)	1.70	0.40	1.60
(*Terminalia tomentosa*)	1.39	0.40	1.80
Subabul (*Leucaena leucocephala*)	3.50	0.48	0.81
(*Xylia dolabriformis*)	1.37	0.30	1.61
Neem (*Azadirachta Indica*)	2.8	0.3	0.4
(*Glyricidia maculeata*)	2.9	0.5	2.8
Sesbania (*Sesbania speciosa*)	2.7	0.5	2.2
Vadanarayanan (*Delonix elata*)	3.51	0.31	0.43
Gulmohur (*Delonix regia*)	2.76	0.46	0.50
Peltophorum (*Peltophorum ferrugenium*)	2.63	0.37	0.50
Compost & farm wastes			
Cattle dung, fresh	0.4-0.5	0.3-0.4	0.3-0.4
Biogas slurry	1.43	1.21	1.01
Cattle urine	0.9-1.2	trace	0.5-1.0
Coirpith compost	1.06	0.06	1.2
Compost	0.5	0.4	0.8
Farmyard manure, dry	0.4-1.5	0.3-0.9	0.3-1.9
Farm litter compost	0.50	0.15	0.50
FYM	1.0	0.5	1.0
Rural compost, dry	0.5-1.0	0.4-0.8	0.8-1.2
Urban compost, dry	0.7-2.0	0.9-3.0	1.0-2.0
Water hyacinth compost	2.00	1.00	1.40
Vermicompost	1.2 - 1.8	0.4 - 0.8	1.0 - 1.8
Vermiwash (ppm)	200 - 500	70 - 400	400 - 1000
Animal wastes			
Animal refuse	0.3-0.4	0.1-0.2	0.1-0.3
Ash, coal	0.73	0.45	0.53
Ash, household	0.5-1.9	1.6-4.2	2.3-12.0
Blood meal	10-12	1.2	1.0

Contd.

Manure	Percentage content		
	Nitrogen (N)	Phosphoric acid (P_2O_5)	Potash (K_2O)
Bone meal	3.5	21.0	-
Fish meal	4-10	3.9	0.3-1.5
Horse dung, fresh	0.5 -0.5	0.4-0.6	0.3-1.0
Horse urine	1.2-1.5	trace	1.3-1.5
Human urine	0.6-1.0	0.1-0.2	0.2-0.3
Poultry manure, fresh	1.0-1.8	1.4-1.8	0.8-0.9
Raw bone meal	3-4	20-25	-
Sheep urine	1.5-1.7	trace	1.8-2.0
Pig dung	0.60	0.50	0.40
Pig urine	1.10	0.10	0.45
Steamed bone meal	4.7	25-30	-
Others			
Ash, wood	0.1-0.2	0.8-5.9	1.5-36.0
Press mud	1-1.5	4-5	2-7
Sewage sludge, activate dry	4.0-7.0	2.1-4.2	0.5-0.7
Sewage sludge, dry	2.0-3.5	1.0-5.0	0.2-0.5
Green manures fresh			
Dhaincha (*Sesbania aculeate*)	3.3	0.7	1.3
Sunhemp (*Crotalaria juncea*)	2.6	0.6	2.0
Wild indigo (*Tephrosia purpurea*)	2.4	0.3	0.8
Kolinji (*Tephrosia purpurea*)	2.4	0.3	0.8
Pillipesara (*Phaseolus trilobus*)	2.8	0.5	1.15
Manila agathi (*Sesbania rostrata*)	3.30	0.60	1.20
Weeds			
Parthenium (*Parthenium hystorophorus*)	2.68	0.68	1.45
Water hyacinth (*Eichhornia crassipes*)	3.01	0.90	0.15
Sarannai (*Trianthema portulacastrum*)	2.64	0.43	1.30
Aduthoda (*Aduthoda vesica*)	1.32	0.38	0.15
Ipomea (*Ipomoea cornea*)	2.01	0.33	0.40
Calotrophis (*Calotrophis gigantea*)	2.06	0.54	0.31
Cassia (*Cassia fistula*)	1.60	0.24	1.20

(Hand book of Manures and Fertilizers, 1964)

3

ITK Based Special Inputs / Techniques for Plant Growth

Technical Knowledge (ITK) based special inputs / Techniques for plant growth

Growth of the plant has for long been believed to be due to the minerals absorbed from the soil and the food materials synthesized by the palnt. It is recognized that growth of the plant is very much regulated by certain chemical substances known as growth regulators. These substances are formed in one tissue or organ of the plant and are then transported to other sites where they produce specific effects on growth and development. These materials can be produced organically for improving plant growth.

3.1. Growth Promoting Substances

Are those organic solutions (like Panchagavya, Buttermilk & Coconut milk slurry, Amudhakaraisal) prepared using the plant and animal products, decayed fruit and other agricultural wastes. For example

1. Amudhakaraisal
2. Avootam / Panchagavya
3. Themor Karaisal (Coconut + buttermilk)
4. Arappu buttermilk Karaisal (*Albizzia amara* + Buttermilk)
5. Biodigested gas slurry with fermented fruits
6. Biodigested gas slurry
7. Microbial consortia (Tholluyiri) Anaerobes [Those micro organisms live in the absence of air (O_2)]
8. Activated biofertilizer slurry (Uyir Urakkaraisal)

Activated BGS (Uyir urakkaraisal)

Biogas slurry -75 litres

Water - 75 litres Mix well

Ferrous sulphate (Annabetti salt) – 100 g

or

Cowdung - 50 kg

Water - 100 litres Mix well

Pour this slurry mix in a plastic drum of 200 litres capacity.

Simultaneously in an 35 litres capacity plastic can add 100 gm of yeast, palm sugar or indigenous sugar – 3 kg, castor oil – 250 ml and water – 20 litres. Mixed well and stir at every 15 minutes for 3 hours so that castor oil dissolves in the slurry mixture. Add this to the first container with 200 litres capacity and make up to 200 litre with water and ensure air tight condition inside. Allow it for a week time (7 days).

After that digestion period, use @ 1 litre of this microbial consortia (Tholluyiri) with 4 litres of water and foliar sprayed or mixed with irrigation water. About 200 litres of Tholluyiri is sufficient for an acre. If mixed with Avootum (5-10 litres) (or) Amudha karaisal (30 – 50 litres) (or) Themor karaisal (5-10 litres) (or) Fish aminoacid (3 litres) (or) a mixture of *Pseudomonas* (200 gm), *Trichoderma viride* (-200 gm) and Basilomyces (-200 gm) applied through irrigation water.

Follow all the above possible combinations one after another will in prove the soil and enhance its productivity.

3.1.1. Fish amino acid

This input is mostly used in organic farming in Japan, Korea etc. It is prepared in the following way:

1 kg fish not suitable for consumption along with 1 kg of indigenous sugar kept inside an air tight glass contained with lid. Water should not be added during the process and kept for 21 days for anaerobic fermentation. Filter on 22nd day to get approximately 300 – 400 g honey like extract spray @ 1% (1ml / litre of water) which produces greening effect on crops.

3.1.2. Activated fertilizer slurry Uyir Urakkaraisal (For paddy)

Fish waste - $\frac{1}{2}$ kg

Waste fruits - 2 kg All in cement tank covered with moist gunny bag

Cow dung - 1 kg

Cows urine - 3 litres ↓

Country sugar - kg $\frac{1}{2}$ Stirred twice a day

Azospirillum - 200 gm ↓

Phosphobacteria - 200 gm

Trichoderma viride - 200 gm Use after 10 days by mixing

Water - 15 litre it with irrigated water for paddy crop.

Fermented butter milk - 2 litres

Ref: Experience by farmer Mr. Bharani komal, village kothangudi Nagapattinam District, Ph 94862-78569

3.1.3. Fermented fruit mixture (For vegetable and fruits) (Pazhakkadi)

Ripened Poovan Banana - 25 nos (2 kg) Made as a paste

Country sugar - 5 kg ↓

Cowpea - 2 kg Mix with 5 litres of water

Pumpkin - 2 kg ↓

Rice - 2 kg Add 4 litres of fermented curd

↓

Stirr two times a day for a week

↓

Supernatent liquid used for foliar spray (1 litre of pazhakkadi in 12 litres of water)

3.1.4. Amudhakaraisal

Cowdung - 1 kg Put in a drum and allow for 24 hours

Cow's urine - 1 litres ↓

Palm sugar. - 1/4kg In 10 lit of water, after 24 hours used as 10% foliar spray

It act as a growth promoter pest repellent, antifungal enhances flowering in crops.

3.1.5. Aavootam or panchagavya

It is the organic solution prepared using the cow's products viz., dung, urine, milk, clarified butter/ghee and curd. Sugarcane juice, tender coconut water and banana are also added to the mixture to facilitate fermentation and microbes build up. It is foliar sprayed (as and when the crops looks sicky) @ 3%.

3.1.6. Themore karaisal

Butter milk -5 litres - Leave it for 15 days

Coconut milk (milk extracted from 10 coconuts) mixture fermentation

↓

Soil applied @ 10 litres / acre (Annuals /Vegetables) and also 10% as foliar spray to all crops.

3.1.7. Buttermilk + Aarapputhool karaisal

Fermented Butter milk - 10 litre Fermented for a week, filter and take 2 Usilai marathool (*Albizzia amara*) - 5 kgs kg of Usilai leaf & ground as paste with water to get 5 kg karaisal

Spray on foliar at 10% concentration. (Contains GA3)

3.1.8. Elumichai muttai karaisal/ Lemon egg solution

Keep 10 eggs in a plastic bucket

↓

Pour lemon juice / Extract until the eggs submerged well

↓

Then make air tight this bucket for 10 days

↓

Take the eggs after 10 days and squeeze it well.

↓

Add ¼ kg Jaggery to the lemon juice + squeezed egg and mix it well

↓

Close it well to maintain air tight for 20 days.

3.1.9. Coconut milk + Butter milk solution

Milk of 10 coconuts and 5 litres butter milk

Soak it for 7 days in earthern pot

10% as foliar spraying – 1 litre is 10 litre of water

3.1.10. Arappu more karaisal

Arappu leaves (2 kg) – Grind it with water (3 litres)

Add 5 litres of Butter milk to the arappu ilai extract of 5 lit

Keep it for 7 days

For spraying, 1 litre of this in 10 litres of water.

3.1.11. Gunabajalam

Take 2 or 3 kg of meat waste in an earthen pot

Pour cow's urine until the meat waste submerge well

Cover the pot with cloth for 20 days

Can be used in the ratio of 1:20 with water

3.1.12. Sanjivak

In a 500 litres closed drum

Cow dung - 100-200 kg

Cow urine - 100 lit

Jaggery - 500 g

Water - 300 litres

Ferment the above for 10 days. Dilute 20 times with water and sprinkle in 1 acre. Three applications are needed on before sowing, second after twenty days of sowing and third after 45 days of sowing.

3.1.13. Jeevamrut

In a 100 lit water barrel

Cow dung - 10 kg

Cow urine - 10 lit

Then add

Jaggery - 2 kg Mix well – keep for fermentation – 5 to 7 days

Any pulse flour - 2 kg ↓
(preferences moonadhal)

Shake the solution regularly 3 times a day

Three applications are needed on before sowing, second after twenty days of sowing and third after 45 days of sowing.

Usage & proportion

- 50-200 litres of Jeevamrut is required for one acre land.
- Jeevamrut has to be applied once in 15 days compulsorily during vegetative stage, flowering stage and grain filling stage (@ 250 ml/plant)
- Jeevamrut can be applied to the paddy crop along with irrigation.
- Jeevamrut can be administered to other varieties, either directly to the roots or the rows in the field.

3.1.14. Amrit pani

In a 100 litres water barrel

Cow dung - 10 kg

Honey - 500 g

Then add

Desi cow ghee - 200 g

- Mix at high speed to form a creamy paste
- Dilute with 200 litres water.
- Sprinkle this suspension in one acre over soil or with irrigation water
- After thirty days apply second dose in between the rows of the plant or through irrigation water

3.1.15. Cows urine

- Dilute 1 litres of cow urine with 100 litres of water and use it as foliar spary
- For 1 acre – 200 litres of dilute suspension is sufficient

3.1.16. Vermi wash (Manpuzhu kuliyal neer)

- This Manpuzhu kuliyal neer acts as a growth promoting substance
- Pour water drop by drop into a pot or tank which contains earthworms
- This water will wash the worms and helps to collect the hormones available on the earth worm's body.
- The texture of the water will look like black tea.
- This can be used for foliar spray

3.1.17. Payir valarchi ooki (Plant growth promoters)

In a mud pot take

Cow's urine - 5 litres allow to ferment for 10 days

Neem oil - 5 litres

Nayuruvi - ½ kg

Fermented solution is transferred into another mud pot and small hole is made on the bottom of the pot. This pot is kept in the entrance of the irrigation channel.

3.1.18. Fruits solution

- At evening, collect rhizosphere soil of 10 well grown plants from the root and pour into the plastic container.
- Add 3 kg papaya fruit, 3 kg pumpkin, 2 desi bird egg and ½ kg Indigenous sugar as into the container
- Add water until the container is filled
- Close the lid and air tight a container
- Kept under shade for 20 days
- Daily just open and close the lid
- Use 50ml of this solution in 10 litres of water.

3.1.19. Arisi kanchi thirami

- 1 kg of rice must be boiled well and make kanchi out of this
- Pour this into a mud pot or a plastic container and air tight it well

- Keep this buried under shade for 7 days
- Take out this on 8 th day and add 1 kg naattu sakkarai then again burry for 7 days
- 15th day it will be ready for usage
- Add 20 litres of water into this solution
- Then spray it by using hand sprayer (1 litre of solution for 10 litres of water/tank)

3.1.20. Bhabut amrit pani

- Mix 10 kg. cow dung with 250 gm. desi ghee and 500 gm. honey.
- This material is mixed with 200 litres of water and spread in the field after sowing a crop.

3.1.21. Pitcher khad

- This is a fermented preparation made from 15kg. of cow dung, 15 litres of cow urine, 15 litres of water and 250 gm of jaggery.
- All this is mixed in a container and covered with a cloth or gunny bag. The material is fermented for 4-5 days.
- The fermented mixture is mixed with 200 litres water and sprayed over the crop in one acre area. Two to three sprays are sufficient for short duration crops

3.1.22. Cow urine

- Dilute 1 litre cow urine in 10 litres of water in as plastic drum or earthern pot.
- Spray the dilution
- For 1 acre – 200 litres of dilute suspension is sufficient

3.1.23. Sasyamrutha- liquid fertilizer

Indigenous cow dung – 30-40 kgs

Indigenous cow urine 3-5 litres.

Plants having latex (Milk hedge, Calatropis, Jatropa, Sweet scented oleander) – 3-5 kgs

Plants having strong aroma (Parthenium, Lantana, Eupatorium, Cassia, Datura) – 3-5 kgs

Oil cake (Groundnut, Pongamia, Neem) - 1-2 kgs.

Jaggery – 1 kg

Ash of Agnihotra – 500 gms

Water - 200 litres

A container with a capacity of more that 200 litres.

- Chop all the plant materials into finer pieces.
- Clean the drum well and fill it with the finely chopped plant -materials.
- Add 200 litres of water and all other -ingredients into this drum like indigenous cowdung, and cow urine, oil cake, Agnihotra ash, Jaggery in the mentioned- quantities to the drum.
- Stir well and close the drum with a lid and allow it to ferment for one week. Daily stirring of the solution for aeration should be done.
- After a week the whole mixture must be filtered and sasyamrutha- has to be collected in another drum.
- The filtered leaves and green should be collected in a separate- container and can be used in the fields along with water.
- Filtered sasyamrutha liquid manure is -given through irrigation water @ 200 litres per acre

3.1.24. Fish medicine

- Chop the fish (1 kg) into small pieces, powder the jaggery (1 kg) and mix the two ingredients in a plastic bucket having a lid. Cover the lid and keep it aside for 15-20 days.
- Stir this mixture everyday to avoid foul smell and odour upto 15 to 20 days.
- This mixture gives a pleasant fruity odor after complete fermentation
- Add 6 ml fish medicine to 1 litre water and sprinkle this on vegetable plants once a week.

3.1.25. Egg lime : (Muttai rasam)

- Take 20-25 limes and squeeze the juice into a bucket. Take about 250g of jaggery and mix it well with the lime juice to form a solution.
- Then take about 10-15 chicken or duck eggs and place it in the bucket containing the solution in such a way that all the eggs are well immersed inside the solution.

- Close the bucket with an air tight lid and keep it in the shade for about 10 days. On the 10th day, the eggs along with the shells inside the solution would have become -rubbery, like a rubber ball.
- Thorough Mix – Use your hands to mix the eggs (along with the shell) with the lime and jaggery solution-. After thorough- mixing, add jaggery solution again in equal measure- to the lime jaggery solution. Eg. If 2 litres of solution is prepared, then add 2 litres of -jaggery -solution. Then close the bucket tightly for about 10 days. After the 10th day the formulation can be used as spray for the crops. About 10-15 ml of the -formulation can be diluted in one litre of water and sprayed.
- The concentration varies according to the area to be sprayed. This formulation can be sprayed for any crops such as -paddy, wheat, banana, vegetables, greens and fruit trees. Spraying should be done either in the morning or late evening. It can also be mixed with panchagavya, vermiwash and sprayed. The solution can be stored for about six months. Hence it is advisable to drill small holes in the lid to facilitate escape of gases which emits from the solution.

3.1.26. Herbal plant growth regulator

1. Sea weed (Sargassum) 20 kg
2. Illuppai seednut cake (*Bassia latifolia*) 1 kg
3. Vasambu (*Acorus calamus*) 1 kg
4. Adadhoda (*Justicia adodhoda*) 500 gm
5. Fish – jaggery mixture 20 ml
6. Alcohol 50 ml
7. Sothukathalai (*Aloe vera*) 1 kg
8. Soap solution 100 gm

- Grind all the leaves and mix along with other items
- It has to be applied 250 ml for one acre at twice or thrice.

3.1.27. Fish Jaggery mixture as organic liquid fertilizer

Fresh fish procured from ponds ("Kendai" fish) or from sea water – 25 kg,

Jaggery – 25 kg,

Water – 100 litres

Papaya fruits – 6 number.

- Mixes jaggery and fish together and soak in water for 15 days. On 16th day add papaya fruits after removing all the seeds from the fruits and crush the fruits along with the existing solution.
- After 4 days (totally 20 days of initial soaking) of putting papaya fruits add 50 litres of water and finally filtered.
- This has to be sprayed at 1 liter per acre or 100 ml per 10 litres of water. This has to be sprayed 2 times for paddy, 4 times for sugarcane and 2 times for onion starting from 15 – 20 days after planting at 15 days interval.

3.1.28. Starter solution

- 20 kg of cow dung, 20 litres of cow urine, 3-4 kg of Jaggery are to be mixed in 200 litres of water.
- After 24 hours, the solution is applied to crops by mixing it with irrigation water (1:10).
- This solution could convert the infertile soil into a productive one.
- The solution is applied 2-3 times for 3-4 months old crop and for long duration crops, this applied twice in a month. Along with this solution, water may also be mixed and is sprayed over the crops

3.1.29. Amara leaf extract

- 1 kg of leaf is to be mixed with 5 litres of buttermilk and is kept for 7 days.
- 1 litre of this solution is mixed with 10 litres of water and can be applied to the crops as foliar spray

3.1.30. Fermented buttermilk & coconut milk solution

- 5 litres of fermented buttermilk and 5 litres of coconut milk are mixed and kept in a mud pot or plastic drums for 1 week during which period it has to be stirred often.
- Then the solution is mixed with water @ 1 litre solution in 10 litres of water. This also is applied as foliar spray.

3.1.31. Liquid manures

i. Green manuring plants – *Sunnhemp, Daincha, Sesbania, Erythrina* etc. and other leguminous plants can be used

ii. Tree leaves – Leaves of other local medicinal trees.

iii. Tender stems – *Artemesia* sp., *Eupatorium* sp. and local plants.

iv. Weeds – *Parthenium,* Stinging nettle, and other weeds before flowering.

A mixture of different plants results in good quality liquid manure.

Materials required

i. Cow dung 3kg
ii. Cow urine 3lts
iii. Green materials / Weeds 3kg
iv. 20lt. water bucket

Method

The basic principle is to allow the materials to ferment over a certain period.

1. A plastic water bucket of capacity of 20 litres or more is taken.
2. The plant materials for liquid manures are chosen based on the availability.
3. The plant materials are shredded into small pieces and 3 kg of the plant material is put into the bucket.
4. Cattle dung 3 kg and cattle urine 3 litres are added to the bucket.
5. The bucket is filled with good quality of water and covered with a gunny sack.
6. The contents are stirred for minute's everyday for 7 days. To hasten the fermentation the contents are stirred once a week for the next three weeks.
7. The biomass will get fully fermented and the liquid manure would ready in a month for use.

Use

The liquid manure has to be diluted with water prior to application. The standard dilution is one part liquid manure in ten parts water and sprayed on the foliage.

3.1.32. Dung brew

Materials required

i. Cattle dung - 5 kg
ii. Cattle urine – 15 litres
iii. Plastic bucket of 20 litres capacity

Method

1. Add 5 kg of cattle dung and 15 litres of cattle urine in a bucket and mix it well.
2. Cover the bucket with a gunny sack or any material that provides sufficient aeration.
3. Stir the materials every alternate day for 15 days.
4. The dung brew is ready for spray after 15 days and can be stored for 2 months.

Use

One litre of dung brew is diluted in 10 litres of water and mixed thoroughly. The solution is sieved with a muslin cloth and then put into the knapsack sprayer and sprayed on the foliage of crops. Crops respond to this spray very well.

3.1.33. Compost tea

Materials required

i. well matured compost
ii. water

Method

1. Good quality compost has to be prepared by adopting heap, aerobic or vermicomposting methods.
2. One part of compost is immersed in 10 parts of water in a bucket or barrel depending on the quantity required.
3. The contents are stirred 5 minutes every day for 5-7 days.
4. The compost tea is ready to be used as a foliar spray after 5-7 days.
5. Before the application the solution is filtered and sprayed using a knapsack sprayer.

Other Method of Preparing Compost Tea

Green leaves of legumes: 5 to 7.5 kg

Cow dung or droppings of goat, sheep or poultry: 5 to 7.5 kg.

Drum: One

Gunny bag: One

Water: 200 lit

Molasses (gur): 200-250 g

- The cow dung or droppings of goat, sheep or poultry are mixed in the green leaves of legumes.
- These materials should be kept in the gunny bag which is kept in the drum having 200 lit of water.
- A brick is placed on the gunny bag to keep it on the bottom of drum.
- The water of the drum should be stirred for a period of five minutes every day and this is to be continued up to 20 days.
- The manures become ready within three weeks.
- When the liquid manure is fully prepared, no bad odour should be smelt out from the drum and colour of water of the drum should be as like of tea.
- Now, gunny bag is taken out from the drum. 200-250 gms of molasses (gur) is added to the mixture to increase the quality of the liquid manure.
- Equal amount of water should be added with the mixture before use.
- The quality of manure remains good for a period of 4-6 weeks, if it is preserved in cool and shady place.

3.1.34. Stem paste

Materials required

i. Cattle dung – 50 kg
ii. Sieved sand – 10 kg
iii. Soil from the field – 10 kg
iv. Cattle urine – 25 litres

Method

1. Mix cattle dung, sand and soil with cattle urine to form a thick paste. Add sufficient cattle urine till the consistency is reached.
2. The stem of the tree is cleaned with a gunny cloth or brush with sobristles.
3. The paste is plastered over the stem and branches of the tree.
4. The paste can be used as healing the cut end of the tree after pruning.

3.1.35. Neem seed and cow dung spray

- Grind 5 kg neem seed and dilute with water. Filter out neem seed. Mix 25 kg fresh cow dung with neem seed filtrate.

- The resulting product is sufficient for spraying 1 hectare of land, provided the required quantity of water is added. The neem seed and cow dung spray prevents excessive flower drop

3.2. Panchagavya – An Effective on Farm Input

Fig. 8: Products used in making of Panchagavya

Panchagavya is a bio promoter with a combination of five products obtained from the cow, which includes cow dung, cow's urine, milk, curd and ghee. All the five products are individually called gavya and collectively termed as panchagavya. Panchagavya has the potential to play the role of promoting growth and providing immunity in plant system. The Panchagavya acts as growth promoter (75%) and immunity booster (25%) and exactly fills the missing link to sustain the organic farming without any yield loss. It plays an important role in providing resistance to diseases, pests and in increasing the overall yield due to the immunostimulant activity of the cow dung and it can be prepared by the farmers themselves with the materials available on the farm. It was prepared by a slightly modified method as mentioned in Vrksayurveda and it was standardized by innovative farmers under the expert guidance of Dr. K. Natarajan, President, Rural Community Action centre, Kodumudi, Erode Dt, Tamil Nadu, India

3.2.1. Panchagavya

- Cow dung slurry 4 Kg
- Fresh cow dung 1 Kg
- Cow Urine 3 lit
- Cow milk 2 lit
- Curd 2 lit
- Cow deshi ghee 1 kg

Method of Preparation

Mix all the ingredients thoroughly and ferment for 15 days with twice stirring per day. Dilute 3 litres of Panchagavya in 100 litres water and spray over soil. 20 lit panchagavya is needed per acre for soil application along with irrigation water. Panchagavya can also be used for seed treatment. Soak seeds for 20 minutes before sowing.

Table 5. Panchagavya spray schedule for various crops

Crops		Time schedule
Rice	:	10,15,30 and 50th days after transplanting
Sunflower	:	30,45 and 60 days after sowing
Black gram	:	Rainfed : 1st flowering and 15 days after flowering
	:	Irrigated : 15, 25 and 40 days after sowing
Green gram	:	15,25,30,40 and 50 days after sowing
Castor	:	30 and 45 days after sowing
Groundnut	:	25 and 30th days after sowing
Bhendi	:	30,45,60 and 75 days after sowing
Moringa	:	Before flowering and during pod formation
Tomato	:	Nursery and 40 days after transplanting, seed treatment with 1% for 12 hrs
Onion	:	30,45 & 40 days after transplanting
Rose	:	At the time of pruning & budding
Jasmine	:	Bud initiation and setting
Vanilla		Dipping setts before planting
Fruit trees		
First time	:	One month before flowering
Second time	:	15 days after flowering
Third time	:	When the fruit is in peas size
Fourth time	:	After harvest once
Leafy crops		
Greens, curry leaf, tea, coffee,	:	2 % weekly once

3.2.2. Modified panchagavya

Panchagavya, by name constitutes five products evolving from cow. Rural community Action Centre (NGO) has modified this Panchagavya by adding a few more ingredients and this modified version has a lot of beneficial effects on a variety of crops and livestock. The present form of Panchagavya standardized by TNAU is a single organic input which can act as a potentiator. It is essentially a product containing the following:

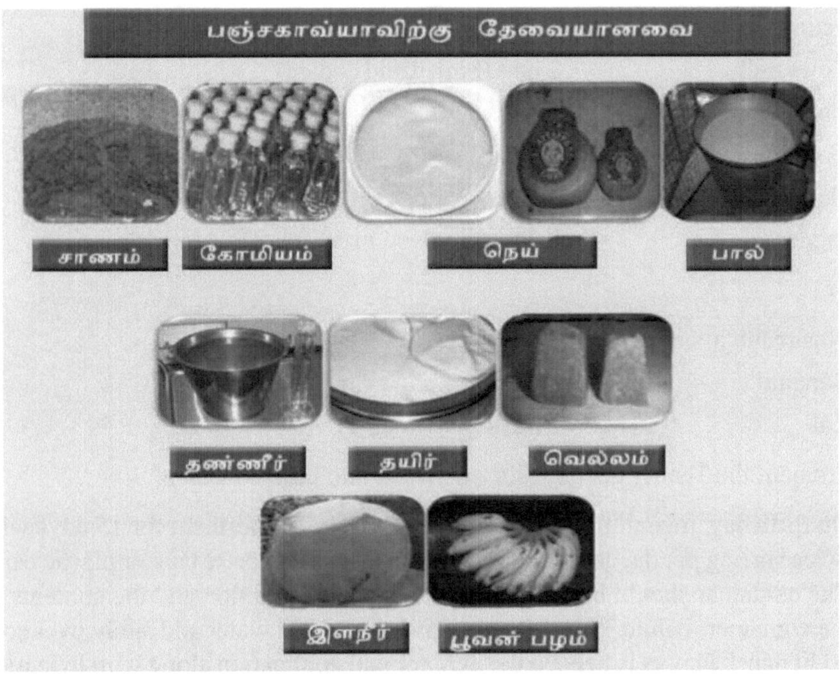

Fig. 9: Products used in making of panchagavya

Table 6: Ingredients of Modified panchagavya

Sl.No.	Ingredients	RCAC	TNAU Standard
1.	Cow dung slurry (from Gobar gas plant)	4 Kg	-
2.	Fresh cow dung	1 Kg	5.0 kg
3.	Cow's urine	3 litre	3.0 lit
4.	Cow's milk	2 litre	2.0 lit
5.	Cow's curd	2 litre	2.0 lit
6.	Cow's ghee	1 Kg	1.0 kg
7.	Sugarcane juice	3 litre	3.0 lit
8.	Tender coconut water	3 litre	3.0 lit
9.	Banana (ripe)	12 Nos	1.0 kg
10.	Toddy (if available)	2 litre	-

The cost of preparation may be approximately Rs.50.00 per litre

3.2.2.1. Method of preparation

First day

Fresh cow dung - 5 kg

Cow ghee - ½ kg

Mix well these two and keep under shade for three days. Stir it well twice a day

Fourth day

Cows milk - 2 litre (boiled and cooled)

Curd - 2 litres

Cow's urine - 3 litres

Tender coconut - 3 litres

Jaggery - ½ kg (dissolved in 3 litres of water)

or

Sugarcane juice - 3 litres

Banana - 12

Kal - 2 litres (if available)

Ferment the Tender coconut for one week and take 2 litres

On forth day, mix all the ingredients thoroughly and ferment for 15 days with twice stirring per day in a wide mouthed mud pot, concrete tank or plastic cans. The container should be kept open under shade condition and the contents of the container. Dilute 3 lit of Panchagavya in 100 lit water and spray over soil. 20 lit panchagavya is needed per acre for soil application along with irrigation water. Panchagavya can also be used for seed treatment. Soak seeds for 20 minutes before sowing.

The products of local breed of cow is said to have more potency than exotic breeds and care should be taken not to mix buffalo products.

3.2.2.2. Methods of application

The panchagavya can be applied in different ways like foliar spray through irrigation and as seed / seedling treatment. A 3% solution has been found to be the most effective in increasing the production and consumption of organic produces.

As foliar spray

Investigations prove that 3% solution was found to be the most effective compared to the higher and lower concentration. Three litres of panchagavya to every 100 litres of water is the ideal recommended dosage for all the crops. The power sprayer of 10 litres capacity may need 300 ml/tank. When sprayed with power sprayer, sediments are to be filtered and when sprayed with hand operated sprayers, the nozzle with higher pore size has to be selected.

Through Flow / irrigation system

In the case of flow or irrigation systems the solution of panchagavya can be mixed with irrigation water at 50 litres per hectare either through drip or flow irrigation water at 50 litres per hectare either through drip or flow irrigation.

As seed / seedling treatment

3% solution of panchagavya can be used to treat the seeds and seedlings by soaking and dipping before planting. Soaking for 20 minutes is sufficient. Rhizomes of turmeric, ginger and sets of sugarcane can be soaked for 30 minutes before planting. In the case of seed storage 3% panchagavya solution can be used to dip the seeds before drying and storing them.

For seed storage

Panchagavya solution of 1 % can be used to dip the seeds before drying and storing them.

Table 7: Frequency of spray

Pre flower phase (20 days after planting)	Once in 15 days, two sprays depending upon duration of the crops.
Flowering and pod setting stage	Once in 10 days, two sprays
Fruit / pod maturation stage	Once during pod maturation

3.3. Enriched Panchagavya (for one acre-quantity required is 200 kg)

i) Dung (cow/buffalo/ox) : 40 kg
ii) Urine (cow/buffalo/ox) : 40 kg
iii) Butter/mustard oil : ½ litres
iv) Milk : 2 ½ litres
v) Lassi/diluted curd : 8 Ltrs.
vi) Gram flour and Methi flour : One kg each
vii) Old Jaggery (gur) : 2 kg
viii) Rotten Bananas : 1 kg
ix) Mustard (de-oiled cake) : 2 kg
x) Banyan tree & Banana soil : 2 kg each

Application procedure/Implementation:

- Put all the ingredients in drum.
- Stir it thirty times clock-wise and thirty times anti-clock wise (twice daily).

- Keep the drum closed.
- Enriched panchagavya would become ready on 7th day.
- Mix 100 litres of water and stir it properly.
- Mix the mixture with vermi-compost or ash or fertile soil and spread it in the field before sowing.

During irrigation, make a hole in the drum (4" above the base) and set it so that dripping from drum starts and whole field will be nourished. By its application at the time of sowing and during irrigations, soil can maintain remain fertile.

3.4. Dasagavya

Dasagavya is an organic preparation made from ten products in the form of panchagavya and certain plant extracts. "Gavya" is the term given to cow's products comprising of cow dung, cow urine, cow's milk, curd and ghee, which have miraculous effects on plant growth when suitably mixed. These are commonly available weed plants in the district, found abundantly along roadsides and in wastelands. The plants recommended for the tropical areas are neem (*Azadirachta indica*), erukam (*Calotrophis*), kolingi (*Tephrosia purpurea*), notchi (*Vitex negundo*), umathai (*Datura metal*), katamanaku

Fig. 10: Leaves used in making of dasagavya

(*Jatropha curcas*), adathoda (*Adathoda vasica*) and pungam (*Pongamia pinnata*). Since management of these can be made best use of in agriculture, as effective agents against certain pests and diseases.

Method of Preparation

The plant extracts are prepared by separately soaking the foliage in cow urine in 1:1 ratio (1 kg chopped leaves in 1 litre cow urine) for ten days. The filtered extracts of all the plants are then added @ 1 litre each to 5 litre of the panchagavya solution. The mixture is kept for 25 days and stirred well, meanwhile, to ensure thorough mixing of panchagavya and the plant extracts.

Mode of Use

The dasagavya solution is filtered to avoid clogging of sprayer nozzles and is recommended as foliar spray at 3 % concentration. Soaking of seeds or dipping the roots of seedlings in 3 % solution of dasagavya for 20 minutes before planting enhances seed germination and root development.

Periodicity

Weekly sprays during crop growth for all vegetables and plantation crops.

3.5. EM – Technology in Organic Farming

EM or effective microorganisms are a consortium culture of different effective microbes, commonly occurring in nature. Most important among them are: N_2-fixers, P-solubilisers, photosynthetic microorganism's, lactic acid bacteria, yeast, plant growth promoting rhizobacteria and various fungi and actinomycetes. In this consortium, each microorganism has its own beneficial role in nutrient recycling plant protection and soil health and fertility enrichment.

How to use EM

- Procurement of primary EM available in market.
- Preparation of secondary EM - to be carried by the farmer by using the primary EM.

3.5.1. EM-1 formulation

This is used for seed treatment, soil enrichment and spray in field after emergence of seedlings.

- Dissolve 5 kg jaggery (chemical free) in 100 lit of water.
- Add 5 lit of EM.

- Mix thoroughly and pour into a plastic carboy. Seal the carboy and allow to ferment for 7 days.
- Dilute this solution in a ratio of 1:100 and spray over soil or crop residue. For seed treatment soak the seeds in this dilute solution.

3.5.2. EM-5 formulation for control of insects and pests

- Dissolve 100g jaggery (chemical free) in 600 ml of water
- Add 100 ml each of natural vinegar, wine or brandy and EM
- Mix thoroughly and transfer the contents in a plastic bottle or carboy. Seal the container and allow fermenting for 5-10 days under shade
- To increase the potency few cloves of garlic and chilly paste can also be added to this suspension before sealing the container
- Release the gas daily
- Within 10 days the EM solution will be ready for use. This can be stored upto 3 months at normal room temperature in a cool and dry palce
- Dilute the contents in a ratio of 1:1000 and apply as foliar spray.

3.5.3. Fermented plant extracts (FPE)

- Grind 2.3 kg of fresh green weeds to a coarse paste
- Dilute with 14 litres of water
- Add 420 ml of EM
- Transfer the contents to a plastic drum and with the help of a thick plastic sheet cover the drum and tie with a rope
- The drum should be filled up to the top, leaving very little space for air.
- Fermentation and gas formation process will strat slowly
- Mix the contents at repeated intervals
- Finished FPE having a pH of 3.5 with pleasing smell will be ready in 5-10 days time
- Filter the solution through a cloth and collect the filtrate
- For spraying on soil dilute the FPE in a ratio of 1:1000 with fresh water
- For spraying on crops dilute the FPE in a ratio of 1:500 with fresh water
- Sparing should be done after germination of seeds in early morning hours once or twice a week.

3.5.4. Application of EM formulations

At the time of land preparation – Dilute 5-10 lit of simple EM solution in 50-100 lit of water and sprinkle/spray over 0.1 ha of land, when soil is wet a day before sowing.

For seed treatment – Soak seeds for 5-6 hrs in 1 : 100 fold diluted EM solution and sow immediately.

As foliar/ soil spray – After seedling emergence, 1 : 1000 diluted EM solution or FPE should be sprayed at the rate of 500 lit per ha, 4-5 times at an interval of 7-10 days. In fast growing crops such as vegetables, spraying should be done twice a week. In transplanted crops 1 : 500 diluted FPE can be sprayed after 5 days of transplanting @ 750-1000 lit per ha.

For soil enrichment – For every 0.1 ha mix 100-150 kg Bokashi with crop residue and mix with soil just before sowing. Simple EM solution @ 5-10 lit can also be used as spray over this residue-Bokashi mix. Spraying the soil with 5-10 lit of FPE mixed in 500-1000 lit of water per ha also add to the fertility of the soil

3.5.5. EM- Bokashi

It is a type of compost prepared by fermentation of waste organic matter with the help of EM which is mainly used for improving the fertility status of the soil and for enhancing the degradation of crop residue.

- Collect sufficient quantity of organic matter (such as rice bran, fish meal, animal waste etc.,) equivalent to 150 lit drum volume
- Mix 150 g of jaggery and 50 ml of EM in 15 lit of water
- Mix this solution with organic waste thoroughly in such a way that entire contents get uniformly moistened
- Transfer the contents in plastic bag and seal the bag
- To ensure the anaerobic conditions put this bag into another polythene bag and seal
- Allow the contents to ferment for 3-4days in a cool shade place
- Bokashi will ready afire 4 days and stored up to 6 months in a air tight plastic bags
- For 0.1 ha mix 100-150 kg bokashi with sufficient quantity of finely chopped crop residue and mix with soil just before sowing. Spraying of 5-10 lit of 1:500 dilute simple EM solutions over this mixture can further boost the degradation process.

3.6. Gloria Biosol an Effective Homa Biofertilizer

Gloria Biosol is a bio-fertilizer which can be produced in Homa atmosphere. Biosol liquid can be used for foliar application to nourish plants and soil. Agnihotra Ash has a significant positive effect on all the materials used and makes the Biosol rich in macronutrients.

Materials required to make the Biosol are:

- Fresh cow dung
- Vermicompost
- Cow urine
- Agnihotra ash
- Water

Materials are mixed in a large tank (200, 500 or 1000 litre). One copper disc is placed in the tank. The tank is then sealed and kept for 30 days. After digestion is complete, the slurry can be removed. Biosol is used diluted with Agnihotra ash water solution in the ratio of 1:10. For one hectare of agricultural area, 200 litres of Biosol solution are required. Biosol solution can be sprayed on any type of crop at an interval of fifteen days. The application of the Biosol solution should be made before sunrise or after sunset. If we preserve Biosol liquid in air tight cans it will last longer, say about six months. Left over solid Biosol which is having maximum macro nutrients should be mixed with any type of organic manure at a ratio of 1:5.

3.7. Farmer's Effective Microorganism

- Pumpkin 3 kg
- Banana 1 kg
- Papaya 3 kg
- Jaggery 3 kg
- Egg 5 numbers
- Non-chlorinated water 10 litres

Preparation

Cut the vegetables into small pieces. Transfer these pieces into a clean plastic container. Mix jaggery in 10 litres of nonchlorinated water till it dissolves well and add the eggs to it. Mix all the contents. Close the container with airtight lid. Open lid after 10 days to release the air. Mix well again. Keep the set up closed for 45 days.

Collection of FEM

After 45 days there will be three layers in the container. The upper thin layer is in white color, which indicates successive fermentation. The middle layer will be pure brown colored liquid and the lower layer will be the semi solid formed by the dissolved vegetables. Open the tap fixed at the bottom of the container to collect the semi solid portion in one container. The upper and middle portions are collected in another container.

Applications of FEM

- 2% to 5% concentration in water can be used as foliar spray on any crop.
- It also acts as a weedicide if it is used with goat's urine.
- It may act as a pesticide if used after being fermented with neem, papaya & nerium leaves.

3.8. Aattottam

- Goat dung (soaked in water overnight)
- Goat urine (fresh)
- Green gram after grinding (soaked in water overnight)
- Goat milk
- Curd from goat milk
- Banana (ripened)
- Tender coconut water
- Fermented coconut water
- Sugarcane juice.

Preparation

Aattottam is prepared in a wide mouthed clean plastic container. Metal containers should be avoided. Fresh goat dung (soaked in water overnight) and goats urine are to be added into the container and mixed thoroughly. The other ingredients are then added one by one to the container and again mixed thoroughly. This is to be stirred twice a day in clockwise and anticlockwise directions. Aattottam stock solution is ready for use after 14 days. Keep in shade and cover with a thin cotton cloth to prevent insects from laying eggs and formation of maggots on the surface of the solution. The solution can be stored for a period of six months, provided it is stirred twice every day. If the solution is too thick to stir over time, water or tender coconut water can be added in required quantities.

Application

Apply 2% in water as spray during dawn or dusk on any crop. The solution should be filtered properly before pouring into the hand sprayer. For best results, spray at the time of branching, before flowering and fruit setting.

3.9. Other ITKS

3.9.1. Groundnut cake extract spray

- 25 kg of groundnut cake powder is mixed with 150 litres of water and keep it for overnight in a closed vessel.
- Next day, only the cleared filtrate is collected and use as a spray solution.
- About 1 lit from this solution is taken and mixed with 9 litres of water to fill a tank.
- This 10 lit is enough for spraying l acre of tomato field.
- Spray this once in 15 days after tomato transplantation for higher production.

3.9.2. Asafoetida spray in Moringa

- Asafoetida powder (50g) with 1 lit of water.
- This solution is prepared freshly and then poured to individual tree at the root zone region.
- Reduces the dropping of flowers and enhances the fruit formation.

3.9.3. Salt application to coconut trees

- To control flower shredding in coconut applied common salt (Nacl) (1/2 kg per tree) on the apical portion of the flower buds.

3.9.4. Neerium leaf extract spray (Arali saru thelithal)

- 1 kg of nerium seeds was powdered well and mix with 5-litres of curd and 5 litres of milk in an earthen pot and left for 3 days for fermentation.
- After the fermentation period, the solution is filtered using cotton cloth and 1 litre of the filtrate solution is mixed with 9 litre of water and used as a spray solution for 1 acre of the cropped field.

3.9.5. Vermiwash

Ingredients

- Plastic drum
- Earthworms
- Small pebbles or brick pieces
- Sand particles
- Cow dung
- Fertile soil
- Waste grass.

Procedure

Take a plastic drum with 250 kg. capacity and make a small hole at the bottom of the container. Fill the container at the bottom with small stones/pebbles to a height of 1 ½ inches. Add water to test if the water flows out through the hole. Above the stones, fill the container with sand (1 inch) and waste materials like manure, fertile soil, vegetable peels (30-40 cms.) as the food for earthworms. Add one kg of fresh cowdung into the drum and release 200-300 earthworms in the drum. Take a five litre capacity bucket and make a small hole at the bottom. Fill the bucket with water and place it on the drum so that water drips into the drum. Fill this bucket with water as it turns empty. Collect the vermiwash in a bottle that trickles slowly from the bottom of the drum.

Uses

- Mix one litre of vermiwash with five litre of water and use it for crops. It can be sprayed directly to the crop or applied along with irrigation.
- The vegetable plants become very healthy and strong. Controls deflowering and increases the yield of a crop.
- Vermiwash mixed with cow's urine and water in the ratio of 1:1:6 is sprayed to crops for effective control of caterpillars and aphids in vegetable crops.

3.9.6. Activated effective microorganism (AEM)

Ingredients

Standard Ingredients to prepare one litre AEM solution: 850 ml Water, 100 grams Jaggery, 50 ml EM Solution.

Procedure to prepare ten litres AEM

Take a ten litre capacity container having a lid. Take 8.5 litres of water, 1 kg jaggery (dissolved Jaggery) and 500 ml EM in the container. Stir the solution well with a stick and close the lid. Store the mixture for a week in the container. Stir the mixture well twice a day. The solution is ready for use when pH is between 3.5-4.0.

Uses

Strengthens the organic manure. Can be used to prepare fermented plant extract. Disease causing bacteria and unnecessary microorganism can be controlled.

3.9.7. Egg amino acid (EAA)

Ingredients:

- Egg – 10 numbers; Lemons – 10-15; Jaggery – 250 g

Procedure

Place eggs in a jar and pour lemon juice in it until eggs are completely immersed. Keep it for ten days. Smash the eggs and prepare the solution. Add equal quantity of thick jaggery syrup to it and set aside for ten days. The solution will then be ready for spraying. It boosts plant growth.

3.9.8. Pitcher khad

This is a fermented preparation made from 15kg. cow dung, 15 litres cow urine, 15 litres water and 250 g jaggery. All this is mixed in a container and covered with a cloth or gunny bag. The material is fermented for 4-5 days. The fermented mixture is mixed with 200 litres water and sprayed over the crop in one acre area. Two to three sprays are sufficient for short duration crops.

3.9.9. Archaebacterial solution (Plant Growth Promoting Rhizobacteria -PGPR)

Ingredients

20 kg dung, 200 litres water, 3 kg jaggery, 100 gms kadukkai powder, 10 gms adhimadhuram.

- Mix dung, jaggery and water well in a container.
- Add kadukkai powder to it and mix well.
- Boil the adhimadhuram powder in 250 ml water and let it cool.
- Add the cooled adhimadhuram solution to the above solution.

- Fill the rest of the container with water so that there will be no air left inside and close it air-tight.
- Methane will be formed inside the container. Let the air out of it once in a while by slightly unscrewing the cap for a moment.
- The solutions will be ready in ten days.
- The solution will be light brown in colour and having a pleasant smell.

Usage

Mix a litre of it with ten litres of water to spray. Or, for use in irrigation, mix 200-300 litres per acre of this solution with one of the following solutions: (a) 30 to 50 litres concentrated Amudham Solution, (b) 5-20 litres Panchagavya, (c) 5-10 litres coconut-buttermilk /arappu-buttermilk / soap nut-buttermilk solution, (d) 3 litres fish extract (given in 4.7). With this mix the following microorganisms for disease control depending on the prevailing diseases.

- To control deseases which damage the rhizosphere/ rhizomes and to control fussarium wilt, we have to use beneficial fungi like *Pseudomonas fluorescence*, *Trichoderma viride*, *Trichoderma harzianum* and *Bacillus subtilis*;
- To control root knot nematodes use *Paecilomyces lilacinus*;
- To control root grubs use *Beauveria brongniartii* and *Metarrhizium anisopliae*.

4

Techniques for Weed Management in Organic Farming

Weeds can be considered a significant problem because they tend to decrease crop yields by increasing competition for water, sunlight and nutrients while serving as host plants for pests and diseases. Since the invention of herbicides, farmers have used these chemicals to eradicate weeds from their fields. Using herbicides not only increased crop yields but also reduced the labor required to remove weeds. Today, some farmers have a renewed interest in organic methods of managing weeds since the widespread use of agro-chemicals has resulted in purported environment and health problems. It has also been found that in some cases herbicides use can cause some weed species to dominate fields because the weeds develop resistance to herbicides. In addition, some herbicides are capable of destroying weeds that are harmless to crops, resulting in a potential decrease in biodiversity on farmers. It is important to understand that under an organic system of weed control, weeds could never be eliminated but only managed. In organic agriculture weeds are the resources as they are mostly native to the field/areas of existence (Children's of the soil)

Weed management methods

Classified as

I. Preventive
II. Breeding
III. Agronomical (cultural)
IV. Eradicative
V. Physical
VI. Mechanical
VII. Thermal
VIII. Biological

4.1. Preventive Methods: Preventing or denying establishment or entry of a new weed species into an area

- Avoid using crop that are infested with weed seeds for sowing.
- Avoid feeding screenings and other material containing weed seeds to the farm animals.
- Avoid adding weeds to the manure pits.
- Clean the farm machinery thoroughly before moving it from one field to another. This is particularly important for seed drills.
- Avoid the use of gravel sand and soil from weed-infested.
- Inspect nursery stock for the presence of weed seedlings, tubers, rhizomes, etc.
- Keep irrigation channels, fence-lines, and un-cropped areas clean.
- Use vigilance. Inspect your farm frequently for any strange looking weed seedlings. Destroy such patches of a new weed by digging deep and burning the weed along with its roots. Sterilize the spot with suitable chemical.
- Quarantine regulations are available in almost all countries to deny the entry of weed seeds and other propagules into a country through airports and shipyards.

4.2. Breeding Competitive Crops

There is an inverse relationship between the crop growth habit and weed growth. So varieties should

- emerge faster
- grow taller
- have wide canopy to prevent light from reaching the weeds
- produce higher yield

4.3. Cultural or Agronomical Methods

4.3.1. Field preparation

- The field has to be kept weed free. Flowering of weeds should not be allowed. This helps in prevention of build up of weed seed population in the fields.
- Irrigation channels are the important sources of spreading weed seeds. It is essential, therefore, to keep irrigation channels clean.

- Deep ploughing in summer, exposes underground parts like rhizomes and tubers of perennial and abnoxious weeds to scorching summer sun and kills them.
- Conventional tillage which includes 2 to 3 ploughings followed by harrowing decreases the weed problem.
- Running blade harrows cuts weeds and kills them.
- In lowland rice, puddling operation incorporates all the weeds in the soil which would decompose in course of time.

4.3.2. Pasturing

In perennial crops like coffee, mangoes, avocadoes or cocoa, the use of sheep and goatsto reduce rampant weed growth is becoming common. In case of cattle, broadleaf weeds tend topredominate due to the cattle preference for grasses. Therefore, it is necessary to rotate with sheepand goats which prefer broadleaves to overcome this selective grazing.

- Prevent dissemination of weeds by eliminating them before seed dispersal.
- Avoiding the introduction of weed seeds into the fields through tools or animals; and by using only weed free seed material.

4.3.3. Sowing time and density

Optimum growing conditions enhance the optimum crop plant development and their ability to compete against weeds. Proper crop spacing will ensure that minimum space is available for the growth of weeds and will minimize competition with weeds.This will effectively restrict weed development. In order to apply this approach, the limiting weeds must be known and the seasons in which they occur. A weed calendar of the area or region, if available, might be of help. It will be used to manage weeds in a targeted fashion with proper timing and effect.

4.3.4. Planting in narrow rows

Keeping all other factors constant, narrow spaced crops smother weeds more efficiently than wide spaced crop. Wide spaced crop allows more light to reach weeds. Hence, weed growth is more.

4.3.5. Planting direction

Crops sown in North-South direction suppress weed growth better than East-West direction. East-West crop allows sunlight to reach weeds throughout the day, whereas, North-South sown crop utilizes more light, shades weeds and does not allow light to reach weeds.

4.3.6. Plant density

In the crop-weed competitive situation, higher number of plants gain, comparative edge, more plant result in broader canopy cover and thus higher weed suppression capacity.

- Proper time of planting
- Crop rotation

Weeds get familiarized with canopy cover, growth habit and cultivation practices adopted for a particular crop. Weed management is difficult in cropping systems where the same crop is grown year after year. On the other hand with a change in cropping sequence, cultivation practices vary and so also the canopy cover and growth habit of the crop.

- Inter cropping

4.3.7. Planting method

1. Sowing of clean crop seeds without weed seeds should be done. It is a preventive method against introduction of weeds.
2. Sowings are taken up one to three days after rainfall or irrigation depending on soil type. Weeds already present in the soil start geminating within two or three days.
3. Sowing operation with seed drill removes some of the germinating weeds.
4. Transplanting is another operation which reduces weed population. Since, the crop has an additional advantage due to its age.

4.3.8. Stale seedbed

A stale seedbed is one where initial one or two flushes of weeds are destroyed before planting of a crop. This is achieved by soaking a well prepared field with either irrigation or rain and allowing the weeds to germinate. At this stage a shallow tillage or non- residual herbicide like paraquat may be used to destroy the dense flush of young weed seedlings. This may be followed immediately by sowing. This technique allows the crop to germinate in almost weed-free environment.

- Crops germinate in weed free environment
- Highly effective in controlling weeds in crops planted in rain less period with the help of irrigation.

4.3.9. Cover crop

Cover Crop means plants or a green manure crop grown for seasonal soil protection or soil improvement. Cover crops help control soil movement and protect the soil surface between crops. Cover crop reduces wind erosion by shielding the soil with vegetation and anchoring the soil with roots. In India, green manure crops like sunnhemp, cowpea, daincha, kolingi, clover, lupins etc. are more commonly used. Legume cover cropping in grape, mango, guava and other fruit crops is becoming a common practice in the management of orchards. Cowpea and French beans grow well under guava and sapota tree. In some places to prevent soil erosion, certain permanent cover crops like *Calapogonium muconoides, Centrosema pubescens* and *Peurariapha-seoloides* are raised in the alley spaces especially in Coconut gardens. They are leguminous crops, establish in a short period, dry up during summer to conserve moisture. With summer showers they come up again because of their profuse seeding habit and spread themselves as a vegetative mat by the time the heavy monsoon starts pouring in. Such permanent cover cropping is a common feature in rubber plantations of Kerala and Kanyakumari district of Tamil Nadu.

4.3.10. Crop rotation

Crop rotation involves alternating different crops in a systematic sequence on the same land. It is an important strategy for developing a sound long term weed control program. Weeds tend to thrive with crops of similar growth requirements as their own and cultural practices designed to contribute to the crop may also benefit the growth and development of weeds. Monoculture, that is growing the same crop in the same field yea after year, results in a build-up of weed species that are adapted to the growing conditions of the crop. When diverse crops are used in a rotation, weed germination and growth cycles are disrupted by variations in cultural practices associated with each crop (tillage, planting dates, crop competition, etc).

Within a rotation, crop choice will determine both the current and the potential future weed problems that a grower will face. Traditionally, potato (*Solanum tuberosum* L.) was included in the rotation to reduce weed problems before a less competitive crop was grown. For an organic grower, crop choice is complicated further by the need to consider soil fertility levels within the cropping sequence and to include fertility building periods in the rotation. Variations in crop and weed responses to soil nutrient levels can also play an important part in weed management. The inclusion of a fallow period in the rotation in known to reduce perennial weeds. It is best to alternate legumes with grasses, spring planted crops with fall planted crops, row crops with close planted crops and heavy feeders with light feeders.

4.3.11. Choice of crops and varieties

Tall crops and varieties with broader leaves will compete better with late occurring weeds than small varieties with narrow leaves. Some varieties will inhibit and suppress weeds while others will tolerate them. For example, there are witchweed (*Striga* spp.) resistant maize and cowpea cultivars in many countries of Africa, which give better performance at the same level of weeds where other varieties are more affected.

4.3.12. Water management techniques

- Depending on the method of irrigation, weed infestation may be increased or decreased.
- Frequent irrigation or rain during initial stage of crop growth induces several flushes of weeds.
- In lowland rice, where standing water is present most of the time, germination of weeds is less.
- Continuous submergence with 5 cm water results in reducing weed population whereas under upland situation, weed population and weed dry matter is very high.

4.3.13. Pre-germination of weeds

In pre-germination irrigation or rainfall germinates weed seeds just before the cash crop is planted. The newly germinated weeds can be killed by light cultivation or flaming. Pre-germination should occur as close a possible to the date of planting to ensure that changes in weather conditions do not have an opportunity to change the spectrum of weeds (cool *vs.* warm season) in the field.

4.3.14. Planting to moisture

Another technique similar to pre-germination is planting to moisture. After weeds are killed by cultivation, the top 2 to 3 inches of soil are allowed to dry and form a dust mulch. At planting, the dust mulch is pushed away and large-seeded vegetables such as corn or beans can be planted into the zone of soil moisture. These seeds can germinate, grow, and provide partial shading of the soil surface without supplemental irrigations that would otherwise provide for an early flush of weeds.

4.3.15. Buried drip irrigation

Drip tape buried below the surface of the planting bed can provide moisture to the crop and minimize the amount of moisture that is available to weeds closer to the surface. If properly managesd, this technique can provide significant weed control during dry period.

4.4. Eradicative Method (Physical method)

It is the complete elimination of all the live plant parts including seeds from an area. Deep digging and hand weeding are the most useful methods to remove weeds completely. However, these have to be adopted in a limited area.

4.4.1. Deep digging

Digging is labour intensive and slow process to be adopted in a situation where weeds are posing problem because of heavy infestation of root and vegetative propagules deep in the soil. It is having a limited scope but very effective. Deep digging is very useful in the case of perennial weeds to remove the underground propagation parts of weeds from deeper layers of the soil. Digging is accompanied by hand pulling of plant or picking up of roots and vegetative propagules in the soil. This practice allows weed free situation in the first year and impact of weeds in subsequent years is less. Roots of the crop establish gradually gain competitive advantage and thus weed growth is restricted.

4.5. Physical Methods

4.5.1. Hand weeding

Hand weeding is a slow but very useful method. Here as weeds are pulled up than weeds loose anchorage from the soil, fall dry desiccate and subsequently die. Thus weeds do not compete with the crop. This cannot be recommended for a larger area but can be adopted successfully as a last resort to remove weeds, which have become big and problematic in specific situations.

4.5.2. Hand hoeing

- The entire surface soil is dug to a shallow depth with the help of hand hoes, weeds are uprooted and removed.
- After hand hoeing, the field is subjected to drying to avoid re-establishing of uprooted weeds.
- This method is adopted in irrigated upland crops like finger millet, pearl millet, onion etc.

Fig.11: Hand weeding in maize

4.5.3. Digging

- Weeds are removed by digging up to deeper layers so as to remove underground storage organs.
- It is very useful in the case of perennial weeds and it is donewith the help of pick axes or crowbars. *Cynodon dactylon* can be effectively controlled by this method.

4.5.4. Mowing

- Mowing is the cutting of weeds to the ground level.
- Mowing is usually practiced in non-cropped areas, lawns and gardens wherein the grass is cut to a uniform height to improve the aesthetic value.
- The common mowing tools are sickle, scythe and lawn mower.

4.5.5. Cutting

- Weeds are cut above the ground surface leaving stubble. It is most common practice against brush and trees.
- Cutting is done with the help of axes and saws.

4.5.6. Dredging and chaining

- Dredging and chaining methods are used to control aquatic weeds.
- Removing of weeds along with their roots and rhizomes with the help of mechanical force is called dredging.
- The floating aquatic weeds are removed by chaining.
- A very heavy chain is pulled over the water bodies to collect the weeds.

4.5.7. General techniques

- Convert weeds into manures / composts
- Mulching reduces weeds
- Weeds can be used as mulches after drying
- Crop rotation reduces weed load
- Closer spacing and judicious water management reduces weed occurrence.
- Biological control of weeds.

4.6. Mulching

Fig. 12: Trash mulching in sugarcane field

Mulching or covering the soil surface can prevent weed seed germination by blocking light transmission preventing seed germination. Allelopathic chemicals in the mulch also can physically suppress seedling emergence. There are many forms of mulches available.

4.6.1. Living mulch

Fig. 13: Straw mulch

Living mulch is usually a plant species that grows densely and low to the ground such as clover. Living mulches can be planted before or after a crop is established. It is important to kill ad till in, or manage living mulch so that it does not compete with the actual crop. A living mulch of *Portulaca oleracea* from broadcast before transplanting broccoli suppressed weeds without affecting crop yield. Often, the primary purpose of living mulch is to improve soil structure, aid fertility or reduce pest problems and weed suppression may be merely an added benefit.

4.6.2. Organic mulches

Such materials as straw, bark, and composted material can provide effective weed control. Producing the material on the farm is recommended since the cost of purchased mulches can be prohibitive, depending on the amount needed to suppress weed emergence. An effective but labor-intensive system uses newspaper and straw. Two layers of newspaper are placed on the ground, followed by a layer of hay. It is important to make sure the hay does not contain any weeds seeds. Organic mulches have the advantage of being biodegradable. Cut rye grass mulch spread between planted rows of tomatoes and peppers was more economic than cultivation.

Fresh bark of conifers and oak as well as rapeseed straw gave good control of weeds when they were laid as mulches under the trees in apples orchards.

Materials such as black polyethylene have been used for weed control in a range of crops in organic production systems. Plastic mulches have been developed that filter out photosynthetically active radiation, but let through infrared light to warm the soil. These infrared transmitting mulches have been shown to be effective at controlling weeds.

4.6.3. Constraints of Organic Mulching

- As beneficial as organic mulch is, too much mulch can be harmful. A thick layer of organic mulch can be effective in suppressing weeds and reducing maintenance, but it often causes additional problems. Deep mulch can lead to excess moisture (root rot, mould fungus); it can create a habitat for rodents, which damage the plant. Furthermore, thick blankets of fine mulch may prevent penetration of water and air.
- Some organisms can proliferate too much in the moist and protected conditions of the mulch layer. Slugs and snails can multiply very quickly under a mulch layer.
- Ants or termites, which may cause damage to the crops also may find ideal conditions for living. When crop residues are used for mulching, in some cases there is an increased risk of sustaining pests and diseases.
- Damaging organisms such as stem borers may survive in the stalks of crops like cotton, corn or sugar cane. Plant material infected with viral or fungal diseases should not be used if there is a risk that the disease might spread to the next crop. Crop rotation is very important to overcome these problems.
- When carbon rich materials such as straw or stalks are used for mulching, nitrogen from the soil may be used by microorganisms for decomposing the material.Thus, nitrogen may be temporary not available for plant growth (risk of N-immobilisation).
- The major constraint for mulching is usually the availability of organic material. Its production or collection normally involves labour and may compete with the production of crops.

4.6.4. Pebble mulch

Here small pebbles like stones are placed on the soil surface. This mulching will be successful in dryland fruit tree culture. The pebbles placed on the basins of trees not only reduce evaporation but also facilitate infiltration of rain water into the basin.

Fig. 14: Pebble mulch in Mango

4.7. Mechanical Methods

Mechanical removal of weeds is both time consuming and labor-intensive but is the most effective method for managing weeds. The choice of implementation, timing, and frequency will depend on the structure and form of the crop and the type and number of weeds. Cultivation involves killing emerging weeds or burying freshly shed weed seeds below the depth from which they germinate. It is important to remember that any ecological approach to weed management begins and ends in the soil seed bank. The sol seedbank is the reserve of weed seeds present in the soil. Observing the composition of the seedbank can help a farmer make practical weed management decisions. Burial to 1 cm depth and cutting at the soil surface are the most effective ways to control weed seedlings mechanically.

Mechanical weeders include cultivating tools such as hoes, harrows, tines and brush weeders, cutting tools like mowers and stimmers, and dual-purpose implements like thistle-bars. The choice of implement and the timing and frequency of its use depends on the morphology of the crop and the weeds. Implements such as fixed harrows are more suitable for arable crops, whereas inter-row brush weeders are considered to be more effective for horticultural use. The brush weeder is mainly used for vegetables such as carrots, beetroot, onions, garlic, cerely and leeks. The optimum timing for mechanical weed control is influenced by the competitive ability of the crop and the growth stage of the weeds.

Hand hoes, push hoes and hand-weeding are still used when rouging of an individual plant or patch of weed is the most effective way of preventing the weed from spreading. Hand-weeding may also be used after mechanical inter-row weeding to deal with weeds left in the crop row.

Blind, 'over-the top' cultivation controls very small weeds, just germinated or emerged, before and sometimes after planting. The entire surface of the fields is worked very shallow using flex-tine cultivators (e.g. Lely weeder or rotary hoes, Inter-row cultivations with a rotary hoe in pinto beans (*Phaseolu svulgaris* L.) gave adequate weed control without reducing plant stand or injuring the crop.

The hoe-ridger is specifically designed to achieve intra-row control in sugar beet, Thistle-bars are simple blades used to undercut perennial weeds with minimal soil disturbance. The brush weeder, or brush hoe, is used primarily for inter-row weeding of vegetable crop.

Shallow between-row cultivators such as basket-weeders, beet-hoes, or small sharp sweeps are used to cut off and uproot small weeds after the crop is up. These can get very close to the crop when it's small, without moving much soil into the row, and may be the only tools used on delicate crops like leafy greens, As vigorous crops grown, soil can be thrown into the row to bury in – row weeds using rolling cultivates (e.g. Lilliston), spyder wheels (e.g. Bezzerides), large sweeps or hilling disks. Some of these tools can be angled to pull soil away from the row when plants are small and later turned around to throw soil back on the row during subsequent cultivators.

4.8.Thermal Weed Control Methods

4.8.1. Flamers

Flamers are useful for weed control. Thermal weed control involves the use of flaming equipment to crate direct contact between the flame and the plant. This technique works by rupturing plant cells when the sap rapidly expands in the cells. Sometimes thermal control involves the outright burning down of the weeds. Flaming can be used either before crop emergence to give the crop a competitive advantage or after the crop has emerged. However, flaming at this point in the crop production cycle may damage the crop. Although the initial equipment cost may be high, flaming for weed control may prove cheaper than hand weeding.

Propane – fuelled models of flamers are the most commonly used. Flaming dose not burn weeds to ashes; rather the flame rapidly raises the temperature of the weeds to more than 130 °F. The sudden increase in temperature causes the plants cell sap to expand, rupturing the cells walls. For greatest flaming efficiency, weeds must have fewer than two true leaves. Grasses are difficult to impossible to kill by flaming because the growing point is protected underground. After flaming, weeds that have been killed rapidly change from a glossy appearance to a duller appearance. Flame weeders can be used when the soil is too moist for mechanical weeding and there is no soil disturbance to stimulate further weed emergence.

Flaming can be used prior to crop emergence in slow-germinating vegetables such as peppers, carrots, onion, and parsley. Onions have some tolerance to flaming and flame weeding has been successful in both pre and post-crop emergence conditions and after transplanting. Transplanted cabbage has some tolerance to heat, allowing band flaming to be used along the crop row. Damage can occur when the treatment is applied too early, but the crop usually recovers. In a young pear orchard, where treatments were started on a clean soil after cultivation, flaming kept weed growth in check. In an established apple orchard, there was insufficient control of perennial weeds. Best results are obtained under windless conditions, as winds can prevent the heat from reaching the target weeds. The efficiency of flaming is greatly reduced if moisture from dew or rain is present on the plants. Early morning and early evening are the best times to observe the flame patterns and adjust the equipment.

4.8.2. Stubble burning in specific occasions

Stubble burning is now banned because of the smoke and other hazards, but this traditional form of thermal weed control was used to reduce the number of viable weed seeds returned to the soil after cereal harvest. Soil surface temperatures under the burning straw reached in excess of 200 ^{0}C for 10 -30 seconds and reduced the viability of freshly shed wild oat (*Avena fatua*) and blackgrass (*Alopecurus myosuroides*) seed by up to 30% and 80% respectively. Current methods of thermal weed control use a variety of energy sources to generate the heat needed to kill weed seeds and seedlings.

4.8.3. Freezing

Freezing would be advantageous only where there is an obvious fire risk from flaming. Liquid nitrogen and solid carbondioxide (dry ice) can be used for freezing weeds. Various test systems using electrocution, microwaves and irradiation have also been evaluated for weed control purposes, but high energy inputs, slow work rates and the safety implications for operators have hampered developments. Lasers have been shown to inhibit the growth the *Eichornia crasispes* (water hyacinth) but did not kill the weed completely. Weed control using ultraviolet light has been patented but remains at an experimental stage.

4.8.4. Infrared weeders

Infrared weeders are a further development of flame weeding in which the burners heat ceramic or metal surfaces to generate the infrared radiation directed at the target weeds. Some weeders use a combination of infrared and direct flaming to kill the weeds. In general, flame weeders are considered to be more effective because they provide higher temperatures, but burner height and plant stage are important too. Infrared weeders cover a more closely defined area than those of the standards flame weeder, but may need time to heat up.

4.8.5. Soil solarization

Solarization in agriculture would include thermal, chemical and biological changes caused by solar radiation when covered by clear plastic films especially when the soil has high moisture content. The possible mechanisms of weed control by solarization

* Killing of germinating seeds
* Killing of seeds stimulated to germinate in the moistened mulched soil
* Indirect microbial killing of seeds weakened by sub-lethal heating and
* Direct killing of weeds due to heat.

In general, solarization increases heat. This increased heat of temperature induces weed seeds to germinate and get killed.

Fig. 15: Soil covered with plastic films as a soil solarization method

4.9. Biological Methods

It involves the deliberate use of living organism like insects, fish, disease causing organisms or competitive plants to limit weed infestation.

Biological control would appear to be the natural solution for weed control in organic agriculture.

Table 8: Use of biocontrol agents for weed control

Name of the weed	Bioagent
Cyperus rotundus	*Bactra verutana*
Ludwigia parviflora	*Halticacynea* (Steel blue beetle)
Parthenium hysterophorus	*Zygrogramma bicolarata*
Lantana camara	*Crocidosema lantana, Teleonnemia scrupulosa*
Opuntia dilleni	*Dactylopius tomentosus, D. Indicus* (cochineal scale insect)
Eichhornea crassipes	*Neochetina eichhornea, N. Bruchi* (hyachinth weevil) (hyancinth moth)
Salvinia molesta	*Crytobagus singularis* (weevil) *Paulinia acuminate* (grass hopper), *Sameamutiplicalis*
Alternanthera philoxaroides	*Agasides hygrophilla* (flea beetle) *Amynothrips andersoni*
Tribulus terrestris	*Microlarinus lypriformis, M. lareynii*
Solanum elaegnifolium	*Frumenta nephalomicta*

Fig. 16: *Zygrogramma bicolarata* feeding the *Parthenium* spp.

Table 9. Use of fish & competitive plants for weed control

Name of the weeds	Fish
Lemma, Hydrilla, Potamogeton	Grass carp or white amur
Algae	Silver carp, common carp
Competitive plants for weed control	
Parthenium hysterophorus	*Cassia sericea*
Typha sp.	*Brachiaria mutica*

4.10. Bioherbicide approach

Using microbial plant pathogens for weed control. Fungi are more useful for this and hence the term mycoherbicide is also used to refer bioherbicide.

Table 10: Commercial mycoherbicides

Trade name	Pathogen	Target weed
Devine	*Phytophthora palmivora*	*Morreria odorata* (Strangler vine) in citrus
Collego	*Colletotrichum gleosporoides f.*sp. *aeschynomene*	*Aeschynomene virginica* (northen joint vetch) in rice and soyabean
Biopolaris	*Biopolaris sorghicola*	*Sorghum halepense* (Johnson grass)
Biolophos	*Streptomyces hygroscopius*	General vegetation (non-specific)
LUBAO 11	*Colletotrichum gleosporoides f.*sp. *Cuscuttae*	*Cuscutta* spp. (Dodder)
01	*Alternaria cassiae*	*Cassia abtusifolia*
ABG 5003	*Cercospora rodmanii*	*Eichhornea crassipes* (water hyancinth)
Biochon	*Chondrostereum purpureum*	*Prunus serotina*

4.10.1. Naturally occurring bioherbicides

Using secondary plant products (allelochemicals) and microbial forms as natural pesticides.

4.10.2. Allelopathy in bio-control programmes

Use of cover crops for bio-control

- Parthenium- incorporated into soil-reduction in growth of *Cynodondactylon*
- Dry plants of Menthi (Cumin)-Leachets controls weeds
- Velvet bean- Suppress purple nut sedge

Use of allelopathic chemicls as natural herbicides

- Xanthotoxin-Inhibits germination & growth of *Lactuca sativa*

AAL toxin- *Alternaria alternate lycopersicii* (Pathogen) - Effective against dicots at low concentration.

4.10.2.1. What other factors might need to be taken into account?

Before using allelopathy in weed management programmes there are a number of other factors that might be important in any given situation.

Varieties

There can be a great deal of difference in the strength of allelopathic effects between different crop varieties.

Specificity

There is a significant degree of specificity in allelopathic effects. Thus, a crop which is strongly allelopathic against one weed may show little or no effect against another

Autotoxicity

Allelopathic chemicals may not only suppress the growth of other plant species, they can also suppress the germination or growth of seeds and plants of the same species. Lucerne is particularly well known for this and has been well researched. The toxic effect of wheat straw on following wheat crops is also well known.

Crop on crop effects

Residues from allelopathic crops can hinder germination and growth of following crops as well as weeds. A sufficient gap must be left before the following crop is sown. Larger seeded crops are effected less and transplants are not affected

Environmental factors

Several factors impact on the strength of the allelopathic effect. These include pests and disease and especially soil fertility. Low fertility increases the production of allelochemicals. After incorporation the allelopathic effect declines fastest in warm wet conditions and slowest in cold wet conditions

Allelopathic weeds?

Several weed species have been reported to show allelopathic properties. They include couch grass, creeping thistle and chickweed. Where they occur together they may have a synergistic negative effect on crops

4.11. ITK's in Weed Management

- To control *Cyanodon dactylon*, harvested dried stalks of cumin crop are spread in the field. As the stalks decompose and mix with the soil, the weed is destroyed. (Farmers in Gujarat)
- Farmers change the variety of paddy crop in each season to control the weeds like 'dhakura' (Farmers in Uttaranchal)
- Manure made with mango leaves is applied in the fields to control *Cyperus rotundus* (Farmers in Tamil Nadu)
- Use of multivariate seeds (MVS) for mixed sowing navathaniyam as intercrops.

4.11.1. Navathaniyam

- A multivarietal seed mixture of nine crops viz., two cereals, two pulses, two oilseeds, two spices and condiments and a N fixing green manure ploughed *in situ* as green manure (Somasundaram, 2003).

Fig. 17: Navathaniyam for effective weed control

4.11.2. Control of nut grass weed in the crop field

- Nut grass is locally called as 'Korai'. If the weed occurs then it will affect the growth of the field crop and it also spread to the entire garden.
- In order to prevent the nut grass ploughing the land has to be performed by using neem plough (plough made of neem wood). Besides, the application of neem cake as organic manure is believed to control nut grass weed in the field efficiently.

4.11.3. Recent approaches

Use of rice bran and application of tamarind seed powder in paddy ecosystem

Weeds composting

Relay sowing etc.,

4.11.4. Herbicide / Kalaikolli

Cow's urine 250 ml + salt 100 g

↓

Mix it and spray

↓

Useful for weed control out side of the field

5

Pest Management Techniques in Organic Farming

Insect pests cause huge losses ranging from 5 to 80% of even up to 100%. Acute food shortage following world war –II and Bengal famine (1943) due to failure of rice crop due to a paddy disease indicate the severity of the loss, caused by the pests and diseases. The insects in storage on an average consume and spoil an additional 4 million tones of grains every year. All this indicates the importance of plant protection by which we can save millions of tones of food grains which are otherwise eaten away by different pests. Losses due to insect pests in Indian agriculture are 23.3 per cent. One of the practical means of increasing crop production is to minimize the pest associated losses.

A pest is any organism which occurs in large numbers and conflict with man's welfare, convenience and profit.

Pests are often controlled with man made chemicals which have many harmful effects, for example:

- Artificial chemicals kill useful insects which eat pests.
- Artificial chemicals can stay in the environment and in the bodies of animals causing problems for many years.
- Artificial products are very simple chemicals and insect pests can very quickly, over a few breeding cycles, become resistant to them and can no longer be controlled.

Components of organic pest management

The following components may be included in organic method of pest management

1. Preventive practices
2. Habitat diversification
3. Mechanical methods of pest management
4. Use of plant products / botanicals

5. Use of insect pheromones
6. Biological control of pests
7. Use of synthetic organics permissible for use in organic agriculture
8. Using farmers wisdom in organic farming
9. Physical methods of pest management.

Plant protection: Two categories of insects

i. Insects which feeds on plants (Vegetarians)

ii. Insects which feeds on insects (Non – Vegetarians)

Vegetarians insects are posing problems to farming as pets. Non – vegetarian are beneficial components of organic farming stated by successful organic growers.

5.1. Preventive Practices

5.1.1 Selection of adapted and resistant varieties

- Choose varieties which are well adapted to the local environmental conditions (temperature, nutrient supply, pests and disease pressure), as it allows them to grow healthy and makes them stronger against infections of pests and diseases.

5.1.2. Selection of clean seed and planting material

- Use safe seeds which have been inspected for pathogens and weeds at all stages of production.
- Use planting material from safe sources.
- Planting physically sound seed is also important. In crops such as flax, rye and pulses, a crack in the seed coat may serve as an entry point for soil-borne micro-organisms that rot the seed once it is planted.

5.1.3. Use of suitable cropping systems

- Mixed cropping systems: can limit pest and disease pressure as the pest has less host plants to feed on and more beneficial insect life in a diverse system.
- Crop rotation: reduces the chances of soil borne diseases and increases soil fertility.
- Green manuring and cover crops: increases the biological activity in the soil and can enhance the presence of beneficial organisms (but also of pests; therefore a careful selection of the proper species is needed).

5.1.4. Use of balanced nutrient management

- Moderate fertilization: steady growth makes a plant less vulnerable to infection. Too much fertilization may result in salt damage to roots, opening the way for secondary infections.
- Balanced potassium supply contributes to the prevention of fungi and bacterial infections

5.1.5. Organic matter addition

- Increases micro-organism density and activity in the soil, thus decreasing population densities of pathogenic and soil borne fungi.
- Stabilises soil structure and thus improves aeration and infiltration of water.
- Supplies substances which strengthen the plant's own protection mechanisms.

5.1.6. Application of suitable soil cultivation methods

- Facilitates the decomposition of infected plant parts.
- Regulates weeds which serve as hosts for pests and diseases.
- Protects the micro-organisms which regulate soil borne diseases.

5.1.7. Good water management practices

- No water logging: causes stress to the plant, which encourages pathogens infections.
- Avoid water on the foliage, as water borne disease spread with droplets and fungal disease germinate in water.

5.1.8. Conservation and promotion of natural enemies:

- Provide an ideal habitat for natural enemies to grow and reproduce.
- Avoid using products which harm natural enemies.

5.1.9. Selection of optimum planting time and spacing:

- Most pests or diseases attack the plant only in a certain life stage; therefore it's crucial that this vulnerable life stage doesn't correspond with the period of high pest density and thus that the optimal planting time is chosen.
- Sufficient distance between the plants reduces the spread of a disease.
- Good aeration of the plants allows leaves to dry off faster, which hinders pathogen development and infection.

5.1.10. Use of proper sanitation measures

- Remove infected plant parts (leaves, fruits) from the ground to prevent the disease from spreading.
- Eliminate residues of infected plants after harvesting.
- Incorporating the residue into the soil hastens the destruction of disease pathogens by beneficial fungi and bacteria. Burying diseased plant material in this manner also reduces the movement of spores by wind.

5.1.11. Crop Rotations

Crop rotation is central to all sustainable farming systems. It is an extremely effective way to minimize most pest problems while maintaining and enhancing soil structure and fertility. Diversity is the key to a successful crop rotation program. It involves:

- Rotating early-seeded, late-seeded and fall-seeded crops.
- Rotating between various crop types, such as annual, winter annual, perennial, grass and broadleaf crops; each of these plant groups has specific rooting habits, competitive abilities, nutrient and moisture requirements. (True diversity does not include different species within the same family - for example, wheat, oats and barley are all species of annual cereals).
- Incorporating green manure crops, into the soil to suppress pests, disrupt their life cycles and to provide the additional benefits of fixing nitrogen and improving soil properties.
- Managing the frequency with which a crop is grown within a rotation
- Maintaining the rotation's diversified habitat, which provides parasites and predators of pests with alternative sources of food, shelter and breeding sites.
- Planting similar crop species as far apart as possible. Insects such as wheat midge and Colorado potato beetle, for example, are drawn to particular host crops and may over winter in or near the previous host crops. With large distances to move to get to the successive crop, the insects' arrival may be delayed. The number that find the crop may be reduced as well.

5.1.12. Forecasting

Producers should pay attention to the forecasts for various pest and disease infestations for each crop year. Maps of these forecasts are usually available for many of the major destructive insects such as grasshoppers and wheat midge, as well as some diseases. Agro meteorological warning and forecast can help in this way.

5.2. Habitat Diversification

Habitat diversification makes the agricultural environment unfavourable for growth, multiplication and establishment of insect pest populations. The following are some approaches by which the pest population can be brought down.

5.2.1. Intercropping system

Intercropping system has been found favourable in reducing the population and damage caused by many insect pests due to one or more of the following reasons.

- Pest outbreak less in mixed stands due to crop diversity than in sole stands
- Availability of alternate prey
- Decreased colonization and reproduction in pests
- Chemical repellency, masking, feeding inhibition by odours from non-host plants.
- Act as physical barrier to plants.

The following table gives a few examples of intercropping system where reduction in damage level was noticed.

Table 11: Effect of intercropping system on pest levels

Sl.No.	Crop		Pest reduced
	Sole crop	Intercrop	
1.	Sorghum	Red gram	Earhead bug
2.	Sorghum	Cowpea	*Chilo partellus stem borer*
3.	Pigeon pea	Sorghum	Leaf hopper *Empoasca kerri*
5.	Green gram	Sorghum	*E. kerri* Leaf hopper
5.	Ground nut	Sorghum	*E.kerri* Leaf hopper
6.	Pigeon pea	Sorghum	*H. armigera*
7.	Chickpea	Wheat, mustard or Safflower	*H.armigera*
8.	Sugarcane	Greengram, Balckgram	Early shoot borer

Interplanting maize in cotton fields increased the population of Araneae, coccinellidae and chrysopidae by 62.8-115.7% compared with control fields. Maize also acted as a trap crop for *H. armigera* reducing the second generation eggs and damage to cotton. Intercropping pulses in cotton reduced the population of leaf hopper on cotton and Lablab bean in sorghum reduced the sorghum stemborer incidence. Hence it is highly important that appropriate intercropping systems have to be evolved where reduction in pest level occurs.

5.2.2. Trap cropping

Crops that are grown to attract insects or other organisms like nematodes to protect target crops from pest attack. This is achieved by

- Either preventing the pests from reaching the crop or
- Concentrating them in a certain part of the field where they can be economically destroyed

Table 12: List of successful examples of trap crop

Sl.No.	Main Crop	Trap crop	Pest
1.	Tobacco/ cotton/groundnut	Castor	*Spodoptera litura*
2.	Maize	Sorghum	Shootfly, Stem borer
3.	Cotton	Onion / Garlic	*Thrips tabaci*

Growing mustard as trap crop 2 rows per 25 cabbage rows for the management of diamondback moth. First mustard crop is sown 15 days prior to cabbage planting or 20 days old mustard seedlings are planted. Growing castor along the border of cotton field and irrigation channels act as indicator or trap crop for *Spodoptera litura.* Planting 40 day old African tall marigold and 25 day old tomato seedlings (1:16 rows) simultaneously reduces *Helicoverpa* damage.

Growing trap crops like marigold which attract pests like American bollworm to lay eggs, barrier crops like maize/jowar to prevent migration of sucking pests like aphids and guard crops like castor which attracts *Spodoptera litura* in cotton fields was reported.

Fig. 18: Castor as a trap crop in the field

A. Based on characteristics of trap crop-

1. **Conventional trap crop-** It is most common practice. Growing of trap crops next to a higher value crop is naturally more attractive to a pest as either a food source or oviposition site than is the main crop, thus preventing or making less likely the arrival of the pest to the main crop and/or concentrating it in the trap crop where it can be economically destroyed. Ex: Castor and marigold in ground nut crop, Alfalfa as a trap crop for Lygus bugs in Cotton.

2. **Dead end trap cropping-** Trap crops which are highly attractive to insects but they or their offspring's can't survive. Dead-end trap crops serve as a sink for pests, preventing their movement from the trap crop to the main crop later in the season Ex: Indian mustered for Cabbage diamond back moth, Sun hemp for Bean pod borer.

B. Based on spatial distribution-

1. **Perimeter trap cropping-** Growing trap crops around the borders of the main crop. For example borders of early-planted potatoes have been used as a trap crop for Colorado potato beetle, which moves to potato fields from overwintering sites next to the crop, becoming concentrated in the outer rows, where it can be treated with insecticides, cultural practices.

2. **Sequential trap cropping-** Growing trap crops earlier or later than the main crop to attract the pest. Ex. Indian mustard as a trap crop for diamondback month in cabbage.

3. **Multiple trap cropping-** Planting of several species simultaneously as trap crops with the purpose of either managing several insect pests at the same time or enhancing the control of one insect pest by combining plants for attracting pests. For ex. use of a mixture of castor, millet, and soybean to control Groundnut leaf miner and the use of corn and potato plants combined as a trap crop to control wireworms in sweet potato fields.

5. **Push – Pull trap cropping-** Growing combination of trap crop and repellent crops. The trap crop attracts the insect pest and, combined with the repellent intercrop, diverts the insect pest away from the main crop. Ex. marigold and onion in chilli. A push-pull strategy based on using either Napier or Sudan grass as a trap crop planted around the main crop, and either Desmodium or Molasses grass planted within the field as a repellent intercrop, has greatly increased the effectiveness of trap cropping for stem borers.

System of planting some trap crop: - Ex.

- Planting Indian mustard as a trap crop for management of Diamond Back Moth. Sowing of two rows of bold seeded Mustard in every 25 rows of Cauliflower/Cabbage.
- Planting Cow pea as intercrop for management of *Heliothis* sp. Sowing of one rows of cowpea in every 5 rows of cotton.
- Planting Tobacco as trap crop for management of *Heliothis* sp. Sowing of two rows of tobacco in every 20 rows of cotton.
- Planting African marigold as trap crop for management of fruit borer. Sowing of two rows of marigold in every 14 rows of Tomato.
- Planting Coriander or Fenugreek as trap crop for management of shoot and fruit borer. Sowing of one rows of Coriander or Fenugreek in every two rows of Brinjal.
- Planting Coriander or Marigold as a trap crop for management of Gram pod borer. Sowing of one rows of coriander or marigold in every 4 rows of gram.
- Planting Corn as trap crop for management of *Heliothis* sp. Sowing of two rows of Tobacco in every 20 rows of Cotton.

5.2.3. Planting dates and crop duration

Planting dates should be so adjusted that the susceptible stage of crop synchronizes with the most inactive period or lowest pest population. The plantings should be also based on information on pest monitoring, as the data varies with location. Crop maturity also plays an important role in pest avoidance. The following table (Table 14) shows the importance of planting dates on pest population and damage

5.2.4. Planting density

Plant nutrient status, interplant spacing, canopy structure, etc., affect insect behaviour in searching food, shelter and oviposition site. It also affects natural enemy population. The effect of plant density on pest population is shown in Table 15.

Table 13: Some examples of trap crops used in main crops

Main crop	Trap crop	Method of planting	Controlled pest
Horticultural crops			
Cabbage	Raddish	Planted in every 15 rows of cabbage	Flee beetle
Cabbage	Indian Mustard	2 rows planted in every25 rows of cabbage	Diamond back moth
Cabbage	Nasturtium	Row intercrop	Aphids, Flea beetle, Cucumber beetle
Garlic	Marigold	Border crop	Thrips
Carrot	Onion and garlic	Border crop	Carrot fly
Tomato	Marigold	2 rows planted in every 14 rows of Tomato	Tomato fruit borer and Root knot nematodes
Field beans	Chrysanthamum	Border crop	Leaf minor
Cauliflower /Cabbage	Sesamum	Border crop	Diamond back moth
Potato	Marigold	Intercrop	Nematodes
Potato	Horse radish	Intercrop	Colorado potato beetle
Brinjal	Coriander/ Fennel	1 rows planted in every2,rows of Brinjal	Shoot and Fruit borer
Agronomical crop			
Sunflower	Mari gold	Intercrop	*Heliothis* sp.
Sunflower	Castor	Border crop	Tobbacco catter pillar
Cotton	Marigold	Intercrop	*Heliothis* sp.
Groundnut	Cow pea	Intercrop	Leaf folder
Redgram	Soya bean/ Green gram	Border crop	Thrips
Bengalgram	Marigold	Intercrop	*Heliothis* sp.
Cotton	Alfa alfa	Strip intercrop	Laygus bug
Cotton	Castor	Border crop	*Heliothis* sp.
Cotton	Sunflower/ Tobacco	Border crop	*Heliothis* sp.

Contd.

Main crop	Trap crop	Method of planting	Controlled pest
Cotton	Cow pea	1 rows intercrop, planted in every 5 rows of cotton	*Heliothis* sp.
Cotton	Chick pea	Intercrop	*Heliothis* sp.(Caterpillar)
Cotton	Corn	1 rows intercrop, planted in every 20 rows of cotton	*Heliothis* sp.
Corn	Beans and other legumes	Row intercrop	Leaf hopper, Leaf beetles, Stalk borer
Arhar	Sorghum/Maize/Marigold	Row intercrop	Gram caterpillar
Soybean	Green beans	Row intercrop	Mexican bean beetle
Soybean	Sunflower/Castor	Border crop	Tobbacco catter pillar
Corn	Napier grass/ Sudan grass	Intercrop, border crop	Stem borer
Corn	Desmodium	Row intercrop	Stem borer
Cotton	Okra	Border crop	Bollworms
Groundnut	Castor/ Sunflower	Intercrop	Leaf eating caterpillar, Leaf minor

Table 14: Role of planting dates on pest population and damage

Sl.No.	Host plant	Insect	Response
1.	Rice	Leaf folder	Early palnted rice (upto 3rd week of June) suppressed population
2.	”	BPH	Planting in end of July in Kharif and Early in Rabi escapes attack in AP
3.	”	Gallmidge	Lowest incidence iof planted in Aug or Oct
4.	Sorghum	Shootfly	Advancing sowing date (Sept - Oct) decreased incidence
5.	Cotton	Leafhopper	Higher incidence in late sown crop
6.	Chickpea	*H. armigera*	For every 10 day delay in sowing 5.02% increase in pod damage
7.	Tomato	Whitefly *(B.tabaci)*	Incidence less if planted within Jul- Nov
8.	Chillies	Thrips	Late planted crop severely affected by thrips and leaf curl virus

Table 15: Effect of plant density on pest population

No.	Crop	Spacing/ density	Insect	Response
1.	Rice	Dense planting	Leaf folder, BPH	High incidence
2.	Chickpea	Dense plant population	*H.armigera*	High incidence
3.	"	Less dense population	*Aphis craccivora*	High incidence
4.	Sugarcane	Dense seed rate	Topshoot borer	Low incidence
			Early shoot borer	High incidence

5.2.5. Destruction of alternate host plants

Many insects use a wide range of cultivated plants especially weeds as alternate hosts for off season carry-over of population. Weeds around the crop can alter the proportion of harmful and beneficial insects that are present and increase or decrease crop damage.

Table 16: Alternate hosts to be removed to reduce damage by pests

Crop	Pest	Alternate host to be removed
Groundnut	Thrips *(Caliothrips indicus)*	*Achyranthus aspera*
Rice	Gallmidge	Wild rice *(O.nivara)*
	GLH	*Leersia hexandra*
		Echinochloa colonum
		E.crusgalli
		C.dactylon
Rice	WBPH	*Chleres barbata*
Sorghum	Earhead midge	Grassy weeds

Destruction of off types and volunteer plants, thinning and topping, pruning and defoliation and summer ploughing are other cultural methods which can reduce pest load in field.

5.2.6. Water management

Availability of water in requisite amount at the appropriate time is crucial for proper growth of crop. Hence, water affects the associated insects by many ways such as nutritional quality and quantity, partitioning of nutrients between vegetative growth and reproduction etc. The following table shows the effect of irrigation on pest population / damage.

Table 17: Effect of irrigation on pest population / damage

Sl.No	Crop	Insect	Response
1.	Rice	Mealy bug	Continuous ponding of 5cm water reduced incidence
2.	Rice	Caseworm and BPH	Draining of water to field capacity reduces incidence
3.	Fruit tree nursery	Termite	Copious irrigation reduces incidence
4.	Groundnut	Aphids	Copious irrigation increased incidence

5.2.7. Crop rotation

Sustainable systems of agricultural production are seen in areas where proper mixtures of crops and varieties are adopted in a given agro-ecosystem. Monocultures and overlapping crop seasons are more prone to severe outbreak of pests and diseases. For example growing rice after groundnut in garden land in puddled condition eliminates white grub.

5.2.8. Organic manure

Application of press mud in groundnut @ 12.5 t/ha had a better influence on leaf miner with lower leaflet damage at 38.84 per cent and 2.48 larval numbers per plant during summer 1991. It was 34.93 per cent and 2.72 numbers during kharif, 1991. Farm yard manure, *Azospirillum* and Phosphobacteria has no significant influence on the control of leaf hopper and fruit borer in bhendi. The incidence of paddy plant and leafhopper was low in *Azospirillum* combined with farmyard manure. Application of organic manure lowered the rice gall midge incidence (5.28%)

5.2.9. Depth and timing of seeding

Optimum seeding depth is also important. Deep seeding in cold soils may result in seedling blights and damping-off, especially in pulses and small-seeded crops. Seeding depth should generally be no deeper than required for quick germination and even emergence. Variables include seed size, soil type and moisture conditions. If the soil is loose before seeding, a packing operation will firm up the soil and bring moisture closer to the surface.

For most crops, seeding should ideally be done when the soil is warm enough for rapid germination. Seeds that remain ungerminated in cool soil are more susceptible to damageby insects such as wireworms.

5.3. Herbal Insecticides

Assumptions: Following are pest repellants

1. All non-edible foliages by ruminants: (Leaves Adathoda, *Nochi vitex* spp. etc.).
2. Latex yielding herbs and shrubs;
3. Bitter taste plant products, salty, / astringent Jatropha etc plant materials
4. *Calotropis, Datura, Neem, agave* Nerium etc.
5. Seeds of bitter/ salty plants Neem seed kernal Ettikottai etc. All medical herbal plants are pest repellants.

For example, *Calotropis* + *Clerodentron* + *neem* + *vitex* + *Lantana camera* soaked in cowdung, Urine mixture.

Any 5 herbs belonging to the above assumed category are powdered and soaked in urine (2 kg each), cow dung and100 – 200gm kernel (3:1) slurry for 10-15 days and used @10% foliar spray

For quick extraction, put any of the herbs belonging to the above assumed category in water and boil for 3 hours. After 12 hours, filter it to get 10 litres for immediate use and spray it along with one litre of turmeric powder (10% as foliar spray)

Mode of action: Repellant and antifeedant; detergent activity on copulation, egg hatching, etc.,

5.3.1. Ponneem (Organic insecticide)

Neem oil - Rs. 45

Pungam oil - Rs. 45

Soap - Rs. 10 (soaking gel)

For 10 litre of water 30ml of Ponnem can be added for spraying 1 acre.

5.3.2. Veppangottai karaisal

5 kg Neem seed kernel powder in 6 litre of water – 5 days soaking

5.3.3. Pungai ilai kasayam

Dried neem seed - 10 kgs

Cow's urine - 20 litres

Asafoetida - 200 g
Chewing tobacco - 1 kg
Datura plant - 3
Green chillies - ½ kg

Grind all soak it in urine
↓
Cover it for 5 days (stir it once in 2 days)

Usage

For 10 litres of water ½ liter of kasayam is sufficient after filtered

5.3.4. Neemastra

Neem leaves - 5 kgs
Cow's urine - 5 litres
Cowdung - 2 kg

Ferment for 24hrs with intermittent stirring
↓
Filter the extract and dilute to 100 litres (1 acre)

Useful against sucking pests and mealy bugs

5.3.5. Brahmastra

Neem leaves - 3 kgs
Cow's urine - 10 litres
Custard apple leaf - 2 kg
Papaya leaf - 2 kg
Pomegranate leaf - 2 kg
Guava leaf - 2 kg

Mix all and boil for 5 times at some interval till it becomes half
↓
Keep for 24 hrs then filter the extract

Usage

Dilute 2-2.5 of extract to 100 litre of water for1 acre.

Useful against sucking pests and pod/fruit borers

5.3.6. Agniasthra

Ipomea leaf - 1 kg
Hot chilli - 500 g
Garlic - 500 g
Neem leaves - 5 kg
Cow's urine - 0 litre

Boil all suspension 5 times till it becomes half
↓
Filter the extract & store glass or plastic bottles

Usage

Dilute 2-3 of extract to 100 litre of water for1 acre.

Useful against leaf roller and stem/pod/fruit borers

5.3.7. Formulation 1

In a copper container Ferment all for 10 days

Crushed neem leaf	- 3 kg ↓	
Neem seed kernel powder	- 1 kg	Boil all suspension 5 times till it becomes half
Cow's urine	- 10 litre	

Suspend 500 g garlic paste and 250 g chili paste in 1 litre of water separately and keep over night.

Next day mix all the solutions and filter

Usage

Dilute to 200 litre of water for1 acre and used as foliar spray.

For wide range of leaf eating and sucking pests

5.3.8. Formulation 2

In a 200 litres drum

Chopped neem leaf	- 5kg	Fill with water and ferment all for 10 days
Neem seed kernel powder	- 5kg	↓
Chopped Ipomea leaf	- 5 kg	Boil all suspension 5 times till it becomes half
Pungam seed powder	- 1 kg	↓
Cow's urine	- 0 litre	Distill the suspension

Usage

Distillate can be diluted to 200 litre of water for1 acre and used as foliar spray.

For wide range of leaf eating and sucking pests

5.3.9. Neem leaf extract

- 1 kg of green neem leaf soaked overnight in 5 litres of water, crushed and the extract is filtered. Add 10 ml of emulsifier (neutral pH adjuvant)
- This is beneficial against leaf eating caterpillars, grubs, locusts and grasshoppers.

5.3.10. Neem cake extract

- 100 g of neem cake is soaked overnight in 1 litre of water in a muslin pouch, crushed and the extract is filtered. Add 1 ml of emulsifier (neutral pH adjuvant) per llitre of water

5.3.11. Neem oil spray

- 15-30 ml neem oil is added to 1 litre of water and stirred well. To this emulsifier is added (1ml/1 litre).

5.3.12. Agrochem an herbal insecticide (Innovator- Sri Dhyaneshwar Patil, Shindkheda, Dhule)

To prepare a paste A

Datura leaves	- 50 g	
Chandrajoyti (*Jatropha grandulifera*)	- 50 g	
Ipomea fistula leaves	- 50 g	
Tobacco	- 50 g	
Varul (*Garuga pinnata*)	- 50 g	Mix all & crushed well to a paste A
Neem seeds	- 50 g	
Neem leaves	- 50 g	
Jowar tillers	- 100 g	
Congress grass	- 100 g	

To prepare a paste B

Water	- 3litres	Heat till it boils
Vidanga (*Embelia ribes*)	- 50 g	↓
Kerosene	- 150 ml	Add 200 ml of cows urine
Soap powder (Nirma)	- 5 teaspoon	

- Both A and B mixed and kept for 5 days in a vessel.
- After 5 days, the mixture is filtered and stirred well for 5 minutes to get the final formulation.
- Dilute to 1:100 with water and then spray.

5.3.13. Use of Cycas flowers

The flower of Cycas (sannampu) are cut into pieces, wrapped in straw and placed in the field @ 25–30 pieces /hectare. The odour that is emitted from this flower prevents the entry of earhead bugs for two weeks. By this time, the milky stage is over and the grain matures without any interruption.

5.3.14. Five leaf extract

This extract is prepared using the leaves of five different plants. Leaves with the characteristics describeded below can be used for the purpose:

In a mud pot, any five of the below mentioned plant leaves pound well

Plants with milky latex – e.g., *Calotropis, Nerium, Cactus* and *Jatropha.*

Plants which are bitter – e.g., *neem, Andrographis, Tinospora* and *Leucas.*

Plants that are generally avoided by cattle – e.g., *Adhatoda, Ipomea fistulosa.* 1kg each

Aromatic plants – e.g., *Vitex, Ocimum.*

Plants that are not affected by pests & diseases – e.g., *Morinda, Ipomea fistulosa.*

Then add

Water - 10 litre Tie the mouth of the pot tightly with a cloth

Cow's urine - 1 litre

Asafoetida - 100 g mixed well daily for a week

- It can be used after filtration. Cow urine is used for disease control and asafoetida prevents flower dropping, enhancing the yield.

5.3.15. Jatropha leaf extract

In a mud poof, pound

Jatropha leaves - 12.5 kg

To this add

Water - 12.5 litres

- Allow to ferment for 3–7 days.
- Filter and use the extract for spraying (after diluting with 10 parts of water) for one hectare area.

5.3.16. Turmeric rhizome extract

- Shred one kg of turmeric rhizomes. To this, add four litres of cow urine, mix well and filter. Dilute with 15–20 litres of water.
- For every litre of the mixture, add 4 ml of khadi soap solution. This helps the extract stick well to the surface of the plant.

5.3.17. Cow dung extract

- Mix one kg of cow dung with ten litres of water and filter using a gunny cloth.
- Dilute the solution with five litres of water and filter again. The result can be used for spraying.

5.3.18. *Andrographis* or *Sida kashayam*

Collect *Andrographis paniculata* or *sida acuta* leaves and wash first with tap water

Cut the materials into small pieces

Transfer the cut leaves to brass vessel and add 4 parts (four times the quantity of leaves) of water.

Boil them on a low flame until the contents are reduced to one-fourth the original volume (approximate time of boiling to get 500 ml of Kashayam will be about two hours and 30 minutes)

Pour the solution into a glass jar and allow it to cool for half an hour

↓

Filter the solution using khadi cloth add 1% sodium benzoate solution as preservative

- About 500 ml of this solution is mixed with 100 ml of khadi soap solution and diluted in 9.4 litres of water for field spraying.

5.3.19. Eucalyptus or *Lantana* leaf extract

- Boil tender leaves of Eucalyptus or Lantana (1 kg in 1.5 litres of water).
- Cool the solution, filter the next day and dilute with 20 litres of water.
- About 350–450 litres of this solution can be used directly for spraying in a one hectare field area.

5.3.20. Dashparni extract

Crush following plant parts in a 500-litres drum

- Neem leaves 5 Kg
- *Vitex negundo* leaves 2 Kg
- Aristolochia leaves 2 Kg
- Papaya (*Carica Papaya*) 2 Kg
- *Tinospora cordifolia* leaves 2 Kg
- *Annona squamosa* (Custard apple) leaves 2 Kg
- *Pongamia pinnata* (Karanja) leaves 2 Kg
- *Ricinus communis* (Castor) leaves 2 Kg
- *Nerium odorum* leaves 2 Kg
- *Calotropis procera* leaves 2 Kg
- Green chilly paste 2 Kg
- Garlic paste 250 gm
- Cow dung 3 Kg
- Cow Urine 5 litre
- Water 200 litre

Crush all the ingredients and ferment for one month. Keep the drum in shade and covered with gunny bag. Shake regularly three times a day. Extract after crushing and filtering. The extract can be stored up to 6 months and is sufficient for one acre.

5.3.21. Fermented curd water

In some parts of central India fermented curd water (butter milk or *Chaach*) is also being used for the management of white fly, jassids aphids etc.

Take the distasteful butter milk and add equal amount of water in it. Keep it for two days in a semi-shaded place. Now take it and add 40 litres of water in 10 litre of the extract to form 50 litres of the solution. Spray this solution in 1 hectare of field in such a way that all plants get bath in the fogging spray in the early morning time.

5.3.22. Herbal pesticide formulation

- 500g neem seeds, 1000g tobacco, 100g *Acorus calamus*, 250g Asofoetida and 50g *Sapindus emarginata* seeds are ground and the extract is sprayed for one acre cotton to control pests.

5.3.23. Neem-cow urine extract

- 5 kg of neem leaves, 5 litres of cow urine, 2 kg of cow dung, 100 litres of water.
- Crush all ingredients and ferment for 24 hours with intermittent stirring, filter and squeeze the extract and dilute to 100 litres of water.
- Use this extract to fill in the spray machine and spray it over one acre of the crop.

5.3.24. Mixed leaves extract

- 3 kg of neem leaves, 10 litres of cow-urine, custard apple leaves 2 kg, 2 kg papaya leaves, 2 kg pomegranate leaves, 2 kg guava leaves.
- Crush all of ingredient and add 5 litre of water in it. Boil the above mixture 5 times after some intervals of time till the mixture becomes half of the initial.
- Keep it for 24 hours, filter and squeeze the extract. This can be stored in bottles for 6 months.

5.3.25. Lantana leaf extracts

- 1 kg of lantana leaves are cut into small pieces and ground with 250 ml of water to make it into a paste.
- Filter the material by adding another 250 ml of water. Add a suitable emulsifier and stir well. Then dilute in 10 litre of water, and spray.
- This solution is effective against insect pests, and diseases caused by fungi, virus, termites, nematodes and bacterial.

5.3.26. Gliricidia leaf extract

- 500 gm of *Gliricidia sepium* leaves (Gliricidia) are crushed with a mortar-pestle, and soaked overnight in 10 litre of water, then filter and add another 10 litre of water to the filtrate. Spray on infested crops.
- 500 gm of Gliricidia leaves together with 7 pods of chilli and 3 onion bulbs are chopped and ground with a mortar-pestle. Then soak them overnight in 10 litre of water. Strain the mixture; add another 10 litre of water to the solution. Spray on the infested plants. This solution is effective against insects and acts as a repellent and antifeedant.

5.3.27. Onion bulb extract

- Boil 1 kg of chopped onion bulbs in 1 litre of water and keep for 24 hrs. Then dilute the filtrate with 10 litre of water and spray on infested plants. This extract will be helpful in spraying against leaf eating pests like caterpillars, aphids and some diseases.

5.3.28. Calotropis leaf extract

- 1 kg of leaves is crushed with mortar-pestle. Add 1 litre of water to this paste, and filter. 10 litre of water is added before spraying. *Calotropis* solution is active against insect pests and act as a repellent and antifeedant.

5.3.29. Pongamia leaf extract

- Soak 1 kg of *Pongamia pinnata* leaves in 5 litre of water for 12 hours. The following dry leaves are crushed and grounded, and the extract is filtered through a cotton cloth.
- Then add a soap solution as an emulsifier to help the extract to stick well to the leaf surface. This extract is beneficial against the leaf eating caterpillar.

5.3.30. Marigold extract

- Mix ½ to ¾, plastic bucket (volume 15 litre) full of flowering plants (with leaves and stems), with water and leave for 8 days for fermentation. Strain the mixture with about 2-3 litre of water, then stir with a suitable emulsifier.
- To increase effectiveness it can be mixed with tomato leaves. This solution is effective against pests, nematodes, and diseases.
- Mixture of marigold and chilli, garlic and onion extracts can also be used. This solution is active against eggs and larvae of insect pests.

5.3.31. Datura leaf extracts (Attana)

- Chop 1 kg of datura leaves into small pieces and then ground with 250 ml of water to make a paste. Then filter through the cotton cloth with another 250 ml of water. Dilute in 5 litre of water.
- Datura seed can also be used. This solution is active against insect pests, diseases and mites etc.

5.3.32. Tulsi extract

- Soak 100g tulsi in sufficient quantity of water for overnight.
- Grind next day and add 2 litres of water.
- This is sufficient to spray 2 cents of land.

5.3.33. Effective microorganisms fermented plant extract (EM-FPE)

Preparation of Activated Effective Microorganisms

To prepare AEM, EM solution (300 ml) is mixed in jaggery solution (3%) to activate the microorganisms.

Procedure to prepare 10 litres solution of FPE

Take 20 litres capacity container. Fill the container with finely chopped leaves used to prepare FPE in equal proportion (1/2 kg each) to 35% capacity. Add 5.9 litres of water (59%) to this container. Add 300ml (3%) AEM and 300gms (3%) Jaggery dissolved in the water to the container. Stir the mixture thoroughly with a bamboo stick and close the lid. Stir the solution twice a day for a week. After fermentation, the pH must be in the range of 3.5-4. This solution should be used within 6 months.

Use

- Keeping in view the intensity of the pest attack,10-20 ml FPE solution is mixed in 1 litre water or 100-200 ml FPE solution is mixed in 10 litres of water and sprayed to the crops.
- Depending on the intensity of pest attack, FPE is sprayed more than three times at an inter-val of 10-12 days if necessary. As a precautionary measure, it has to be sprayed during flowering and earhead emergence stage.

5.3.34. Chilli-garlic-ginger extract

Green chilli 0.5 kg,

Ginger 0.5 kg,

Garlic 1 kg,

Tobacco leaves 0.5 kg,

Neem oil 200 ml

1 packet of shampoo

- Crush green chillies, ginger, garlic separately with -required quantity of water. Soak tobacco leaves in 1 litre hot water overnight and take the solution for use.
- Mix 200 ml neem oil with one packet of shampoo (use of shampoo is to mix thoroughly neem oil and water).

- Then mix chilli, garlic and ginger extract and tobacco leaf extract and neem oil in a mud pot or plastic bucket and close the lid. Keep this solution for 1 or 2 days and apply to the crops. This solution should be used within 3 days of preparation.
- 10 ml of the solution is diluted in 1 litre of water and sprayed to crops for effective control of pests.
- Spraying is done regularly at an interval of 10-12 days during flowering and fruiting stage.

5.3.35. Garlic brew

i. Garlic 200 gm

ii. Cattle urine – 20 litres

iii. 20 litre capacity bucket

- Crush or grind the garlic into a paste and put into the bucket containing cattle urine.
- Stir the materials for 5-10 min and cover with a gunny cloth or cotton cloth which can provide sufficient aeration.
- Allow the materials to ferment for 5-7 days.
- After 7 days sieve the solution before spraying to avoid clogging of the nozzles of the sprayer.
- The solution is diluted 10 times with water and sprayed on the foliage in the evening hours.
- Care should be taken that the concentrated solution should not be sprayed in the crop.

5.2.36. Garlic decoction

Garlic – 1 kg

Water – 10 litres

- Boil 1 kg of garlic in 10 litres of water separately.
- Allow the solution to cool and sieve the solution with a muslin cloth.
- Dilute one part of the solution with 10 parts of water and spray on the crop during evening hours.

5.4. Herbal Insecticides for Individual Crops and Insects

5.4.1. Nematode control in turmeric (Innovator- K.M. Chellamuthu, Tamil Nadu)

Ginger - 200 g

Chilli - 200 g

Nirgudi leaves (*Vitex negundo*) - 1 kg

Garlic - 500 g Grind all into a fine paste & mix with 150

Aloe vera - 500 g liter water

Neem seeds - 1 kg ↓

Cleodendron inerme - 1 kg Apply over soil 120 days after planting for 1 acre

5.4.2. Insect control in paddy (Innovator- K.M. Chellamuthu, Tamil Nadu)

Nirgudi leaves (*Vitex negundo*)- 1 kg Grind all into a fine paste & mix with 100 l water

Aloe vera - 1 kg ↓

Neem seeds - 1 kg

Cleodendron inerme - 1 kg Foliar spray over one acre

5.4.3. Control of sucking pest in cotton, castor and green leafy vegetables (Innovator- Rajinikant Bhai Patel, Gujarat)

- Crush 3 kg fresh leaves of Black Veldi (a cotton sp) in 20 litre of water and boil till the volume reduced to 5 litre.
- Filter and use as foliar spray.
- Spray 3 to 4 times with a gap of 10 days.

5.4.4. Control of army worm, aphids, cotton semilooper, green leaf hopper, mites, powdery mildew, pulse beetle and rice weevil

- 1 kg Turmeric + 4 liters of cow's urine with 20 litres of water.

5.4.5. Control of american boll worms, aphids, pulse beetle, white fly, etc.

- Make 500 g garlic paste in 100 ml kerosene + 100 g chilli paste in 50 ml water + 100 g ginger paste.

- Add all the paste in to 30 litre of water along with emulsifier.
- Spray in the field over half acre.

5.4.6. Control of aphids and beetles

- Custard apple seed powder 500 g in 2 litre water and boil till 500 ml solution remains. Mix it with 15 litres of water and spray.
- 2 kg custard apple leaves fresh juice in 500 ml water + 500 g of chilli water extract + 1 kg neem seed extract in 2 litre. Dilute it with 60 litre.

5.4.7. Termite control

- Wood ash heap around the base of the trunk prevent termite infestation in coffee bushes and datepalms.
- Repeated pouring of cattle urine dilute at 1:6 with water in termite holes helps in keeping their spread under control.
- Mixture of cattle dung and red coloured clay with water is coated on the trunk and large twigs at the onset of monsoon when termite damage is sever. Fresh and young grafts are can alsobe coated with it to protect them from termites.
- Grow *Aloe vera* around the tress

5.4.8. Supli (*Mundulea suberosa*) – Control of leaf eating, sucking and fruit/shoot borers of vegetables

- Crush 1 kg green leaves of supli in a pot with 10 litre of water.
- Boil the mixture till it is reduced to half.
- After cooling dilute it to 100 litres with water for spraying over 1 acre.

5.4.9. Tamarind and lemon for hairy caterpillar in castor (Innovator-Banidanbhai, Tamil Nadu)

- Mix 500 ml juice of tamarind with 500 ml juice of lemon in 15 litres of water.
- Spray the mixture over 0.25 ha

5.4.10. Phytopalm-Herbal pesticide against coconut mites (Innovator- Louis, Director of centre for Innovation and Transfer of Technology, Kanyakumari, Tamil Nadu)

The extract of 10 herbs

1. Kolingi (*Tephrosia purpurea* (L.) Pers)
2. Nochi (*Vitex negundo)*

3. Veppa ilai (Neem leaves) (*Azadirachta indica*)
4. *Vinca rosea* – Nithya Kalyani
5. Punga mara ilai (*Pongamia glabra*)
6. Unni chedi (*Lantana camera*)
7. Garlic (*Allium sativum*)
8. Ginger (*Zingiber officinale*)
9. *Cassia auriculata – Avarai*
10. Turmeric powder (*Curcuma longa*).

5.4.11. Mukkadaka decoction to control hoppers in paddy (Innovator- B.S. Dinesh, Karnataka)

- To 1 kg of mukkadaka leaves add 10 litres water and boil.
- The solution is filtered and diluted with water in 1:10 ratio and sprayed twice, once during nursery stage and after transplanting.

5.4.12. Caterpillar control in cotton (Innovator from Vill: Choryana, Tal: Savli, Baroda Gujarat)

- Latex of Akda (*Calotropis gigantia*) when dilute with 15 parts of water and sprayed over crop.

5.4.13. Use of *Clerodendrum, Aristolochia, Azadirachta* and *Enicostemma* to control cotton pest (Innovator : Shri Jadubhai Savaliya, Bhavnagar)

Formulation comprises of

Arni (*Clerodendrum phlomidis*)
Kidimari (*Aristolochia bracteata*)

Leaves are crushed in water and juice is extracted ↓

Mamejavo (*Enicostemma littorale*)

Filter it through fine cloth

Neem (*Azadirachta indica*)

200-500 ml of this extract is diluted to 15 litres water for spraying. 30 litres of the extract is required of 800 m^2 area

5.4.14. Termite management (Innovator- Vill: Khagiyali, Tal: Sihor Bhavnagar, Gujarat)

- Akada plant material 8-10 kg is soaked in water for atleast 24 hours then filtered.

5.4.15. Pest Management

- To control pod borer (*Leucinodes orbonalis*) in vegetables, *Anethum sowa* plants are sown in rows along with the vegetables. The strong odour of the plant repels the insects (Farmers in Karnataka, India).
- Mint and marigold are planted in the farm to repel insects (Farmers in Philippines).
- To project the plants from the white ants, *Vetiveria zizanoides, Euphorbia tirucalli* and *Calotropis gigantea* are planted at intervals in the field (Farmers in Maharashtra).
- A mixture is made with 1 litre. neem oil, 3 kg of fine sand and 3 kg. of cow dung and heaped in shade covering with a moist sack for 3 days. On the fourth day, the mixture is dissolved in 150 litre. of water and sprayed to control all sucking pests (Farmers in Tamil Nadu).
- About 10 kg of dried cow dung is ground in to fine powder and mixed with ash (obtained from brick kiln) and dusted in the early morning, to control pests and diseases (Farmers in Tamil Nadu).
- Garlic acts on a wide spectrum of organisms in unrelated crop plants singly or in combination with neem products, chilli, asafoetida etc. Besides garlic is effective against bacteria, fungi and nematodes (Farmers in Tamil Nadu).

5.4.16. Aphid attack on mustard

- Cow urine can be sprayed on infected planst.
- Ash obtained from the firewood can also be used as broadcast for control of aphids.

5.4.17. Control of rice pests

- Raw cow dung is mixed with water to prepare a suspended solution which is sprayed in the rice field – control the trips.
- Erection of bamboo branches or top, or the other stick in the rice field controls rice stem borer.
- Steam decocotion of neem, seed and leaves, and the extract must be sprinkled on rice crop for the control of *Scirpophaga incertulus*. The extract is prepared by mixing 1 to 3 g of ground neem seed/leaf in 11 of water (0.1-0.3 % concentration) for 12 hrs.
- Rope, dipped I kerosene stirred in the water of rice field will control of rice case worm.

- Spreading of raw cow dung in the field water will control rice case worm.
- Placing chopped leaves of Indian rhododentron or few branches of fern or citrus peels in paddy field will reduce the population of *Scirpophaga incertulus*.
- Mixing cowdung with water in the rice field will control *Dicladispa armigera*.
- Spreading of goat dung in the rice field will control rice insect pests.
- Garlic 1 kg + dried tobacco leaf 200 g + washing soap 200 g in 5 litres water @ 150 litre water/ acre for the control of Gandhi bug.
- Neem kernel 5 kg in 100 l water, keep it for 8 hrs. add 100 ml teepol @ 500 l water/ha for the control of green plant hopper.
- Smoking with the help of dried cow dung in one end of the field will reduce the pests in field.
- One kg of chill powder is mixed with 1 litre of kereosne. This should be sprayed early in the morning.
- 5 kg of neem cake is taken in gunny bag and kept in the main channel of the rice field. During irrigation, water passes through neem cake and takes essence from neem cake which will control the root borne diseases and pest.
- Cow urine is taken in a mud pot or any container and boiled well until 10 litres cow urine comes to 5 litres i.e. half of the quantity. With this neem oil 2 litres cow urine 5 litre added. Again boil the contents well until 12 litre comes to 6 litrs then cooled. 1 lit/ac (100 ml/ 10 li water).
- Rice bran (5-10 kg) and neem oil (1 lit) are mixed well and sprinkled overhte rice crop at early morning to control Brown Plant Hopper.

5.4.18. For cotton boll worm

Ginger	- ½ kg	
Garlic	- 1 kg	Pound well and make in to a paste
Green chilli	- 1 kg	

In a buried mud pot, put this paste and Cow urine (10 litres) and keep for fermentation for10 days. Mouth of the mud pot is covered with waste cloth. It has to be stirred well daily in the morning. Dosage: 3.5 litre/ac (0.5 litre/ 10 litre). Spraying can be done weekly intervals.

- Fresh cow urine must be avoided

5.4.19. Low cost practice for control of case worm in paddy (Shri Bhupen Singh, West Bengal Mobile: 08101215689)

In this method a rope of desired length is fully soaked into kerosene and run by two persons from both ends of the paddy field over the standing crop with adequate care that almost all the plants are shaken by the rope. Rice case worms are fallen down in the stagnant water and immediately the water is drained out. These worms are washed out of the paddy field with the flow of water. Kerosene oil acts as repellant to drive away the pests.

5.4.20. Amulya amrit' for pest and disease management (Shri Surya Prakash Bahuguna, Uttarakhand Mobile: 09412147702)

A mixture of cow urine (5 litre), cow milk (0.5 litre), curd (0.5 litre), honey (200 g), green banana (5), coconut paste (1 coconut) and Ghee (50 g) are kept in sealed container. This mixture is kept in shade covered with wet gunny bag for three days. After three days, the gunny bags are removed and the container is opened to release the gas and stirred with stick. After stirring and releasing the gas for another 3-4 days, the fermented solution is filtered through muslin cloth. This solution is named as "Amulya Amrit" and used effectively against foliar fungal diseases and Lapidopterous borers in rice, mango, litchi, wheat, etc. The farmer is using one litre mixture in 5 litre of water in above crops for last 5 years.

5.4.21. Cutworms

- Mix equal quantities of hardwood sawdust, bran, molasses and enough water to make the solution sticky. Spread around the base of the plants in the evenings. The molasses attract the cutworms and as they try to pass through it they get stuck. The substance dries out in the sun and the pest dies.
- Mix 100 grams (g) of bran, 10g of sugar, 200g of water, 5 g of pyrethrum powder. Spread around the base of the plants. The cutworms eat the substance and die.

5.4.22. Fruit fly

Traps need to be put in place before an attack is likely to start. For fruit fly, the traps should be baited 6 to 8 weeks before the fruit ripen.

- Make a small hole in the bottom of a plastic bottle or container. Seal the top of the bottle with a lid or stopper. Fill one quarter of the bottle with the bait. Hang the bottle upside down from trees around fields or gardens. The flies are attracted to the bait through the small hole. They are then trapped and drown in the bait.

- Cut the top of a plastic bottle off. Pour some bait in the bottom half of the bottle. Turn the top half of the bottle upside down and place in the bottom half. Again the flies are attracted into the bottle and drown in the bait.

Here are two different baits for fruit fly that can be poured into the traps:

- Mix 1 litre of water, 250 millilitres (ml) of urine, a few drops of vanilla essence, 100g of sugar and 10g of pyrethrum powder.
- Mix 1 teaspoon of pyrethrum powder, 250g of honey, a few drops of vanilla essence, 250g of orange or cucumber peel or pulp and 10 litres of water.

5.4.23. White grub

- One kg of table salt has been broadcasted in a land.
- Other method is to leave the land fallow for one season, which breaks the life cycle.

5.4.24. Preparations containing Baculovirus

Initially harvest approximately 20 infected cabbage looper caterpillars (*Trichoplusia*) and crush them thoroughly with a drop of water. The homogenate thus obtained is transferred using a clean cloth. Finally, the homogenate is diluted until a sufficient volume of spray to treat approximately half a hectare is obtained. After 3 to 4 days, the caterpillars become sick and die. This preparation must be applied as soon as possible, when the caterpillars are still young, in order to prevent them from causing serious damage. Good results have been obtained with the cabbage looper caterpillars and the cotton bollworm (*Helicoverpa armigera*).

5.4.25. Tobacco-based preparations

Tobacco contains nicotine, which is a highly toxic, organic poison. Its leaves and stems are used to prepare sprays that can be applied to the plants to control the number of insects (aphids, caterpillars, flea beetles, thrips, leaf miners) and mites.

Preparation of a tobacco-based insecticidal spray

Method 1

Crush or grind 1 kg of tobacco stems and leaves, and mix the homogenate thus obtained with 15 litres of water and a small quantity of soap. Leave the mixture to stand for 1 day, and then filter carefully in order to remove any plant particles. Treat with the solution thus obtained using a sprayer fitted with a nozzle to ensure very fine sprays.

Method 2

Mix 250 grams of tobacco (e.g. taken from a collection of cigarette butts) with 4 litres of water. Add 30 grams of soap (e.g. so-called "Marseille" soap). Gently boil the mixture for 30 minutes and then filter carefully. Add 16 litres of clean water to the filtrate and treat with the solution thus obtained using a sprayer fitted with a nozzle to ensure very fine sprays.

Method 3

About 1 kg of tobacco leaf is crushed in to paste and soaked in 10 litre of fresh cows urine and 1 litre kerosene for 15 days. Then the solution is filtered using cotton cloth. For spraying, 1 litre of this filterate is mixed with 9 litres of water. When the pest incidence was severe another spray was done after 15 days interval.

Method 4

Take 1 kg tobacco snuff (*Nicotiana tabacum*) and 2 kg. Aloe (*Aloe Vera*) Wash both plants properly and chop it finely and boil with 6 litre water at moderate and constant heat till it remains half. Mix 4 litre supernatant of buttermilk. Spray 150 ml. of this formulation per pump. About 6-8 pumps per acre are required to control the -infestation. Spray after 10-12 days if infestation is still there. For best result, use this formulation within 6 months of preparation.

5.4.26. Insects and pest management in coconut tree

To protect coconut tree from insect and pest, a formulation made by mixing salt and sand in the ratio of 1:1 is tied at the second leaf of coconut tree. This helps in destroying the rhinoceros beetle and controls white mites.

5.4.27. Control of rice ear head bug

- About 10 kg of the rice husk was mixed with 2 litre of kerosene

5.4.28. Aloe vera for control of termite attack on trees

- Grow *Aloe vera* around the trees.
- These *Aloe vera* plants are believed to possess some property which suppressed the termite attack in trees.

5.4.29. Kerosene spray in gingelly field

- About 1 litre of kerosene is mixed in 20 litre. of water and used as spray solution.

5.4.30. Paddy pests' control

- After 25 days of paddy sowing, the leaves of erukku, *Calotropis gigantea* are cut into pieces (using an indigenous tool aruval) and put in the field.

5.4.31. Aphids management in groundnut

- Fresh cow's urine is mixed with water in the ratio of 1:10 (1 litres of cow's urine with 9 litres of water).
- Spray in Groundnut field at Vegetative phase during early morning /late evening hours.
- Cow's urine produce substance which has antibiotic activity suppressed aphids and other leaf sucking pests.

5.4.32. Pest control in brinjal

- About 2 kg of Neem Seed Kernal is powdered and soaked in 15 litre of cold water in an earthen pot (20 litre capacity).
- At least 10 earthen pots are to be placed in the field of 1 acre at 20-25 feet distance covering the entire field.
- The filled pots are tied to the mouth with a cotton cloth to avoid houseflies breeding.
- This cloth cover also acted as a screen avoiding dust particles, provided good aeration and released the neem odour uniformly over the field and made unpreferrable to most of the pests in brinjal.

5.4.33. Neem leaf extract for pest control

- About 10 kg of green neem leaf was soaked in 10 litres of water.
- The leaves were soaked overnight in water.
- The next day the leaves alone separated and grounded (using an indigenous milling tool called 'Ural').The grounded leaves were then mixed in the same water and left for 3 days. The extract was then filtered using a cotton cloth.
- The extract was then mixed with emulsifiers like teepol, sandovit, soap oil or kathi soap cake powder (2g/litre). This emulsifier (1 ml / litre) would help the extract to stick well to the leaf surface. This neem leaf extract (10 litre) was sprayed (with Knapsack sprayer) in early morning or late evening hours in the cropped field7.

5.4.34. Paddy husk and Kerosene spray

- In this practice, mix 10 kg of rice husk with 2 litres of kerosene for ear head bug control.

5.4.35. Tobacco leaf extract spray in groundnut

- About 1 kg of tobacco leaf was crushed in to paste and soak in 10 litres of fresh cows urine and 1 litre of kerosene for 15 days.
- Then the solution is filtered using cotton cloth.
- For spraying, 1 litre of this filterate is mixed with 9 litres of water and spray the solution in the early morning hours.
- When the pest incidence is severe another spray has to be given after 15 days interval.

5.4.36. Sweet flag extract for crossandra thrips

- About 1 kg of the dried Sweet flag is powdered and soak it in 10 litres of water for 4-5 days and keep closed in an earthen pot.
- After 5 days the solution was filtered with the help of cotton gada cloth.
- About 1 litre of this filtrate is mixed with 10 litres of water for foliar spraying.

5.4.37. Erukku leaves in paddy field

- 25 days after paddy sowing, the leaves were cut into pieces (using an indigenous tool aruval) and put in the field.

5.4.38. Aphids management in groundnut

- In this practice fresh cow's urine is mixed with water in the ratio of 1:10 (1 litres of cow's urine with 9 litre of water).
- Spray the solution in groundnut field at vegetative phase during early morning /late evening hours.

5.4.39. Neem leave extract for field pest control (*Veppam saaru*)

- Soak about 10 kg of green neem leaf in 10 litres of water for overnight.
- The next day separate the leaves alone and ground it. The grounded leaves were then mixed in the same water and left for 3 days. The extract was then filtered using a cotton cloth.
- The extract was then mixed with emulsifiers like teepol, sandovit, soap oil or kathi soap cake powder (2g/litre). This emulsifier (1 ml / litre) would help the extract to stick well to the leaf surface.

- Spray the neem leaf extract (10 litre) in early morning or late evening hours in the cropped field.

5.4.40. Curd + neem leaf paste spray for bollworm control in cotton

- In this method, about 15 litres of curd is first mixed with 15 litres of water in an earthen pot.
- Then add 5 kg of crushed neem leaf paste to the pot. This mixture was stirred in the morning and evening hours thoroughly with the help of neem stick and closed with a lid.
- After 15 days of fermentation this solution is filtered using a cotton cloth.
- Spray the filtered extract to the cotton crop at the early stage of square formation.
- These 30 litres of solution is enough for spraying a hectare of the cropped field.

5.4.41. Cow's urine spray for bollworm

- About 10 litres of fresh cows urine was mixed with 9 litres of cold water and sprayed to the 45 days old cotton seedlings.
- Like that 10 tanks of the spray solution is needed for spraying 1 acre of cotton field.

5.4.42. Pungam seed extract for sucking pests

- Pound 20 kg of pungam using water.
- Soak the crushed material in water for overnight and next day the solution is filtered using gada cloth.
- The filterate obtained was mixed with water in (1:9) ratio and spray using knapsack/hand sprayer during early morning or late evening hours.
- Pungam seed extract spray is first done to 20-25 days old crop and repeated after 1 month based on the pest incidence.

5.4.42. Ash application in groundnut field

- About 2-8 bags of capacity (20 kg) in the 35 days old groundnut field.
- The fact is that the ash filled the intergranular spaces in insect body and thereby restricts insect movement.

5.4.43. Tobacco leaf extract for brinjal pests

- Mix 2 kg of tobacco leaf dust wit10 litres of cow's urine in an earthen pot and keep for 5 days.

- Stir the solution thoroughly with the wooden stick during morning and evening hours.
- After filtering the solution using cotton cloth, about 1 litre of this filtrate is mixed with 9 litres of water and spray an acre of cropped field during early morning hours.

5.4.44. Smoking in paddy field

- In this practice, an earthen pot with holes in the bottom is taken and 100g of pig dung was burned to create smoke.
- These pots were placed on the sides of bunds so that smoke would enter the field.
- This practice done in early morning or late evening with 10 days interval during milky stage reduced ear head bugs.
- For an acre 8-10 such earthen pots were placed in the field.

5.4.45. Neem seed kernal and sweet flag extract spray

- Mix 15 kg of neem kernels and half kg of Sweet flag and grind it.
- Then the grounded, materials were soaked in 100 litre of water for overnight. Then next day the solution was filtered using cotton cloth.
- This filtrate was used for spraying in 25-30 days old cotton crop with knapsack sprayer.
- Repeat the spray once in 15 days.

5.4.46. Management of mites

- About 1kg of omathai leaf (*Datura fastuosa*), Erukku leaves (*Calotropis gigantea*), Thulasi (*Ocimum* sp.), *Castor* (*Ricinus communis*) each - soak in 9 litres of cow's urine for 3 days.
- After filtering, the extract is sprayed to the trees like lemon, mango etc., and to the crops like paddy. This spray was done once in 6 months interval to the perennial trees and once in a month in case of paddy crop.

5.4.47. Dusting ash and cow dung to manage pests and diseases

- About 10 kg of dried cow dung is grounded into fine powder and mix with 10 kg of ash (obtained from brick kiln). This mixture was dusted in early morning hours to control wide range of pests and diseases.
- Dust this mixture usually in 1 month old crop (vegetative phase) for effective pest and disease control.

5.4.48. Grasshopper control in beetroot

- Within 10 days of beetroot seed sowing, karpooravalli leaves are scattered in the field.

5.4.49. Ash application in avarai

- Take ash in a basket, (1 basket of 20kg capacity) and dust as foliar application using a bunch of neem leaves. Ash was applied in such a manner that it spreaded on both upper and lower surface of the leaves to control sucking pests

5.4.50. Whitefly control in cotton

- About ½ litre of castor oil is taken in a vessel and about 50 ml of shampoo was added and stirring thoroughly using a wooden stick. Then the solution was mixed up with 7 litres of water.
- About 1 litre of this solution was enough to mix with 9 litres of water.
- One kg of Jaggery is dissolved in 10-12 litres of water and filtered through a thick cotton cloth. Approximately 5-6 litres solution is sprinkled in one acre with the help of sprayer pump gently on the foliage.
- To get rid of white flies, papers pasted with castor oil/grease are hung at five or six places in the cotton field. Then air is blown using a sprayer over the crop. The disturbed adult flies (white fly) come out of the plant and stick to the oily paper. By this method 90% of white fly can be successfully reduced.

5.4.51. Control of aphids

- Application of tobacco decoction mixed with soap emulsion is done to control aphids in pulse crops.
- Freshly cut branches of erukku (*Calotripis gigantea)* and places in irrigation channels to control aphid infestation in Lucerne (*Medicago sativa).*
- A little quantity of castor oil is added during irrigation in the water channel for reducing aphid population.

5.4.52. Insect control measures in cotton

- Concentrated solution of sugar (500 g in a litre of water) is prepared and allowed to ferment. Fermented solution is placed in open dishes at different locations between the rows of cotton. One or two drops of edible oil are added to the dish. The users say that adult insects (which can fly) are attracted towards this solution. The idea behind this practice is to kill the adult population and ultimately reduce the reproduction.

- The latex of Erukku *(Calotropis gigantia*) when diluted with 15 parts water and sprayed on the crop, effectively controls the pest within three days. The new growth after treatment is also free from infestation.
- About 250 to 300 g of *Dhatura'* leaves along with stem are dipped in 1 litre of lukewarm water. After cooling down, 250 to 300 grams of the solution is mixed with 15 litres of water and sprayed on the crop. Pests perish within a period of six to seven hours. Spraying the mixture when the crop is of one-month-old yields better.
- Arali *(Nerium oleander)* seeds are pulverized, soaked in water overnight and filtered. This filtrate is diluted in water and sprayed on cotton fields. This practice provides 70% control.
- About 10Kg pearl millet flour is mixed with 200 litres water in plastic drum and it is kept for fermentation under the heap of compost for 8 days. After 8 days the solution is sprayed on cotton to check infestation of *Helicoverpa.*

5.4.53. Asafoetida (Perungayam)

- Asafoetida or 'hing ' is used by farmers for the control of termites (*Odontotemzes obesus*). About 40-50 gm of asafoetida is tied in a cotton cloth.
- Two to three packs are placed in the irrigation channel at a distance of about 20-30 metres from each other during irrigation.
- The disagreeable odour repels termites and other insects.

5.4.54. Management of insect pests of mustard crops

- Leaf decoction (1 kg) of *Aloe vera* and tobacco powder extract (200 gm) is prepared in 5 liters of water for 3-4 hours to make a 2 litres solution. Neem leaf extract (200 ml) is added after evaporation process and decoction of 50 gm *Sapindus trifoliatus* powder is added to the above solution and mixed thoroughly.
- This is sprayed on the mustard crop at interval of 2-3 weeks.

5.4.55. Control of termites in rainfed paddy nursery

- When paddy nursery raised under rain fed situation in light textured soil like red loam, the termite incidence is more. The termites will move in gangs and devast the young seedlings and eat them away completely, causing total disaster.
- This will lead to re sowing of the nursery bed once again at heavy cost. Besides the optimum season is also missed.

- To prevent this loss of nursery, it is followed traditionally to spread bitter leaves of neem and hairs of goat and human beings in thick layer. The termites that eat them will perish soon.

5.4.56. Traditional method of control of rhinoceros beetle in coconut palm

- Two numbers of earthen pots are buried into the soil so that the top of pot alights at ground level and it is openly exposed.
- Water is filled up to ¾ th of the pot. To each pot 250 gm of powdered castor cake (castor cake obtained after extraction of castor oil) is added and mixed well.
- After 3 days it ferments and typical odour emanates from this attracts the adult rhinoceros beetles which cause severly loss to the Coconut palms leading to poor yield of nuts.

5.4.57. Control of paddy earhead bugs

- The flowers of herbal plant called in tamil "peyurunjan" flower (or) "Sannambu" flower (*Cycas cercinalis*) will release obnoxious odour at the time of flowering.
- These flowers are cut in to small pieces, covered and tied with wet cotton cloth. These cloth bundles are again covered with straw and fixed on the top of the sticks (at 3 to 4 feet height) which are placed in 2 or 3 spots of 1acre standing paddy land.
- Dew fall in the early morning during winter months will help to emit obnoxious odour that will repel the ear head bugs away from the paddy field.

5.4.58. Control of 'pugaian' pest (brown plant hopper) in paddy

- Rice bran – 15 kg mixed with kerosene – 2 litres and apply in the morning hours near 1 acre of paddy crop affected by brown plant hoppers known as Pugaian pest (*Nilaparvata lugens*).

5.4.59. Control of jassids in mango

- Fish oil – 2 litres mixed with water – 200 litres and the solution is sprayed with a hand operated sprayer over the inflorescence of mango during flowering stage, to control green jassid.
- When young nymphs of green jassid are seen on the inflorescence repeat the spraying of fish oil solution at weekly interval.

5.4.60. Making bon fire for control of groundnut pests

- Bon fire is set up with agriculture waste materials near field crops like groundnut to attract female moths of Red hairy caterpillar (*Amsacta albistriga*) pest on ground nut and *Prodenia litura* caterpillar pest on several field crops.
- Bon fire is made during night time. The adult moths of *Amsacta albistriga* are attracted towards the light. The attracted moths will fall and perish in the fire. This method is practiced in Palladam, Kannivadi, Singampunari areas of Tamil Nadu with success.
- Red hairy caterpillar is a devastating pest to rain fed groundnut. Few farmers also make light traps and set them in the night time nearby crop field. By setting up light traps, the emergence of moths and its outbreak can be known in advance and all possible control measures on a war – foot basis can be followed to prevent this major disaster to ground nut.

5.4.61. Control of insect pests of tree crops

- Water is filled in small cement trough or pots.
- This will attract the predatory birds to the garden which will prey on the caterpillar pests damaging the trees incidentally and thus enabling the biological pest control. About 80% of bird species are insectivores and feed on the larval caterpillar.
- Still some pests may persist. To control them the following method is prescribed: In a mud pot neem cake, leaves of Calotropis and Nux-vomica are soaked in water and kept in few places in the garden. The insects are attracted towards the fermented smell of the ingredients and fall into them and perish.

5.4.62. Botanical pesticides to prevent coconut pests

- For preventing the incidence of Rhinoceros beetle and Nephantis caterpillar pests in coconut palm, 1% neem oil is sprayed.
- In between the tender leaves of coconut fronts, the leaves of wild indigo (*Tephrosia purpurea*) are placed intact, which will prevent the incidence of Rhinoceros beetle.
- Waste oil like castor oil / ground nut oil is taken in a small mud pots and tied to coconut tree, lamp post (at 10 feet height)in the coconut garden. The adult rhinoceros beetles are attracted due to the smell of oil and fall into the pots and perish soon.

5.4.63. Control of pests of paddy and pulse crops

- Leaf - caterpillars of paddy are controlled by spreading neem leaves in the field or applying wood ash to crop.
- Similarly top dressing of well decomposed farm yard manure and green leaf manures such as neem, 'poovarasu' (*Thespesia populnea*), 'Vagai' (*Albizia lebbeck*) on 30 the day of planting. Apply powdered neem cake (or) "Pungam" (*Pongamia pinnata*) cake at 1 kg per acre over the foliage of paddy crop.
- When pulse grows like black gram, small pulse (Siru payaru) are stored, small quantity of 'Vasambu' (*Acorus calamus*) is powdered and mixed to prevent the attack of store pest.

5.4.64. Mixed pesticide

- Cow's urine, neem oil and tobacco leaf decoction are mixed together into a solution and sprayed over the foliage of vegetable crops, to prevent the incidence of sucking of pests like thrips, aphids, jassids.

5.4.65. Aphids control in cotton / chillis

- After the distillation of 'Palmarosa' –oil (*Cymbopogan martine*) the residue steam left in the condenser is cooled and the liquid is collected. This liquid is made use of ,in plant protection, to control the sucking pest like aphids in cotton and chilli crops.
- Turmeric rhizome and waste finger rhizomes are dried and finely powdered and sieved. This turmeric powder and wood ash sieved are mixed in 1:1 ratio and used as duster during early morning hours over the foliage of vegetable crops, when dew fall is there, to control the sucking pests like aphids.

5.4.66. Pest control in paddy crop

- Take a mud pot half filled with water. To this 3 kg of short ' kendai ' variety of fish is added and buried in a manure pit to ferment for 3 days.
- Then they take the pot out and add required quantity of neem seed extract. All the contents are thoroughly mixed and diluted with water at 1: 5 ratio.
- This solution is sprayed over the foliage of paddy to control all caterpillar pests.

5.4.67. Control of castor semi looper

- Around 150 ml lemon (*Citrus limona*) juice and 150 ml tamarind (*Tamarindus indica*) juice is mixed in 15 liters of water.

5.4.68. Control of larval pest in pigeon pea

- Take *Capsicum annum* (Chilli powder, Neem leaves and fresh Garlic in proportion of 1:4:1 respectively.
- Boil it with 16 times water and keep half (8 times). Filter this solution and use after recommended dilution.
- Spray 150 ml. of this formulation per spray.
- About 6-8 spray per acre are required to control the -infestation. Spray after 10-12 days if infestation is still there. For best result, use this formulation within 6 months of preparation.

5.4.69. Control of heliothis in chick pea

- Take equal amount of leaves of *Adhatoda vasica* (Vasaka) and *Pongamia pinnata* (Karanja) Wash both plant material properly in water and chop it finely and boil with 16 times water at moderate and constant heat until the final volume remains half.
- Spray 150ml. of this formulation per spray. About 6-8 spray per acre are required to control the infestation. Spray after 10-12 days if infestation is still there. For best result, use this formulation within 6 months of preparation.

5.4.70. Control of cotton bollworm and all type of heliothis

- Take 1.5 kg fresh leaves and tender parts of naffatiyo (*Ipomea fistulosa*) 1 kg gandhati (*Lantana camera*), 1.5 kg tobacco snuff (*Nicotiana tabacum*) and 1.5 kg neem (*Azadirachta indica*) in 10 litre water and boil till it remains half. Filter this solution and use after recommended dilution.
- Spray 150m l. of this formulation per spray. About 6-8 spray per acre are required to control the infestation. Spray after 10-12 days if infestation is still there. For best result, use this formulation within 6 months of preparation.

5.4.71. Termite

- Take 3.5 karanj (*Pongamia Pinnata*) leaves and 3 kg neem (*Azadirachta indica*) leaves. Wash all the ingredients properly and chop it finely and boil with 10 liter water at moderate and constant heat until the final volume remains half. Mix 1 liter castor oil after filtration.
- Add soap nut (*Sapindus emarginatus*) powder to emulsify- oil contents.

- Spray 150m l. of this formulation per spray. About 6-8 spray per ac re are required to control the infestation. Spray after 10-12 days if infestation is still there. For best result, use this formulation within 6 months of preparation.

5.5. Herbal Fungicides

Leaves and flowers of the following herbs/shrub, 3 to 5 kg each is ground in fine powder and soaked in BDGS/ cow's urine and for 12 hours and the extract is used as fungicide (3%) with 1/4kg of Turmeric Powder and 500gm of *Pseudomonas fluroscens*.

Aloe vera 3-5 kg and leaves of any two of the following Bougainvilla + Lantana camera+ Seethpal etc (3-5kg)- pour all these in 15 litres of water and boil it until reduces to 10 litres. Then filter it. Add 1 kg of turmeric powder into that and leave it for 12 hours. Then 500gm of *Pseudomonas fluroscens* 250-500 in 1 litre solution and dilute it with 10 litres of water for spraying

As a holistic approach, following are the organic farming practices in a nutshell followed by majority of Organic growers of Tamil Nadu.

Crops grown Rice, groundnut, sugarcane, banana, cotton, vegetables, turmeric, maize, coconut

5.5.1. Nochi karaisal

Take 5 kg Nochi leaves soak it in 20 litre of water for 6 days

↓

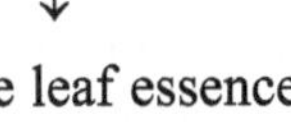

Filter the leaf essence

↓

Take 30 ml of that essence in 1 litre of water and sprinkle it using neem leaves.

5.5.2. Fungal disease control

- A mixture of ash (2-3 kg) and 1 litre of castor oil is spread on a sedd bed of a size of about 100 m^2. The application is repeated 2-3 times at intervals of 7-10 days. This protects against soil borne diseases in tobacco nurseries.
- A mixture of 2 kg turmeric powder and 8 kg wood ash is used as dust over leaves for treatment against powdery mildew.
- Ginger powder at 20 g/litre of water and sprayed thrice at interval of 15 days can be used for powdery mildew and other fungal diseases.
- Handful of slaked lime applied at the base of tomato plant can combat damping off disease.

- Cattle and goat urine have fungicidal properties. 2 cups of cattle urine with 5 ml peppermint oil and 10 litre of water can be used to control fungal diseases on grapes.

5.5.3. Control of leaf curl diseases in brinjal (Innovator- Popatbhai Jambucha, Vill: Mathavada, Gujarat)

In a drum

Dried Spirogyra	- 2 kg	Keep it for 12 hrs
Wood burnt ash	- 3 kg	
Cow's urine	- 5 litres	Filter through muslin cloth and spray
Water	- 40 litres	

- Spraying is taken up at an interval of 10 days as long as crop is in the filed.

5.5.5. Eradication of vectors by anti viral principles (AVP)

Plants are also known to contain some compounds which are inhibitory to virus. They are called Anti-Viral Principles (AVP) or AntiViral Factors (AVF). The leaf extracts of sorghum, coconut, *bougainvillea*, *Prosopis juliflora* and *Cyanodon dactylon* are known to contain virus inhibiting principles.

Preparation of antiviral principles

Dried coconut or sorghum leaves are cut and powdered. Twenty kg of leaf powder is mixed with 50 litres of water and heated at 60^0 C for one hour. It is filtered and volume is made upto 200 litres. This gives 10 per cent extract. Five hundered litres of extract is required to cover one hectare.

The 10 per cent AVP extract is very effective in controlling groundnut ring mosaic virus (bud necrosis). Two sprays are to be given at ten and twenty days after sowing. Similarly 10 per cent leaf extracts of *P. juliflora* and *C. dactylon* effectively reduced the tomato spotted wilt virus in tomato. The leaf extracts are known to contain some proteinaceous substances which induce virus inhibition in the plants

5.5.6. Milk as virus inhibitor

Reduction of viral diseases has been found to occur mostly in infected plants of Solanaceae, Piperaceae and Malvaceae families by milk spray. It was tested that fresh milk spray reduces nearly 73 per cent of TMV infections in tobacco. Fermented buttermilk (1l in 20 l of water) sprayed on tobacco plants cures leaf curl disease. Fresh goat milk sprayed over plants like chilli (*Capsicum annum*), brinjal and species like black pepper also helps control fruit and leaf curl. The neutral to near-neutral pH of milk not only washes off the fungal spores but the adhesive property of milk fat also prevents germination of spores.

5.5.7. Neem and ash

Before transplanting paddy, seedlings are kept in small plots of pounded water mixed with ash and pulverized neem seeds. For a plot of 15 sq. feet, half a kilogram of neem seeds and a kilo of ash are sufficient for mixing with water to accommodate 50 bundles of seedling at a time for a period of one hour. Treated seedling produce a crop free of pest and diseases.

5.5.8. Soil borne diseases in tobacco nurseries

- A mixture of ash (2-3 kg) and 1 litre of castor oil is spread on a seed bed of a size of about 100 m^2.

5.5.9. Powdery mildew

- 2 kg of turmeric powder and 8 kg of wood ash.

5.5.10. Powdery mildew and other fungal diseases

- Ginger can be used at the concentration of 20 gm / litre of water and sprayed thrice at intervals of 15 days.

5.5.11. Fungal/bacterial diseases

- 5 kg of *Lantana camara* + 5 litres of water + cow's urine 10 litres for 3-4 days in earthern pot.

5.5.12. Control of onion tip die back

- In this practice, 10 kg of nerium leaves is mixed along with 100 litres of water in a vessel.
- Then boil this in local earthen store for 30 minutes due to boiling, leaf extract was obtained which was then cooled and filtered using a cotton cloth.
- This filterate of about (10 litre) was sprayed using hand sprayer per acre in the cropped field within a month of crop emergence.
- Repeat the spraying once in 20 days as required depending on disease intensity.

5.5.13. Control of disease in cotton

- When red spots appear on cotton leaves (in July or August), dilute buttermilk with water and sprinkle it on the crop.

- When the application is made during the early stages of the disease, control comes within a week. This practice can also be used as a preventive measure.

5.5.14. Casuarina leaf extract

- 20 kg of leaves are boiled in water for 20 min. After cooling, the solution should be filtered. Then the extract is diluted with water and can be given to control some bacterial and fungal disease.

5.5.15. Baking soda + castor oil

- One large tablespoonful (roughly 20 ml) of baking soda should be mixed with 15 ml of sesame or castor oil and dissolved in 3.5 litres of water.
- Twenty grams of molasses should also be added to the solution before spraying it on plants affected by powdery mildew, brown patch and other fungal diseases.

5.5.16. Control of *Semporian* Disease in Turmeric

Garlic – 1 kg,

Ginger- 500 gm,

Green chillies - 500 gm,

Pepper – 200 gm,

Tobacco 500 gm,

Neem oil 200 ml,

Khadi soap 30 gm.

- First garlic has to be soaked 100 ml of kerosene for over night and this has to be ground along with green chilies and pepper.
- Then tobacco is to be soaked in water for one day and then filtered and has to be mixed in with 200 ml of Neem oil.
- Now all of them are to be mixed together. Finally, soap 30 gm, has to be added and stirred well. Now the total quantity of the mixture will be about 7 litres.
- This entire quantity will be sufficient to spray 1 acre. (Add 700 ml of the fluid in 10 litres of water and sprayed; totally 10 tanks are to be sprayed; tank capacity is 10 litres).

5.5.17. Control of root rot & foot rot diseases in paddy crop

- After planting paddy seedlings and within 20 to 25 days, drain out the standing water from the field. Then apply poultry droppings (40 kg/ac) mixed with neem cake powder (20kg/ac) uniformly. No irrigation for 3 days.
- Three days later irrigation water is allowed to stand for a depth of 5cm. By this method the diseases could be efficiently controlled.

5.5.18. Control of leaf spot disease of paddy crop

- Kandhari chilli (A type of red pepper) is a wild Capsicum variety (said to be more pungent and has more capsicin content) fruits are very small and tiny and it will be pointing upwards direction.
- About 4 kg of Chilli fruits are taken, ground and mixed with water to spray for 1 acre of paddy crop affected with leaf spot disease. This spray gives sufficient protection to paddy crop from all kinds of leaf diseases caused by fungal organisms.

5.5.19. Buttermilk to prevent tobacco mosaic disease

- To prevent the incidence of tobacco mosaic disease, well fermented buttermilk is mixed with water at 1:20 ratio and sprayed over the foliage of tobacco with a hand operated sprayer.

5.5.20. Herbal recipe for control of mosaic virus disease in vegetables

For preventing the spread of mosaic virus, in chillis (Capscicum frutscense) and tomato, spray any one of the following solution.

1) Milk – 1.5 litres
   water – 5 litres
   Bogainvilla leaf – 3kg

All these are boiled, decoction prepared and mixed with 100 litres of water.

2) Butter milk – 4 litres, add maida flour - 1 kg, mix and dilute with 100 litres of water. The tubers of yam (Karunai) – 2kg are cut into small pieces, cooked and mixed with 100 litres of water. Both are mixed together and used for spraying in 1 acre of vegetable crops.

5.5.21. Blight disease in cumin

- Take equal amount of *Azadirachta indica* (Neem) leaves and *Annona squamosa* (Custard apple) leaves.

- Wash plant material properly in water and chop it finely and boil with 16 times water at moderate and constant heat until the final volume remains half.
- Spray 150 ml. of this formulation per pump. About 6-8 pumps per acre are required to control the infesta-tion. Spray after 10-12 days if infestation is still there. For best result, use this formulation within 6 months of preparation.

5.5.22. Turmeric leaf spot

- *Aloe vera* 3 - 5 kg, Bougainvillea, Papaya leaf 3 – 5 kg, Custard apple leaves 3 – 5 kg, *Lantana camera* 3 – 5 kg,– Any two of the above mentioned leaves along with Aloe vera are cut into pieces, mixed with 15 litres of water and is boiled till 10 litres of solution is obtained.
- After filtering the solution, 1 kg Turmeric powder is added and is kept for 12 hours. Then Pseudomonas florescence 250 – 500 gm is added and is sprayed at the rate of 1 litre solution in 10 litres of water.

5.5.23. Mint extract

- Prepare mint leaf extract with 250 gm mint leaf powder in two liters of water. Spray on affected crop until signs of disease disappear (leaf spot, potato blight, anthracnose).

5.6. Pest Repellant Spray Mixture

Neem seed kernal powder	- 5 kg	Put in a drum and allow for 3 days
Cows urine	- 12 litres	↓
Water	- 10 litres	Filter spray as 2 litres of pest
repellant in 11 Lemon/Acid lime fruit	- 4 nos	litres of water

Experienced by Mr.Senthil Kumar, Kulithalai, Karur district. Ph: (097882-03694)

5.6.1. Herbal insect repellants

Herbals available around us most commonly are

1. Thumbai (*Leucas aspera*)
2. Sacred Basil Thulasi (*Ocimum sanctum*)
3. Nochi (*Vitex* sp.)
4. Peenaari (*Peechchangu*)
5. Erukku (*Calotropis gigantea*)

Table 19: Biocontrol module for disease management

Crop	Pest	Biocontrol	Rate of application
Rice	Leaf spot	*Pseudomonas*	**Seed treatment** @ 4-5g/kg seed
	Sheath blight	*Pseudomonas*	**Seed treatment** : Apply the product @ 2.5kg/ha mixed with 50 Kgs of well decomposed farmyard manure (FYM) or sand at 30 days after transplanting.
			Foliar application: Spray the product @ 0.2% concentration (1Kg/ha) commencing from 45 days after transplanting at 10 days interval for 3 times depending on the disease intensity. If there is no disease incidence, a single spray is sufficient.
	Sheath spot	*Trichogramma japonicum*	2.00 lakh eggs/ha
	Neck blast	Nimbecidence + *Trichogramma japonicum*	500 g/acre + 2.00 lakh eggs/ha
Ragi	Blast	*Pseudomonas*	**Seed treatment:** 10g/Kg of seed
			Foliar spray: Spray the product @ 0.2% concentration (1kg/ha) after 50% of the flower emergence and again repeat the spray at 10 days af ter the first spray.
Cotton	Root rot and wilt	*Pseudomonas*	**Seed treatment:** 10g/kg of seed
			Soil Application: Apply 2.5 Kg/ha formulation mixed 50 Kg of well decomposed FYM/sand at 30 days after sowing.
Vegetable crops	Damping off	*Pseudomonas*	**Seed treatment:** The seeds of vegetables have to be treated with formulation at the rate of 10 g per kg of seed 24 hours before sowing to prevent the damping off disease.
			Main field application: Apply talc based formulation *Pseudomonas fluorescens* as soil application at the rate of 2.5 kg/ha mixed with well decomposed FYM 50 kgs.
Cabbage and	Club root disease	*Pseudomonas*	**Seed treatment:** 10g/ha of seed

Contd.

Crop	Pest	Biocontrol	Rate of application
cauliflower			**Soil application:** 2.5 kg/ha *Pseudomonas fluorescens* mixed with 50 kgs of FYM and then applied to the soil before planting. **Seedling root dip:** Seedling dip in solution containing formulation of 5g/litre of water for 30 min.
Mango	Anthracnose	*Pseudomonas*	**Foliar spray** with *Pseudomonas fluorescens* talc formulation at 0.5% concentration with 15 days interval after fruit setting.
Banana	Wilt and anthracnose	*Pseudomonas*	**Soil Application:** 2.5 Kg/ha of formulation+ 50 kg of FYM or Sand. Apply once at the time of planting and repeat it once in 3 months. **Capsule Application:** 50 mg/Capsule/Sucker apply once in three months from 3rd month after planting. **Bunch spray:** spray the formulation @ 0.5% concentration at last hand emergence and then at 30 days interval. Combination of all the above treatments will give effective management against the wilt and anthracnose disease with increase in yield.

6. Adathoda (*Adathoda vasica*)
7. Sottru kathalai (*Aloe vera*)
8. Veppa ilai and kottai (Neem leaves and seed kornels) (*Azadirachta indica*)
9. Kaattamanaku ilai and kottaikal (wild castor leaves and seed kernels) (*Jatropha curcus*)
10. Seethapalam tree leaves and seeds (Custard apple seeds leaves) (*Anona squamosa*)
11. Arali leaves and seeds (*Nerium odorum*)
12. Etti mara ilai and fruits
13. Punga mara ilai and fruits (*Pungamia glabra*)
14. Datura leaves and fruits (*Datura metal*)
15. Papaya leaves and fruits (*Carica papaya*)
16. Veliparuthi leaves
17. Unni chedi (*Lantana camera*)
18. Vilva ilai (*Agle marmalos*)
19. Pirandai
20. Kuppaimeni (*Acalypha indica*)
21. Tobacco leaves (*Nicotiama tabacum*)
22. Garlic (*Allium sativum*)
23. Ginger (*Zingiber officinale*)
24. Green chillies (*Capsicum annuum*)
25. Turmeric powder (*Curcuma longa*)

Select 5 or 6 different above said herbals

If leaves – 2 kgs – seed/fruit kernals- ½ kg take and grind well and put into plastic bucket

Add cow's urine into bucket to until the grinded leave powder is submerged (if needed, small amount of Panchagavya can be added)

Keep on stirring for 2 times a day

So that it can be long last for 2 months

Dosage

- 1 litre herbal insect repellant in 10 litres of water for spraying after proper filtration.
- Should be sprayed morning or evening.

5.6.2. Improved insect repellant or herbal insecticide/veeriya poochi virati or poochi kolli

Needed

1. Leaves which have bitter taste – 2 kg
2. Latex yielding leaves (milk excretion if broken) – 2 kg
3. Odour producing (obnoxious to inhale) leaves – 2 kg

In general, the leaves which are not touched by animals can be used as a insect repellent

Cowdung	- ½ kg	Soak it all for 7 to 10 days
Cows urine	- 5 litres	
Leaves	- 6 kgs in it	

Dose

After 10 days take 500 ml or 1 litre of this extraction/ decoction add 10 litres of water and spray.

To use as insecticide

Add ginger + garlic + chilles + tobacco leaf juice to this improved insect repellant to make this as an insecticide.

5.6.3. Agniasthiram

Garlic	- 10 kg	Put it to an earthen vessel and boil it well
Green chillies	- 10 kg	↓
Tobacco leaves	- 10 kg	cover the earthen vessel with white cotton cloth
Cow's urine	- 100 litres	↓
Neem leaves	- 25 kg	After 48 hrs of shade filter it for further use
Ginger	- 10 kg	

Dose

Take 500 ml from the filtered solution add 10 litre water for spraying at evening and morning.

5.6.4. Brammashthiram : Another form of improved insect repellant

Ingredient		Quantity	Method
Cowdung	-	5 kgs	Grind all the leaves then add cow's urine & dung
Cow's urine	-	5 litres	↓
Neem leaves	-	5 kgs	
Pungam leaves	-	2 kgs	Mix all with 60 litres and keep it for 24 hrs
Adathoda leaves	-	2 kgs	↓
Erukkan leaves	-	2 kgs	
Noni leaves	-	2 kgs	Filter it and use
Datura leaves	-	2 kgs	

5.6.5. Ingi Poondu karaisal – ginger garlic solution

Ingredient		Quantity	Method
Garlic	-	500 g	Grind garlic and soak it in kerosene (for 1night)
Kerosine	-	250 ml	↓
Ginger	-	250 +	Then add ginger green chilli paste and 10 lit water
Green chillies	-	100 g	↓
			Filter it and use

Dose

1. 1 litre solution for 13 litre water / tank

   Add 50 g khadi soap

To this add 500 g of tobacco + 500 g Asafoetida (Perungayam) can be used against paddy stem borer

5.6.6. Control of coconut rhinocerous beetle

1 kg cow dung + ½ kg neem oil cake

↓

Mix well

↓

Keep the mixture under the tree

↓

It will attract all the beetles

5.6.7. Maavu poochi

10 kg of grounded neem seed

↓

Add 20 litre water – soak it for 24 hrs

↓

Then add 100 g khadi soap

↓

Use this solution to control at the initial incidence of mealy bugs (mix with 200 litres of water)

5.6.8. Mooligai poochi viratti (Herbal pest repellant)

Oomathai (*Datura metal*) leaves	-	2 kg
Erukku (*Calotropis gigantean*) leaves	-	2 kg
Peenaarisangu (*Cleodendtron interme*) leaves	-	2 kg
Thumbai	-	2 kg
Water or cow's urine	-	10 litre

- All above said leaves are to be ground well by using a grinder and soak in an earthen or plastic vessel and allow fermentation for 15 days.
- Then it is filtered and added with water for sparying
- Ratio is 1:10

5.6.9. Pasum komiyam (Cows urine) spray

- It is rich in nitrogen and also it's a substitute for urea.
- 1 litre cows urine is dissolved in 10 litres of water and it can be sprayed.
- For 1 acre of crop, 5 litres of cow's urine is sufficient.

5.6.10. Neem formulations

Neem seeds contain more than 100 compounds among which azadirachtin has been found to be biologically most active. The biological effects of neem products are insect growth regulation, feeding deterrent and oviposition deterrent effect.

Commercial Neem formulations are available in market which contain varying levels of azadirachtin (from 0.03% to a maximum of 5%). In India more than 50 firms are manufacturing neem formulations which are available in different brand names.

Table 19: A few examples of neem formulations

Sl.No.	Brand name	Azadirachtin content
1.	Nimbecidine	0.03%
2.	Neem guard	0.03%
3.	Bioneem	0.03%
4.	Jaineem	0.03%
5.	Neem gold	0.15%
6.	Fortune-aza	0.15%
7.	Econeem	0.3%
8.	Achook	0.5%
9.	Neem azal TS	1.0%
10.	Neem azal F	5.0%

5.6.11. Bird attractant

1 kg of rice and 50 gm of turmeric powder is required to treat an acre.

5.6.12. Notchi leaves

- Crush about 10kg of notchi leaves and mix with 10 litres of water. This solution was then filtered using a cotton cloth, and then 1 litre of filterate is mixed with 9 litres of water and used as spray solution for an acre of cropped field (using knapsack sprayer).
- Spray the solution in early morning or late evening hours.
- The best time for spraying is in the early stage of square formation (cotton) and in the first appearance of pod (chilli and pulse crops).

5.6.13. An organic spray

- First mix 15 litres of curd with15 litre of water in an earthen pot.
- Ground 1 kg of neem leaves to paste then add to the same pot.
- After thorough mixing of these three ingredients, the mouth of earthen pots were tied with cotton cloth and kept inside the soil by digging small pits on the corners of field for 15 days.
- In this period the ingredients got fermented and gained the pesticidal property. After filtering, the extract was ready for spraying in 1-month-old crops.
- This organic spray controlled the pests like beetles, leaf sucking pests, boll worms and mite in earlier stage of the crop itself.

5.6.14. Green leaf extract

- Crush about 1 kg each of the green leaves of neem, pungam, nochi, erukku and thulasi and mix with cow's urine (10 litres) and allowed to ferment for 10-15 days in an earthen vessel.
- The solution has to be stirred daily using a wooden stick so that fermentation process would take place uniformly.
- After 10-15 days, the mixture was filtered using a cotton cloth. The leaf extract (10 litres) was used for spraying 1 acre of the cropland on 1 ½ months old crop even before the pest attack. This leaf extract is sprayed on a wide variety of crops like paddy, redgram, blackgram, brinjal, bhendi etc.

5.6.15. Botanical control of pod borer in redgram (Poochi viratti)

Solution 1

About 0.5 kg of garlic were crushed and mixed with half a litre of kerosene and kept in a container overnight and filtered using a cotton cloth.

Solution 2

About 50 gms of green chillies were crushed (with the help of the indigenous milling tool 'ural') and then mixed with 1 litre of water.

Solution 3

About 100 gms of kaathi soap were dissolved in 1 litre of water and soap solution was prepared. These three solutions were mixed thoroughly. About 25ml of the mixture was diluted in 16 litre of water and are sprayed in the cropped field of 1 acre using knapsack sprayer to control pod borer damage.

5.6.16. Herbal pesticide against crop pests

Solution 1

Garlic bulb 500 gm peeled cut in to pieces and soaked in kerosene over night and filtered.

Solution 2

Green chilli(*Capsicum fruitescens*) 50 gm ground into a paste and mixed with water – 1 litre and filtered.

Solution 3

Soap powder 100 gm dissolved in 1 litre of water to serve as sticking agent.

All the 3 solutions are well mixed with 100 litres of water and used as spray fluid in the morning hours over the foliage of affected crops like cotton, redgram, sunflower, pomegranate, grapes etc.,

5.6.17. Liquid manure for pest management

Materials required

i. Plants like *Parthenium, Lantana, Vitex, Eupatorium, Artemesia,* Stinging nettle etc., - 3 kg

ii. Cattle dung - 3 kg

iii. 20 litre capacity plastic bucket

iv. Water – 20 litres

Method

1. Collect 3 kg of plants (leaves and tender parts) of plants which have pesticidal activity. Chop them into small pieces and put into 20 litre bucket.
2. Add 3 kg of cattle dung into the barrel and fill it up with water.
3. The barrel is stirred every day for 7 days and then stirred once a week for the next 3 weeks. The preparation will be ready in 30 days.
5. The concentrated solution is diluted ten times in water and used as a foliar spray.
6. These sprays are very efficient in managing a variety of pests.

Precautions

1. The liquid manure has to be diluted ten times before spraying on the crop otherwise it scorches the plant.
2. The solution has to be sieved through a cloth or gunny bag before spraying to avoid blockage of nozzles.
3. The efficacy of the solution is up to one month and has to be used within that period.

Organic Pest Repellent

Materials required

i. Artemesia (Khempa) leaves or Stinging nettle leaves or Lavender leaves – 200 gm

ii. Dolla Khorsani (or any hot chilli)- 100 gm

iii. Garlic – 100 gm

iv. Egg Yolk – 1 (acts as an emulsifier)
v. Any cooking oil – 100ml.

Procedure

1. Mix the above materials in a container or put them directly in a mixer grinder. These materials are grinded until the contents are turned into a fine paste.
2. Add 5 litres of water to the thick paste and mix thoroughly. This can be used as stock solution.
3. Take one part of the above concentration solution and dilute with 10 times with water. Stir the solution for 5 minutes and it is ready for spraying.
4. Strain the solution with the help of a muslin cloth before pouring the solution into a sprayer to avoid the clogging of the nozzles.
5. Spraying should be undertaken during evening hours.

5.6.18. Marigold extract

Materials required

i. Marigold flowers along with the entire plant
ii. Soapnut or locally made soap (Nepali sabun)

Method

1. Take 5 kg of Marigold plants along with the flowers. Grind into fine paste.
2. Add 100 litres of water and allow it for fermentation for 7 days.
3. Add 100 grams of soap nut powder on the day of spraying which acts as emulsifier.
4. Stir for 15 minutes and spray during evening hours.

5.6.19. Garlic Brew

Materials required

i. Garlic 200 gm
ii. Cattle urine – 20 litres
iii. 20 litre capacity bucket

Method

1. Crush or grind the garlic into a paste and put into the bucket containing cattle urine.
2. Stir the materials for 5- 10 min and cover with a gunny cloth or cotton cloth which can provide sufficient aeration.

3. Allow the materials to ferment for 5-7 days.
4. After 7 days sieve the solution before spraying to avoid clogging of the nozzles of the sprayer.
5. The solution is diluted 10 times with water and sprayed on the foliage in the evening hours.
6. Care should be taken that the concentrated solution should not be sprayed in the crop.

5.6.20. Curd or Yogurt spray

Materials required

i. Curd or Yogurt
ii. Water

Method

- Allow the curd to ferment for 3-5 days until one can sense a strong smell of alcohol.
- Take one litre of curd and dilute in 30 litres of water. Stir the solution for 15 minutes and sieve the solution with a muslin cloth.
- Spray the solution during the evening hours.

5.6.21. Garlic decoction

Materials required

i. Garlic- 1 kg
ii. Water – 10 litres

Method

- Boil 1 kg of garlic in 10 litres of water separately.
- Allow the solution to cool and sieve the solution with a muslin cloth.
- Dilute one part of the solution with 10 parts of water and spray on the crop during evening hours.

5.7. Cultural Techniques

5.7.1. Cultural methods

Adjust the time of sowing to modulate growth of the crop. Plant to plant and row-to-row spacing is similarly used to alter the microclimate and reduce risks. There are no standard prescriptions for these and are generally based on the knowledge and experience of the farming community.

5.7.2. Trap crops

Pests are strongly attracted by certain plants and when these are sown in a field or along the border, tend to gather in them, enabling their easy collection and destruction. African marigold, mustard, maize, etc., can be grown as trap crops in cole crops, cotton and vegetables.

Table 20: Trap Crops - an important tool of pest management

Crops	Trap crops	Pest controlled
Cotton, Groundnut	Sunflower, castor, marigold	*Spodoptera* Castor
Cotton, Chickpea	Marigold	*Helicoverpa*
Pigeon pea	Marigold	*Helicoverpa*
Groundnut, Sesamum	Cowpea , Sunhemp	Red hairy Caterpillar
Sorghum	Cowpea	Stem borer
Bhendi	Bitter Gourd (emits a bitter substance momordicin)	All pests
Pulses	Sun hempCastor	Beetles, *Spodoptera* and Hairy caterpillar
Crosandra	Castor	All pests
Cabbage*	Mustard *(1 row of mustard for every 25 rows of cabbage)	Diamond Back Moth*
General	Marigold	Nematodes
Vegetables	Onion (emits allicin compound which irritates the pest)	All pests
Paddy	Live fencing of Vitex negunda	Stem borer & Earhead bug
Red gram	Marigold	Fruit borer
Tomato	Marigold (1 row of marigold for every 16 rows of Tomato)	Fruit borer

Fig. 19: Summer ploughing

5.7.2.1. Marigold raised as Inter crop to minimize tomato fruit borer Incidence

- For every 16 rows of tomato 1 row of "Maattu Sevvanthi" (mari gold – *Calendula officinalis*) is raised as an inter crop found to minimize the incidence of fruit borer in tomato.
- 40 days old seedlings of marigold when planted in tomato field (25 days old tomato seedlings after planting), there will be synchronized flowering of both the crops.
- This will minimize the incidence of fruit borer in tomato.

5.7.2.2. Other traps

- On all the four sides of cultivated crop dry land, tall grown sorghum or bajra are raised thickly in 4 rows which prevent the entry of pests like white fly and aphids into the cultivated crop, as they cannot fly over the height of tall hedges.
- In the bunds of irrigation channel, coriander is planted as intercrop when chilli is grown as main crop.
- To avoid diamond back moth incidence inter cropping with mustard is recommended : 2 rows of mustard for every 25 rows of cabbage is suggested. One row of mustard is sown 10 -15 days in advance and the 2nd row of mustard is sown 25 days after cabbage sowings.

5.7.3. Use of resistant/tolerant varieties

Genotypes showing tolerance or resistance to pest and disease are preferred in organic cultivation. A series of resistant varieties of different crops have been developed in recent years for most climatic conditions.

5.7.4. Summer ploughing

Summer ploughing is an important cultural practice for pest control. When the land is ploughed, the inactive stages of pests like egg masses, larvae and pupae present within 5–10 cms surface of the soil get exposed. They are killed due to the intense heat of summer and are also eaten away by predatory birds.

Keeping bunds clean

Field and field bunds are the favorite egg laying spots of most pests. Hence, wild grasses and weeds found in the field and on the bunds should be periodically removed.

Plastering of Bunds

Weeds found on the bunds should be removed and the bunds should be plastered. By doing this, rat holes found near the bunds can be sealed and rodent damage controlled. Such a procedure also prevents water leakage.

5.7.5. Proper spacing among the seedlings

When paddy seedlings are transplanted to the main field, they should be laid out with proper spacing. For short duration varieties, the inter row spacing should be 5 cm and inter hill spacing should be 10 cm. For medium duration varieties, it should be 20 cm x 10 cm and for long duration varieties, 20 cm x 15 cm. This facilitates penetration of sunlight to the lower portions of the crop and thus prevents pest and disease incidence.

Providing sufficient gaps

While planting seedlings, a one-foot gap should be provided after every eight feet to enable sunlight to reach the lower segments of the plants. This reduces the incidence of pests that are found on the under surface of the crop. Such spacing also helps during the application of manures and the spraying of biopesticides.

5.7.6. Provision for sunlight

Neekal podum murai is a traditional practice that has been followed in the Chengalpet district of Tamil Nadu for a number of years. In this method, women stand in a line in between the plants and walk from one end of the field to the other, pressing the under parts of the plants. This enables sunlight to reach the lower segments of the crop. The practice not only kills the nymphs of the brown plant hoppers sticking to the under surface of the plants but also helps to check the rodent population.

5.7.7. Rope method

The field should be filled with water up to a height of 5 cm. One litre of kerosene should be mixed with 25 kg of sand and strewn in the field. Later, a string should be dragged over the surface of the leaves vigorously so that the caterpillars fall into the water. The caterpillars are killed by the kerosene present in the water. Later, the water should be drained to remove the dead caterpillars. The field should be dried and then freshly irrigated. This method should be used only during the vegetative stages of the crop.

5.7.8. Bird perches

'T' shaped bird perches should be erected in the field at the rate of 15–20 per hectare. They should be placed one foot above the crop canopy. These perches serve as resting places for the birds which feast on the larvae they find in the field. Mix rice with the blood of a chicken, make it into pellets and broadcast these in the field. The smell of blood and rice attracts predatory birds to the perches in the field from where they pick up the swarming caterpillars.

5.7.9. Clipping practice in brinjal

- Remove 1-2 inch of terminal portion along with dried part of the plant. Clipped portion (containing larvae) should be carried away from the field and fired to destroy all the larvae.

5.7.10. Fumigation for disease control

Diseased crops can be sprayed with 10% cow urine solution. On the same day or the following day, fumigation should be carried out in the evening. About 200 gm of vaividanga (*Embelia ribes*) or sweet flag (*Acorus calamus*) and turmeric (*Curcuma longa*) is powdered well, put in a wide mouthed pot with burning charcoal and carried into the field in a direction opposite to the wind. On the seventh day after fumigation, sweet flag rhizome extract should be sprayed. This method controls bacterial and fungal diseases.

5.7.11. Use of effigies

A human-like figure, made of paddy straw and wearing a white dress (@ two effigies per hectare) kept in the field at milky to grainfilling stage, will scare away the birds.

5.7.12. Cultural control methods for rice

- Avoid close planting especially in BPH and leaf folder prone areas or seasons.
- For every 5-6 m leave a spacing of 75 cm (hoppers).
- Grow horse gram, green gram, soy bean on bunds to attract natural enemies.
- Controlled irrigation by intermittent draining (BPH).
- Remove the weeds on bunds that harbour BPH, gall midge GLH, leaf folder, ear head bug.

5.7.13. Cultural control methods for pulses

- Sow good and healthy seeds.
- In stem fly endemic areas use a higher seed rate to the extent of 25-30% to compensate the loss of seedlings.
- Maintain the fields and bunds free from weeds.
- Avoid crops susceptible to some pests either as mixed crops of in crop rotation.
- Provide 'T' shaped bird perches.
- Grow castor along the borders to trap *S.litura*, marigold to trap *H.armigera* and cowpea to trap stem fly.

The plant density should not exceed 30-35/sq.m. If it exceeds it creates favourable microclimate suitable for the multiplication of pests and diseases.

5.7.14. Cultural control methods for cotton

- Growing one variety throughout the area as far as possible.
- Deep summer ploughing on bright sunny days during the months of May or June should be done to expose soil inhabiting or resting stages of insects, pathogens and nematode population. The field should be kept exposed to sunlight for at least 2-3 weeks.
- Growing of less preferred crops like green gram, black gram, soybean, castor, sorghum etc., along with the cotton as intercrop or border crop or alternate crop to reduce the pest infestation.
- Growing two rows of maize or sorghum or cowpea along the border to sustain and enhance the build up of natural enemies such as lady bird beetles, *Staphylinids*, *Chrysoperla carnea*, *Anthocorids*, *Reduviida* etc. Pollen of maize helps in retaining *Chrysoperia* in main cotton field.
- Plant trap crops like marigold or okra or pigeon pea along the border and irrigation bunds to divert American boll worm oviposition from main cotton crop.
- Growing castor along the border and irrigation bunds as trap crop for tobacco cut worm, okra for spotted boll worm and aphid.
- Use neem cake @ 1 t/ha under assured moisture conditions in nematode infested fields.
- Earthing up on 45th day (stem weevil).
- Basal application of FYM 25 t/ha and 250 kg/ha of neem cake (stem weevil).

- Install 15-20 bird perches per acre for the benefit of predatory birds like black drango, king crow, orange myna etc., after 90 days of crop growth. Provide drinking pots with water to them by placing them around the perches.

5.7.15. Cultural control methods for tomato

- Grow simultaneously 45 days old American tall marigold and 25 days old tomato seedlings @1:16 rows so that both will come to flowering at the same time and *Helicoverpa armigera* adults will be attracted to marigold for oviposition.
- Grow 50 castor plants / acre as a trap crop for tobacco cut worm and the egg masses and early instar larvae in clusters on castor crop should be periodically collected and destroyed.
- Remove alternate weed host of whitefly, *Abutilon indicum.*
- Apply press mud @ 5kg/m^2 for nematode diseases complex.
- Installation of T shaped bird perches @ 15-20/ha.
- Deep summer ploughing (tobacco cut worm and nematode).
- Flood irrigation to bring out the hiding larvae (tobacco cut worm).
- Grow sorghum or bajra as border crop (sucking pests and viral diseases transmitted by them).
- Ploughing the nursery area, uniformly spreading paddy husk @ 20 kg/m2 (about 15 cm) thickness) buming it and ploughing back facilities production if nematode free seedlings.
- In nursery apply 200 g of neem cake / sq m (nematode).
- Crop rotation with marigold, gingelly, mustard, maize wheat etc (nematode).

5.7.16. Cultural control methods for brinjal

- Application of press mud at 5 kg/m^2 at the time of sowing.
- Application of 200 kg neem cake / acre as basal (shoot and fruit borer, nematode and ash weevil).
- Maintain the correct spacing (shoot and fruit borer).
- In nursery apply 200g of neem cake/sq m (nematode).
- Crop rotation with marigold, gingelly, mustard, maize, wheat etc (nematode).
- Deep summer ploughing (nematode).
- Transplant healthy seedlings.

- Avoid rationing of brinjal crop since woody stem is preferred by stem borer larvae.

5.7.17. Cultural control methods for chillies

- Raise 2 rows of maize or sorghum for every 5 rows of chilli crop against wind direction (chilli mosaic).
- Crop rotation with maize, soybean, green gram, black gram etc., (*H.armigera* and *S.litura*).
- Deep summer ploughing (*H.armigera* and *S.litura* and root worm).
- Raise 2-3 rows of maize or sorghum as border crop to prevent the attack of *H.armigera* and *S.litura* from surrounding fields and enhances the activity of lady bird beetle.
- Plant the trap crops for *H.armigera* (marigold) and *S.litura* (castor) @ 100 plants/acre and periodical destruction of egg masses and larvae found on trap crops.
- Provisions of bird perch @ 15-20 / ha and also provide water pots around them.
- Application of well composted farm yard manure along with neem cake (root worm).
- Don't crop the chilli in the fields where tomato and brinjal were raised previously.
- Intercrop onion 1 line after 10-12 lines of chilli to enhance the natural enemy population especially, coccinellids and syrphids.

5.7.18. Manipulation of temperature

- Sun drying the seeds to kill the eggs of stored product pests.
- Hot water treatment (50 - 55°C for 15 min) against rice white tip nematode.
- Flame throwers against locusts.
- Burning torch against hairy caterpillars.
- Cold storage of fruits and vegetables to kill fruitflies (1 - 2°C for 12 - 20 days).

5.7.19. Manipulation of moisture

- Alternate drying and wetting rice fields against BPH.
- Drying seeds (below 10% moisture level) affects insect development.
- Flooding the field for the control of cutworms.

5.7.20. Manipulation of light

- Treating the grains for storage using IR light to kill all stages of insects (eg.) Infra-red seed treatment unit.
- Providing light in storage go downs as the lighting reduces the fertility of Indian meal moth.
- Light trapping.

5.7.21. Manipulation of air

Increasing the CO_2 concentration in controlled atmosphere of stored grains to cause asphyxiation in stored product pests

5.7.22. Use of Abrasive dusts

- Red earth treatment to red gram: Injury to the insect wax layer.
- Activated clay: Injury to the wax layer resulting in loss of moisture leading to death. It is used against stored product pests.
- Drie-Die: This is a porous finely divided silica gel used against storage insects.

5.8. Mechanical Methods

5.8.1. Mechanical control methods for rice

- Collection an destruction of rice stubbles from field after harvest as they harbour egg, larvae, pupae of stem borer, gall midge, white tip nematode and root knot nematodes.
- Clipping the tips of the seedlings up to 2 inches prior to transplanting to remove the egg masses of stem borer if nay.
- Collection of egg masses of stem borer and silver shoots from the nursery seedlings.
- Flooding the nursery to make the hiding larvae in the soil to come to the surface and thus they are picked by the birds (army worm).
- A rope may be passed over the young crop for dislodging the larval cases from the tillers and then the water should be drained for eliminating them (case worm).
- Providing bird perches of 2-3 ft height in vegetative stage @ 15-20/acre. They should be removed after seed setting to avoid the bird damage to seeds. Drinking pots with water should be provided around the perches.
- Collection and destruction of egg masses of stem borer and ear head bugs in main field.

- A thorny hedge may be passed over the crop when it is affected by leaf folder to unfold the lead folds and to expose the larvae within to natural enemies and botanical sprays.

5.8.2. Mechanical control methods for pulses

- Remove and destroy stem fly damaged seedlings.
- Pull out plants manifesting symptoms of sterility mosaic, yellow mosaic, leaf curl and leaf crinkle virus disease since they will serve as a source of inculum spread by sucking pests.
- Collect eggs, larvae, pupae and adult of the insects to the extent possible to reduce their population (leaf feeding caterpillars, beetles, weevils, grasshoppers etc).
- Burn the crop residues after harvest.

5.8.3. Mechanical control methods for cotton

- Removal and destruction of crop residues after harvest to avoid the carry over population of American boll worm to next season.
- Removal of terminals of cotton crop (topping) at 80-90 days of growth to reduce *Helicoverpa* oviposition and also to encourage sympodial branching which bears more fruiting bodies.
- Removal and destruction of alternate weed hosts of white fly like *Abutilon indicum*, *Chrozophore rottlari, Solanum nigrum* and *Hibiscus ficulensus* from the fields and neighbouring areas and maintaining field sanitation.
- Collection and destruction of leaves infested with white fly.
- Hand picking and burning of the pink bollworm affected and dropped squares, flowers and fruits and squashing the pink boll worms in the rosettes.
- Removal and destruction of egg masses, early stage larvae found in clusters and hand picking and destruction of grown up caterpillars to minimize heavy build up of future population of tobacco cut worm.
- Uprooting and destroying the weeds like *Sida spp., Abutilon indicum* and *Xanthium spp.,* before sowing of cotton crop to reduce the initial build up of boll worm, whitefly and cotton leaf curl virus (CLCV).
- Rouge the plants infested with CLCV regularly.

5.8.4. Mechanical control methods for tomato

- Collection and destruction of damaged fruits, early instar and grownup caterpillars of *Helicoverpa armigera* and *Spodoptera litura.*
- Collection and destruction of virus affected plants.

5.8.5. Mechanical control methods for brinjal

- Collection and destruction of the beetles, grubs and pupae of Epliachna beetle.
- Removal and destruction of dried and withered shoots of shoot and fruit borer to arrest the spread of the pest.
- Remove the affected fruits and destroy (shoot and fruit borer).
- Remove the little leaf affected plants.
- Collection and destruction of aphid infested twigs.
- Removal and destruction of webbed leaves (leaf webber).

5.8.6. Mechanical control methods for chillies

- Collection and destruction of damaged fruits and grownup caterpillars (fruit borer).
- Collection and destruction of dropped flowers and fruits (midge).
- Collection and destruction of virus affected plants.

5.9. Trap - Behavioral control methods

5.9.1. Yellow sticky trap

An empty tin or a plate painted yellow and smeared with castor oil should be placed one foot above the crop canopy in the field. The adults of sucking pests that are attracted by the bright yellow colour get trapped in the oil smear. These pests should be wiped out every day and oil should be applied afresh.

- Monitoring the activities of the adult white flies by setting up yellow pan traps and sticky traps at 1 foot height above the plant canopy. Locally available empty yellow palmoline tins coated with grease/Vaseline/castor oil on outer surface may also be used.
- Paint yellow colour on plastic drinking water pot, apply castor oil on it and move it on both sides with hand by walking in the field to attract and trap whiteflies.
- Install yellow sticky traps to attract the adult whiteflies in tomato
- Monitor the whitefly with yellow sticky trap @ 12 numbers/ha in brinjal, bhendi and chilli, respectively.

5.9.2. Light trap

Light traps can be used to monitor and trap adult insects, thereby reducing their population. Some formal light traps that could be used are electric bulbs, hurricane lamps and bonfires. Water mixed with kerosene is filled in a large plate or vessel and kept near the light. The trap should be fixed 2–3 ft above the crop canopy and set up in the field between 6 and 9 pm. (If it is kept beyond 9 pm, there are chances that the beneficial insects will also get trapped and killed.) The adult moths, which get attracted by the bright light, fall into the water in the vessel and perish.

- Installation of light traps with incandescent light at 1-2m height @ 4/acre to monitor the population (stem borer, leaf folder, BPH, gall midge and ear head bugs in rice). AT the base of light trap put a tub filled with water to which kerosene was added to kill the trapped insects
- Use of light trap to monitor and kill the attracted adult moths of tobacco cut worm in pulses.
- Installing light traps with incandescent lamp (1-2 / acre) for monitoring of insect activity (American boll worm) and tobacco cut worm). The cotton crop around the light trap may be sprayed with neem oil.

Note

- Use only incandescent light in light traps, as mercury lamp attracts natural enemies in large numbers.
- The light trap should be lighted between 8- 10 pm.

5.9.3. Pheromone trap

About eight traps should be used per hectare. They should be placed two feet above the crop canopy. The level of the trap should be adjusted according to the plant height (1–2 feet above). A chemical called a 'pheromone' – obtained from female moths – is used in the pheromone trap to attract male moths. The latter are pulled into the trap by the pheromone and die.

- Install pheromone traps @ 8 acre (stem borer and leaf folder in rice).
- Set up sex pheromone traps to attract and kill male moths of *Helicoverpa armigera* and *Spodoptera litura*. Set up five traps per acre from floral bud formation and change the septa once in 3 weeks in pulses.
- Use pheromone traps for monitoring American boll worm, pink boll worm, spotted boll worm and tobacco cut worm in cotton. Install pheromone traps at a distance of 50 m @ 5 traps per acre for each insect pest. Use specific lures for each insect species and change it after every 15-20

days. Trapped moths should be removed daily. If the number of trapped adult moths is 10 (American boll worm), 20 (tobacco cut worm), 15 (spotted boll worm) and 8 (pink boll worm) necessary action should be taken.

- Set up pheromone traps @ 12/ha (*H.armigera* and *S.litura*). When the trapped moths are 8/day necessary action may be taken in tomato.
- Installation of pheromone traps at 12/ha (shoot and fruit borer) in brinjal , chilli and bhendi

Points to be remembered while using insect sex pheromone traps

- Different types of traps available viz., funnel trap, wing trap, water trap etc. The selection of trap is depending on which insect you want to manage.
- Pheromone lures are available only for specific pest and select exact lure for targeted insect pest.
- Traps should be placed 40-100 m apart.
- Check traps twice a week, recording trap count and removing insects from the trap.
- Always inspect crop in conjunction with the traps in order to make an assessment of insect activity and damage.
- Replace pheromone lures throughout the season, referring to manufacturer's instructions on how long pheromones last.
- Replace when they become wet or dirty.

5.9.4. TNAU insect probe trap

TNAU Insect traps are excellent insect detection devices in food grains and more effective in the detection of stored grain insects namely *Rhyzopertha dominica* (F.), *Sitophilus cryzae* (L.) and *Tribolium castaneum* (Herbst) in stored food grains both in terms of detection as well as number of insects caught than the standard normal sampling method (by spear sampling).

5.9.5. Lanther pori (Light trap)

- Aricane or lanther light is used as trap,
- Lanther is kept on the stool which is in one corner of the field. Below this structure, water and kersonen mixture is kept in a pan.
- It can be used between 6.00 – 9.00 pm

5.9.6. Poochi pori (Pest trap)

- A mud is filled with 0.5 litr castor oil and 4 litre of water. It is mixed well and buried upto pot's neck inside the field. This solution creates odour which attracts the Rhinocerous beetles and root grub (40 pots per acre).

5.9.7. Yellow sticky pot trap (Laxmidhar Mohanta, Odisha)

Developed yellow sticky pot trap by using locally available materials i.e. earthen pot and mahua oil (*Madhuca Indica*). The outer part of the earthen pot was painted with enamel yellow paint and smeared with mahua oil. The pot was placed with wooden tag in the field @ 20 umbers per ha. The colour attracts insects and sticks to the pot due to stickyness of mahua oil.

5.9.8. Fly trap

Fly traps are large boards measuring about 30cm by 30cm which are painted bright yellow/orange and covered with an adhesive such as oil or glue. Different pests are attracted to different colours so you need to experiment. The flies are attracted to the bright colour of the board and fly onto it. They get stuck in the oil or glue and die.

5.9.9. Parapheromone trap

The capsule is placed on a sticky, glued, (cardboard) base within a delta trap. It must be replaced every 4 weeks together with the glued (cardboard) base. Attracted by the pheromone, the males (in the case where a female lure is used) will remain glued to the base of the trap. Pheromone traps may be used for monitoring population development, to better target control measures, or, in large numbers, to effect control themselves by reducing the numbers of insects in the plot.

In the case of the mango fly, populations of the male *Bactrocera invadens* flies can be reduced in mango orchards by capturing the flies using "parapheromone" traps. Used on a large-scale and in high numbers, they can also halt population development at the start of the season. The technique involves laying, at the start of the season, plates saturated with a specific attractant and treated with a contact insecticide. The traps must be placed in the orchard at

Fig. 20: Capsule placed on a cardboard base in the field

least one month before the fruit starts to attract insects. It is also advisable to place these traps in orchards where other susceptible fruit trees are growing (e.g. citrus orchard).

5.9.10. Castor beans trap

- Castor beans are well roasted in heat, powdered and put in flat mud vessels. To this water is added, mixed and allowed to soak for 24 hours; When a typical odour is developed, 4 -5 such mud vessels (traps)are put inside and around cultivated field crops.

5.9.11. Colour and water traps

Colour and water traps can be used to monitor adult thrips. In some cases thrips can even be reduced by mass trapping with coloured (blue, yellow or white) sticky traps or water traps in the nursery or field. The colour spectrum of the boards is important for the efficacy of the sticky traps. Bright colours attract more thrips than darker ones. Sticky traps with cylindrical surfaces are more efficient that flat surfaces. They are best placed within a meter of crop level. Traps should not be placed near the borders of fields or near shelter belts.

Water traps should be at least 6 cm deep with a surface area of 250 to 500 cm, and preferably round, with the water level about 2 cm below the rim. A few drops of detergent added to the water ensure that thrips sink and do not drift to the edges and escape. Replace or add water regularly.

5.9.12. Fruit bagging

Prevents fruit flies from laying eggs on the fruits. In addition, the bag provides physical protection from mechanical injuries (scars and scratches). Although laborious, it is cheap, safe and gives a more reliable estimate of the projected harvest. Bagging works well with melon, bitter gourd, mango, guava, star fruit, avocadoes and banana (plastic bags used).

Recommendations to farmers regarding fruit bagging

Cut old newspapers to fruit size and double the layers, as single layers break apart easily. Fold and sew or staple the sides and bottom of the sheets to make a rectangular bag. Blow in the bag to inflate it. Insert one fruit per bag then close the bag and firmly tie the top end of the bag with sisal string, wire and banana fibre or coconut midrib. Push the bottom of the bag upwards to prevent fruit from touching the bag. For example, start bagging the mango fruit 55 to 60 days from flower bloom or when the fruits are about the size of a chicken egg. When using plastic bags (e.g. with bananas), open the bottom or cut a few small holes to allow moisture to dry up. Moisture trapped in the plastic bags damages and/or promotes fungal and bacterial growth that causes diseased fruits. Plastic also overheats the fruit. Bags made of dried plant leaves are good alternatives to plastic.

5.10. Biopesticides

The use of microorganisms as biocontrol agents is gaining importance in recent years. Biopesticides are living organisms or their derived parts which are used as bio-control agents to protect crops against insect pests. Entomopathogenic viruses of baculovirus group, bacterial insecticides, particularly *Bacillus thuringiensis*, entomo-fungal pathogens, protozoans and insect parasitic nematodes have been found to control important pests of crops. These biopesticides are commercially available and are quite difficult to formulate in field conditions.

5.10.1. Fungal biopesticides

Seed treatment: 10 gm/kg of seed

Nursery bed: 1 kg/100 kg soil mix

Soil drenching: 10 gm/litre of water

Seedlings dip (30 min): 10 gm/litre of water

Soil application: 5 kg/acre with FYM

Foliar spray: 1 kg/acre

5.10.2. Panchagavya-A Biopesticide (Shri Umesh Sood, Himachal Pradesh, 09816254302)

500 ml Panchagavya (milk, curd, ghee, dung and urine of cow) + 150-200 ml extract of Ritha fruits and Neem leaves [50 ml Neem leaves extract + small quantity of gur + fermented wheat flour + 100 ml extract of Ritha] + 300 ml cow urine (collected earlier) to make 1lt biopesticide. Apply 1 lt/100 lt of water 30-45 days after transplanting

5.10.3. Leaves decoction as Biopesticide (Shri Anand Singh Thakur, Madhya Pradesh, Mobile: 09301301901)

The components used were 5 kg leaves of Neem, Pongamiya, Custard apple, Ipomia, and *Calotropsis gigantia* (Commonly known as Madar) [1 kg each] + 250 gm Garlic (mashed) +10 litre water + 10 litre Cow urine. He boiled it till it remains half of the total quantity (approx. 10 litre) and filtered. This decoction of 10 litres is dissolved in 600-700 litres of water and applied (sprayed) for 1 ha.

5.10.4. *Trichoderma viride* (TV) -Biopesticide

Trichoderma viride is a fungus with multipurpose use in agriculture. As a biological pesticide, it is useful in fungal attack like wilt, rusting of leaves and root rot disease. It helps in germination of seed. It can enhance the growth of the plant and partly satisfies the nutrient requirements of the plant. TV is not harmful to either plants or animals. For soil application, it can be used @ 250 ml/ha, after mixing with farmyard manure. It can be used as foliar spary @ 250 ml/ha.

Seed treatment

- Seed must be washed first to get rid of any chemical fertilisers and pesticides.
- TV culture is taken @10 gm per kilo of seeds.
- The culture is mixed with starch to make it sticky.
- Seeds are coated with the paste and dried in the shade.
- The dried seeds are sown the same evening.

Seedling treatment

- 20 gm of TV is mixed in a litre of water.
- The seedlings of brinjal, chili, tomato, cabbage, etc., are immersed in this mixture for five minutes before transplantation
- For nursery treatment, 50 gm of TV culture is mixed with 500 gm of vermicompost or compost and mixed in 65.8 m^2 of land.

5.10.5. *Beauveria basiana*

Spray @ 200 ml/ha by knapsack or any conventional sprayer at 12-15 days interval based on the pest populations. It should not be mixed with fungicides at any stage. Spar in the evening on young larval stages or on sighting of egg-laying.

5.10.6. *Metarhizium anisopliae*

For soil application, mix 200 ml of *Metarhizium anisopliae* in 50 kg FYM and broadcast in one acre, plough in the root zone. For potato and sugarcane, apply before planting and mix thoroughly with the soil by ploughing.

5.10.7. *Verticillium lecanii*

It is applied as sprays @ 5 ml/litre of water. Addition of jaggery or soap solution to the spray solution before spraying on the standing crop improves the results.

Table 21: Bio-pesticides for different crops

S.No	Biological agents	Pest	Crop
1.	*Beauveria bassiana* – 1.0% affects in young age	Hairy Caterpillars, *Helicoverpa, Spodoptera*, Borers, Mites, Scales, etc.,	Vegetables cereals, fruits
2.	*Metarhizium anisopliae* - 0.5 % affects all stages	White grubs, Bettle grubs, Caterpillars, Semiloopers, mealy bugs, BPH	Sugarcane, Cotton, groundnut, Rice, Potato, Cotton, Cereals
3.	*Verticillium lecanii* - 0.5 % affects all stages	All sucking soft body insects	Sugarcane, Cotton, groundnut, Rice, Potato, Cotton, Cereals
4.	*Bacillus thuringiensis Var kustaki* 0.3 – 0.4 %	Hairy Caterpillars, *Helicoverpa, Spodoptera*, Borers, Mites, Scales, etc.,	Vegetables cereals, fruits
5.	Nuclear polyhedrosis virus of *Spodoptera litura* – 200-500 ml/ha 2-3 time at 10 days interval	*Spodoptera litura*	Cotton, Groundnut, Pulse, Cabbage, Chillies
6.	Nuclear polyhedrosis virus of *Helicoverpa armigera* – 200-500 ml/ha 2-3 time at 10 days interval	*Helicoverpa armigera*	Cotton, Groundnut, Pulse, Cabbage, Chillies

Table 22: Commercially important microbial pesticides and biorationals used in India

S.No	Category	Products	Target pest	Major crops
1.	Bacteria	*Bacillus thuringiensis*	Lepidoptera	Cotton, maize,vegetables, soybean, groundnut, wheat, peas, oilseeds,rice
		Bacillus sphaericus	Mosquitoes, flies	
		Bacillus subtilis	Fungal pathogens	
		Pseudomonas fluorescens	Fungal pathogens	
2.	Fungi	*Trichoderma viride* *Trichoderma harzianum* *Trichoderma hamatum*	Fungal pathogens	Wheat, rice, pulses, vegetables, plantations, spices and sugarcane
		Beauveria bassiana *Verticillium lecanii* *Metarrhizium anisopliae* *Paecilomyces lilacinus* *Nomuraea rileyi*	Insect pests such as bollworms, whiteflies, root grubs, tea mosquito bugs and nematodes	Cotton, pulses, oilseeds, plantation crops, spices and vegetables
3.	Viruses	Nuclear polyhedrosis virus (NPV) *of Helicoverpa armigera, Spodoptera* sp.and *Chilo infescatellus*	American bollworm,tobacco caterpillar and shoot borer	Cotton, sunflower, tobacco and sugarcane
4.	Biorationals	Pheromone traps, Pheromone lures, stickytraps and mating disruptants	*Bactocera sp.Chilo* sp. *Dacus* sp. *Earias vittella* *Helicoverpa armigera* *Leucinodes orbonalis* *Pectinophoragossypiella* *Plutella xylostella*	Cotton, sugarcane, vegetables, fruit crops

5.10.5. *Chaetomium globosum*

Spore concentration must be 10^7- 10^8 spores/ml. it must be applied to the crops @ 1 ml suspension to 220 seeds.

5.11. Biocontrol methods

5.11.1. Predators

These are naturally occurring insects who use pests for feeding or multiplication. By introducing large numbers of predators in a greenhouse they will destroy the pest. Predators are single minded and are only interested in their particular prey. If the pest has been destroyed they will disperse in search of more food or die of starvation.

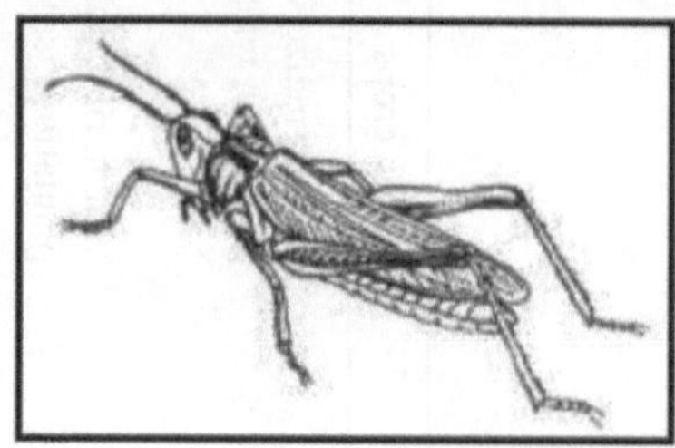

Fig. 21: Predator

- Predators like *Chrysopa* spp, *Menochilus* spp. are highly useful in controlling a wide variety of pests like aphids, whiteflies, cotton bollworms, leaf insects, etc.
- The eggs of these parasitoids are commercially available in the form of egg cards. For example, the egg cards of the parasitoid *Trichogramma brasiliensis* are available commercially.
- Each egg card (e.g., *Trichogramma*) contains 20,000 live parasitised eggs which have 90–96% hatching potential within 7–10 days of parasitisation. These are applied @ 3–5 cards/ha.
- Each egg card cost Rs.20 to Rs.50. *Chrysopa* spp. is available in vials containing 1,000–5,000 live eggs/larvae.
- The standard recommendation for crops like cotton, sunflower, tobacco, groundnut, mustard and vegetables is 5,000–10,000 eggs/larvae per ha. Each vial cost Rs.150 to Rs.200.

5.11.1.1. Plants to attract predators and parasites

Similar to companion planting, which seeks to deter pests from the main crop, attractant plants can be grown to attract predatory insects.

Areas of Natural Habitat

Bushes and trees are a home for many useful insects and birds. They provide resting areas, shelter and food. Areas of natural habitat can be left around the edges of fields where crops are grown. If these areas are destroyed then there is likely to be an imbalance between the populations of predator and pest.

5.11.1.2. Specific plants to attract beneficial insects

There are many plants that can be grown to attract natural predators and parasites, which will help to keep down pests and diseases.

Flowers such as Marigolds (*Tagetes)* Mint (*Mentha*), Sunflower (*Helianthus annus*), Sunhemp (*Crotalaria juncea*) as well as local legumes are useful attractant plants.

However flies, whose larvae feed on greenfly, are attracted to the flowers of herbs and vegetables such as fennel, celery, dill, carrots and parsnips (Umbelliferae family). The nectar and pollen that these flowers provide will help to increase the number of eggs that these insects lay. Umbellifers will also provide food to various parasitic wasps whose young live on aphids and some caterpillars. Red hot pokers (*Kniphofia uvaria*) are used in parts of Africa to attract birds that eat aphids.

Table 23: Predator and their prey

Predators	Prey
Lady Bird Beetles	Aphids, scale insects and mealy bugs
Rove Beetles	Many soil dwelling insects
Tiger Beetles	Various insects on the ground surface
Lace Wings	Aphids
Hover Flies	Aphids
Wasps	Caterpillars of many insects
Spiders	Many flying adult insects
Predatory Mites	Plant feeding mites
Toads	Various soil inhabiting insects
Lizards	Various soil inhabiting insects
Birds	Various soil inhabiting insects

Chandy (2013)

5.11.1.3. Barriers

Barriers are physical structures put in place to prevent a pest from reaching a plant. They keep pests away from a plant but do not kill them. Here are some examples that you can adapt, depending on the resources available.

A ploughed strip should preferably be left around fields. These should not contain any crops or weeds. The strip will act as a barrier for insects that migrate from the uncultivated areas to the fields. A ploughed strip can thus protect crops against armyworms that migrate from other crops. However, this approach involves a great deal of work (ploughing and weeding) for the producer over an area of land that will not produce any yield. Most producers do not, therefore, readily agree to this practice. It should be recommended, therefore, when an invasion seems likely (e.g. when the adjacent crop plots are heavily infested).

To reduce the invasion of bands of grasshopper larvae and make a barrier against these pests, trenches can be hollowed out between the field and the approaching grasshoppers. If the larvae are very young and not very mobile, they will fall into these trenches as they approach the field. It only remains for the larvae trapped in these pits to be buried or destroyed by fire.

Crawling insects

Cut the top off a transparent plastic bottle and place it firmly into the ground, over a young plant. This stops pests such as slugs from reaching the plant.

Climbing insects

To help protect trees from attack by insects, grease bands can be used. Wrap a piece of plastic or a long leaf around the trunk of the tree. Spread any kind of thick grease on top of this. Fold over the top of the foil or plastic to form an overhang to protect the grease from being washed away by rain. Check the grease every week to ensure that the grease is intact. This prevents crawling insects such as ants, fruit fly larvae, slugs, snails, beetles or caterpillars from damaging trees, especially fruit trees, or grain or stored grains.

Termites

Digging a 70-100 cm trench around buildings and nurseries can prevent attack from subterranean species of termites. This is a good method of control however it is hard work. Alternatively, barriers can be built. These should be partially above and below ground and should be made from material that is impenetrable to termites such as basalt, sand or crushed volcanic cinders. Particle size of the material is critical, they should not be too large for the termites to carry away, and not so small that termites can pack the particles to create a continuous passage through which they can move.

5.11.1.4. Red Ants for control of tea mosquito

Red ant colonies are established in cashew garden by bringing the colonies from other trees to cashew and connecting the trees with jute or plastic ropes to facilitate ant movement. Pieces of meat or fish should be kept in the other end of rope in the tree which needs to be invaded. Ant colonies should be kept in the evening time and provide food in the form of pieces of meat or fish for successful colonization (Shri N. Vasavan, Kerala, and cell - 09847871575)

Rearing of red ants

Under the trunk of tree apply 5 baskets of ash near the tree collar region. Clay soil or loamy soil taken from tank. The soil should not be easily dissolved when moist. Both are mixed together. Above this sprinkle a layer of jaggery powder. This will attract red ants and make nesting nearby. The increased ant population will be beneficial in controlling stem borers or any caterpillar pests in the garden.

5.11.1.5. Wild birds

It has been shown that a pair of blue tits can consume 10,000 caterpillars and one million aphids in a 12 month period. The installation of tit boxes is a worthwhile activity. Wrens, thrushes and blackbirds similarly contribute to the control of garden insects.

5.11.1.6. The black-kneed capsid

The black-kneed capsid (*Blapharidopterus angulatus*) is an insect found on fruit trees alongside its pestilent relative, the common green capsid. It eats more than 1000 fruit tree red spider mites per year. Its eggs are laid in August and survive the winter. Winter washes used by professional horticulturalists against apple pests and diseases often kill off this useful insect. The closely related anthocorid bugs, such as *Anthocoris nemorum*, are predators on a wide range of pests, such as aphids, thrips, caterpillars and mites, and have recently been used for biological control in greenhouses.

5.11.1.7. Lacewings

Lacewings, such as *Chrysopa carnea*, lay several hundred eggs per year on the end of fine stalks, located on leaves. Several are useful horticultural predators, their hairy larvae eating aphids and mite pests, often reaching the prey in leaf folds where ladybirds cannot reach.

5.11.1.8. Ladybird beetle

Almost 40 species of lady bird beetle are predatory among pest in horticultural crops. The red two-spot ladybird (*Adalia bipunctata*) emerges from the soil in spring, mates and lays about 1000 elongated yellow eggs on the leaves of a range of weeds, such as nettles, and crops such as beans, throughout the growing season. Both the emerging slate-grey and yellow larvae and the adults feed on a range of aphid species. Wooden ladybird shelters and towers are now available to encourage the overwintering of these useful predators.

Fig. 22: Predators used in rice field

A worrying development in the last few years has been the rapid spread and increase of the harlequin ladybird from South-East Asia. This species is larger (6–8 mm long) and rounder than the two-spot species (4–5 mm). It has a wider food range than other ladybird species, consuming other ladybird's eggs and larvae, and eggs and caterpillars of moths. Furthermore, it is able to bite humans and be a nuisance in houses when it comes out of hibernation.

5.11.1.9. Carabid beetles

The ground beetle (such as *Bembidion lampros*), a 2 cm long black species, is one of many active carabid beetles that actively predates on soil pests such as root fly eggs, greatly reducing their numbers.

5.11.1.10. Mites and spiders

Predatory mites such as *Typhlodromus pyri* eat fruit tree red spider mite and contribute importantly to its control. The numerous species of webforming and hunting spiders help in a very important but unspecific way in the reduction of all forms of insects.

5.11.1.11. Wasps

The much maligned common wasp (*Vespula vulgaris*) is a voracious spring and summer predator on caterpillars, which are fed in a paralyzed state to the developing wasp grubs. Digger wasps also help control caterpillar numbers and benefit from dead hollow stems of garden plant which they use as nests all year round.

5.11.2. Parasitoids

Parasitoids can be effectively used for the management of pest of various agricultural and horticultural crops. These are host specific when a parasite egg hatches on pest it will stop feeding or die. Parasites are able to attack the eggs, larvae, nymphs, pupae or adult. Depending upon the attacking stages parasites can be divided as egg parasites, larval parasites.

5.11.2.1. *Trichogramma*

Trichogramma is a bio-control agent against lepidopteron insect pests. It is available in the form of cards 20,000 live parasitoid eggs which have 90-96% hatching within 7-days of parasitization.

5.11.2.2. *Chrysoperla*

Chrysoperla is a bio-control agent against most of the insect pests. It is applied to crops @ 5000-10000 eggs/larvae per ha, 3-4 times at 15 days interval coinciding with egg laying young stages of pests. These are applied @ 3 cards per ha, 3-4 times at 15 days interval in various crops.

5.11.3. Biocontrol method for rice

- The crop should be observed from 20 DAT and when stem borer eggs are observed on leaf tips and leaf folder eggs on veins, release *Trichogramma chilonis* (for leaf folder) and *T.japonicum* (for stem borer) thrice @1,00,000/ha each and spray *Bacillus thuringiensis* @ 1.0 kg/ha when the stem borer/leaf folder crosses ETL.
- Release *Platygaster oryzae* parasitized galls @ 1 per 10 m^2 on 10 days after transplanting (DAT) (gall midge)

Table 24: Biocontrol module for pest management

Pest	Biocontrol	Rate of application
Yellow stem borer	*Trichogramma japonicum* BT	2.00 lakh eggs/ha 0.75 kg/ha
Leaf folder	*Trichogramma japonicum*	2.00 lakh eggs/ha
Hoppers	Neem (*Azadirachta indica*)	1500 ppm

5.11.4. Biocontrol method for pulses and cotton

- Spray specific NPV suspensions of *H.armigera* and *S.litura* in the evening hours
- Application of *Helicoverpa armigera* or *Spodoptera litura* nuclear polyhedrosis virus (NPV) @ 250-500 LE/ha (1 LE =6 x 109 POBs) (1LE / litre of water) depending upon the crop growth with jaggery and teepol in evening hours at 7th and 12th week after sowing.
- ULV spray of NPV at 3 x 10^{12} POB /ha with 10% cotton seed kernel extract, 10% crude sugar, 0.1% each of Tinopal and Teepol for effective control of *H.armigera*.

- Inundative release of egg parasitoid, *Trichogramma spp.,* at 6.25 cc/ha at 15 days interval 3 timers from 45 DAS (American boll worm)
- Inundative release of egg-larval parasitoid, *Chelonus blackburni* and predator, *Chrysoperia camea* at 100000/ha at 6th, 13th and 14th weeks after sowing (American boll worm)
- Seed treatment with *Trichoderma spp.* @4g/kg of seed for seed borne diseases

5.11.5. Biocontrol method for tomato

- Spray *Bacillus thuringiensis* @ 2g/litre. (*H.armigera* and *S. litura*).
- Release *Trichogramma chilonis* @ 50000/ha/week (up to 6 weeks) coinciding with flowering time and based on ETL (*H.armigera* and *S.litura*).
- Application of NPV : For *H.armigera* : H.a.NPV 250 LE/ acre
- For *S.litura* : *S.l.*NPV 250 LE/acre along with 1 kg jaggery and 100 ml teepol at 10 days interval.
- Treat the seeds with antagonistic fungi, *Trichoderma viride* @ 4 g/kg seed (nematode).

5.11.6. Biocontrol method for brinjal

- Seed treatment with antagonistic fungi viz., *Trichoderma harzianum* or *T.viride* @ g/kg seed (nematode)
- Apply *Pseudomonas fluorescens* @ 10 g m^2 for nematodes
- Release 1st instar larvae of green lace wing, *Chrysoperia carnea* @ 10,000 per ha (aphid).

5.11.7. Biocontrol method for bhendi

- Release of egg parasite, *Trichoderma* @ 1.0 lakh/ha (fruit borer)
- Release 1st instar larvae of green lace wing bug, *Chrysoperia carnea* @ 10,000/ ha (fruit borer).

5.11.8. Biocontrol method for chilli

- Spray *Bacillus thuringirnsis* at 2 g/l (fruit borers)
- Apply TNAU formulation of VAM (containing 1 spore/g to control root knot nematode in nursery)
- When *H.armigera* is in egg stage, release *Trichogramme* @ 60000/ acre two to three times.
- Spray H.a NPV and S.l./NPV @ 200 LE/ha + 500 g jaggery +100 ml teepol in evening hours.

Table 25: Field Use – Release, Frequency

***Trichogramma* spp**	
Sugarcane borers 1. Early shoot borer (*Chilo infuscatellus)*	*Trichogramma chilonis* sugarcane strain, 4-6 releases @ 50,000/ha at 10 days interval starting from 45 days after planting.
2. Top shoot borer *(Scirpophaga excerptalis)*	*Trichogramma japonicum* sugarcane strain @ 50,000/ha. 4-6 release at 10 days interval with observation of the pest, usually 60th day onwards.
3. Stalk borer (*Chilo auricilius)* Internode borer (*Chilo sachhariphagus indicus)* Gurdaspur borer (*Acigona steniellus*)	*Trichogramma chilonis* sugarcane strain 8-10 release @ 50,000/ha at 10 days interval starting from 90th day onwards.
Cotton bollworms Old world bollworm*(Helicoverpa armigera)* Pink bollworm*(Pectinophora gossypiella)* Spotted/spiny bollworms(*Earias* spp.)	*Trichogramma chilonis* or *Trichogramma achaeae* cotton strain @ 1,50,000/ha. from 45th day onwards 6 weekly releases. Moths can be monitored by pheromone traps.
Maize stem borer *(Chilo partellus)*	*Trichogramma chilonis* @ 75,000/ha. 6 releases from 45th day onwards at an interval of 10 days.
Tomato fruit borer *(Helicoverpa armigera)*.	*Trichogramma brasiliensis* 6 releases @ 50,000/ha. from 45th day onwards at weekly interval
Paddy stem borer*(Tryporyza incertulas)*	*Trichogramma japonicum* @ 50,000/ha / week; with the appearance of the pest or 30 days or after transplantation, 6 release to be made in one season.
***Chrysoperla* spp**	
Cotton Old world bollworm (*Helicoverpa armigera*) Spotted/spiny boll worms (*Earias* spp.) Pink bollworm (*Pectinophora gossypiella*) White fly (*Bemisia tabaci*) Aphid – (*Aphis gossypii*)	*Mallada boninesis* or *Chrysoperla carnea*@ 1,00,000/ha twice during the season with a gap or 15 days.
Tobacco Aphid (*Myzus persicae*) Tobacco caterpillar (*Spodoptera litura*) White fly (*Bemisia tabaci*)	*Chrysoperla carnea* @ 100,000/ha or 6larvae per plant twice during the season with an interval of 15 days.
Sunflower Head borer (*Helicoverpa armigera*) Aphid *Aphis* spp.	*Chrysoperla carnea* @ 100,000/ha twiceduring the season with an interval of 15 days.
Groundnut Aphid (*Aphis craccivora)*.	*Chrysoperla carnea* @ 100.000 1st instarlarvae/ha twice during the season with aninterval of 15 days

Table 26: Biocontrol agents to control pests of different crops

S.N.	Biological agents	Pest	Crop
1.	*Trichogramma brasiliensis* – 1.0 cc/ac. Once in 10 days	Lepidopteran , *Heliothis* spp.	Cotton, Tomato
1.	*Trichogramma chilonis* - 2.0 cc/ac. Once in 15 days	Borers	Sugarcane, paddy, pulses, Vegetables
2.	Nuclear polyhedrosis virus – 100/200 LE/ac	Spodoptera sp & *Heliothis* spp.	Vegetables
3.	*Chrysoperla* spp @ 5000-10000 eggs/ha, 3 – 4 times in 15 days	Prudenia, Caterpillars, Whiteflies, Thrips, Aphids	Vegetables

5.12. Farmers Wisdom in Pest Management

The knowledge of traditional agriculture with millions of farmers should be utilized and modern technology in agriculture should be blended with traditional wisdom. The following are certain practices of farmers which they have been following time immemorial

a) Diluted cow dung slurry sprinkled to hasten paddy germination.

b) Coconut fronds cut into small bits erected as perches in field to attract nocturnal birds which preys upon rats.

c) Chilli mash and garlic juice sprayed to control rice earhead bug.

d) Application of common salt at 1 - 1.5 kg/ palm of coconut gives insect resistance and prevents button shedding.

e) Use of scarecrows to ward off bird pests in day time, which also serve as perches to nocturnal predatory birds.

f) Use of Kavankal where stones are released from slings to scare birds

g) Ploughing of field during Agninakshatra (April-May) when temperature is around 40 - 45 C brings about killing of soil insects, nematodes and pupae of lepidopteran pests.

h) Treating stored pulses with red earth (1 kg of red earth : 1 kg of pulses) to prevent insect damage.

i) Use of Tanjore bow trap, a common traditional gadget to kill rats in rice fields of Cauvery delta.

j) 'Vrikshayurveda", a science of plant health, similar to 'Ayurveda" which is science of human life deals with maintenance of plant health and provides literature on control measures for control of pests and diseases.

Table 28: Farm level practice for insect pest control

Cropping Techniques	Pest Checked
Ploughing	Red hairy caterpillar (*Amsacta albistriga*)
Puddling	Rice mealy bug (*Brevennia rehi*)
Trimming and plastering	Rice grass hopper (*Hieroglyphus daganensis*)
Pest free seed material	Potato tuber moth (*Phthorimaea operculella*)
High seed rate	Sorghum shootfly (*Atherigona soccata Rondani*)
Rogue space planting	Rice brown planthopper (*Nilaparvata lugens*)
Plant density	Rice brown planthopper (*Nilaparvata lugens*)
Earthing up	Sugarcane whitefly (*Aleurolobus barodensis*)
Detrashing	Sugarcane whitefly (*Aleurolobus barodensis*)
Destruction of weed hosts	Citrus fruit sucking moth (*Othreis fullonica*)
Destruction of alternate host	Cotton whitefly (*Bemisia tabaci*)
Flooding	Rice armyworm (*Pseudaletia unipuncta*)
Trash mulching	Sugarcane early shoot borer (*Chilo infuscatellus*)
Pruning / topping	Rice stem borer (*Scirpophaga incertulas*)
Intercropping	Sorghum stem borer (*Chilo partellus*)
Trap cropping	Diamond back moth (*Plutella xylostella*)
Water management	Brown planthopper (*Nilaparvata lugens*)
Judicious application of fertilizers	Rice leaf folder (*Cnaphalocrosis medinalis*)
Timely harvesting	Sweet potato weevil (*Cylas formicarius*)

5.12.1. Community level practices

- Synchronized sowing : Dilution of pest infestation (eg) Rice, Cotton.
- Crop rotation : Breaks insect life cycle.
- Crop sanitation.

a) Destruction of insect infested parts (eg.) Mealy bug in brinjal.

b) Removal of fallen plant parts (eg.) Cotton squares.

c) Crop residue destruction (eg.) Cotton stem weevil.

6

Indigenous Technical Knowledge (ITK) Based Techniques

Indigenous Technical Knowledge (ITK) based inputs (Most of the ITK's needs scientific validation before large scale adoption)

Is the systematic body of knowledge acquired by local people through the accumulation of experiences, informal experiments, and intimate understanding of the environment in a given culture (Rajasekaran, 1993). Learning from ITK can improve understanding of local condition and provide a productive context for activities designed to help the communities. In addition, the use of ITK's assures that the end user of specific agricultural development projects are involved in developing technologies appropriate to their needs'. Yet, ITK is still an underutilized resource in the development activities. It needs to be intensively and extensively studied, and incorporated into formal research and extension practices to make agriculture and rural development strategies more sustainable.

Many of our indigenous practices in agriculture and allied fields have been replaced by the so-called modern technologies and they have become obsolete, especially among the younger generations. Now these indigenous practices are endangered ones and these are a possibility for them to become extinct particularly during this era of globalization, liberalization and commercialization. Now the need has come to re-examine and then gradually re-introduce the effective traditional technologies of crop production.

Therefore, there is a need to systematically document the ITK as they are the unwritten body of knowledge. There is no systematic record to describe what is, what it does, how it does, means of changing it, its operations, it boundaries and its applications. It is held in different brains, languages and skills in as many groups, cultures and environments. Tamil Nadu is also a treasure land of indigenous practices in agriculture. However, only a few attempts have been undertaken by social scientists to document the available indigenous agricultural practices but none has evaluated scientifically. The documented ITK practices are given below.

6.1. Soil and Water Management

- For soil improvement in 'theri' lands of Tuticorin district, 200 tones of tank silt are applied per acre followed by 50 tones per year for the next few years.
- About 10 kg. of neem cake is soaked in 10 lit. of cow urine along with ½ kg. of waste asafoetida and left over night. In the next day, it is sprayed for 1 acr. after dilution as liquid manure.
- Outer shells of tamarind fruits are applied in the field to control *Cyperus rotundus*.
- A mixture is made with 1 lit. of neem oil, 3kg. of fine sand and 3 kg of cow dung and heaped in shade covering with a moist sack for 3 days. On the fourth day, the mixture is dissolved in 150 lit. of water and sprayed to control all sucking pests.
- About 10kg. of dried cow dung is ground into fine powder and mixed with ash (obtained from brick kiln) dusted in the early morning, to control pests and diseases.
- Garlic acts on a wide spectrum of organism in unrelated crop plants singly or in combination with neem products, chilli, asafetida etc. Besides garlic is effective against bacteria, fungi and nematodes.
- *Calotropis* leaf extract is applied at the place of termite attack to control them.
- Consolidation of lands gives better results even they are less fertile.
- Fields which are nearer to rivers will give lesser yield
- Better yield is obtained from the wetlands near water sluice of canals and the dry lands near foot hills.
- Laying stone bunds around the fields across the slope for preventing soil erosion and for conserving moisture.
- Planting vettiver *(Vetiveria zizanoides)* slips across the slope or around the fields to prevent soil erosion.
- To minimize soil erosion perennial vegetation is grown on the field bunds.
- New garden land and old wetland will yield better.
- Intensive care is required for wetland crops as compared to the garden land crops.
- Water logged dry lands are unsuitable for cultivation.
- Soil character decides the choice of crops for cultivation.

- Red soil is suitable for continuous cropping.
- Black soil has more water holding capacity than the red soil.
- Sandy soil is less suitable for the cultivation of many crops.
- Excessive application of farmyard manure (FYM) improves the soil texture.
- Tank silt is applied to increase the soil texture.
- Manures and fertilizers are applied based on soil character.
- If the weed growth is profuse after the rains, it indicates high soil fertility.
- However the low soil fertility is indicated by the growth of the weed Aduthinnapalai *(Aristolochia bracteolata).*
- Addition of red soil to black soil increases the fertility of the black soil and vice versa.
- Practicing sheep/cattle penning during summer season to improve the soil fertility.
- Practicing mixed cropping or inter cropping of legumes in rain fed areas to maintain the soil fertility.
- Cultivating Kolinji *(Tephrosia purpurea)* in between the fruit trees in sloppy lands to prevent soil erosion and to improve soil fertility.
- High moisture content in the soil is identified with the occurrence of 'Nuna' tree *(Morinda tinctoria).*
- For moisture conservation deep ploughing is done during summer.
- Land is well ploughed and powdered to conserve more moisture.
- Application of tank silt (taken from black soil tanks) on the red soil fields to increase the water holding capacity of the red soil.
- Raising and ploughing daincha (*Sesbania* spp.) and sun hemp (*Crotalaria juncea*) in the field before flowering increase water holding capacity at the soil.
- It is better to grow sorghum, finger millet and chilies if the water is saltish.
- Irrigation is given to any crop at a stage that while walking on the fields, our foot should not create any print on the soil, which is taken as the indication.
- Growing 'Poovarasu' *(Thespesia populnea)* tree near the wells reduces water loss through evaporation.

- Wetlands having 'Aarai' weeds (*Marshilea quadrifolia*) and garden lands having 'Arugu'(*Cynodan dactylon*) weeds give better yields.
- Red soils having 'Arugu (*Cyanodan dactylon*) weeds and black soils having nut grass (*Cyperus rotandus*) weeds are the best of their kind.
- Sowing densely the daincha (*Sesbania* spp.) green manure and ploughing in-situ at its ftowering to correct alkaline soils.
- Growing sun hemp (*Crotalaria juncea*) in alkaline soils and ploughing in-situ before flowering to rectify its alkalinity.
- Application of 'Pirandai' (*Cissus quadrangularis*) to reduce alkalinity.
- Neem leaves are applied to correct alkalinity.
- Application of shells of neem seed to reduce salinity in soils.
- Application of neem cake to correct salinity.
- Palmyia (*Borassas flabellifer*) leaves are cut into pieces and applied in large quantity to correct alkalinity.
- To correct alkaline soils, pungam (*Pongami apinnata*) leaves, or outer shells of tamarind fruits are applied.
- Mixing and applying coir waste with compost to correct alkalinity.
- Application of sugarcane bagasse and sediment after extraction of country sugar to correct alkaline soils.
- Putting leaves and branches of Indian gooseberry (*Phyllanthus distichus*) in the wells reduce salinity in water.

6.2. Preparatory Cultivation

- Achieving fine tilth is better than applying manures.
- It is better to plough intensively than extensively.
- It is better to have deep ploughing rather than shallow ploughing for good crop growth.
- Plough four times for garden land and seven times for wetland.
- Summer Ploughing gives good crop in the ensuing season.
- Tying paddy straw around the sole of country plough to make a wider furrow while forming broad bed and deep furrows.
- Garden lands are ploughed deep to conserve more moisture.

6.3. Manures and Manuring

- Commonly used green leaf manures are Kolingi *(Tephrosia purpurea),* Calotropis *(Calotropis gigantea),* Nuna (*Morinda tinctoria*), Pungam *(Pongamia pinnata),* Neem *(Azadirachta indica),* Poovarasu *(Thespesia populnea)* and Adathoda (*Adhathoda vasica).*
- Green manure crops like daincha (*Sesbania* spp.), kolingi (*Tephrosia purpurea*) sunnhemp (*Crotalaria juncea*) etc., are raised and ploughed insitu before their flowering.
- Red gram is also used as a green manure crop which improves the soil fertility.
- Applying water hyacinth plants either as a compost or as burnt ash to the fields for supplying potash.
- Sheep penning results in more crop yields.
- *Goat* manure gives benefits to crops grown in the same season.
- Goat manure is good for the first season and cattle manure and green manure for the second season.
- Poultry manure serves as a good source of crop nutrients.
- Cow urine is more nutritious than cow dung.
- Soil fertility can be better increased by the use of pig dung than by the cow, sheep or goat waste.
- It is better to apply cattle manure for garden land and dry land and leaf manure for wet land.
- Near the irrigation channel, a pit is dug in which, cow dung, foliage of *Calotropis gigantea,* neem cake powder and cow urine are applied, mixed well and allowed to decompose. Then it is allowed to mix with the irrigation water, to supply the nutrients, and to control pests and diseases. Tank silt is applied every year in dry lands for better yields. Termite hills serve as good manure.
- Adding sand from an ant hill to the field gives good yield.
- Foliar spray of manures given on full moon day yields better results.

6.4. Weed Management

- Weeding is not required under dry land condition. If weeding is not done the weed growth is controlled naturally and it also helps to conserve moisture.
- Repeated ploughing will reduce weed population.

- Crop yield will be less in the fields having 'Arugu' *(Cynodon dactylon)* weeds.
- To control "Arugu *(Cynodon dactylon)* grass" in black soils the field is kept fallow for 3 years.
- Cultivating rice once in three years in garden lands to control 'Arugu' *(Cynodon dactylon)* weeds.
- Raising and ploughing the green manure crops tike daincha, *(Sesbania spp.),* kolingi *(Tephrosia purpurea)* in the field before their flowering to reduce weed population.
- Raising *Calotropis gigantea* as a green manure to check the growth of Aarai (*Marshilea quadrifolia)* weed.
- Growing horse gram to control nut grass *(Cyperus rotundus).*
- Growing cowpea as a green manure to control nut grass.
- Allowing swine in the fields to eradicate nut grass.
- Frequently ploughing the fields by wooden plough made up of neem trees and frequent application of neem cake in the soil to control nut grass.
- Dissolving 1 kg of salt and 100g sarvodaya soap in 10 litre water and spraying this solution to control all the weeds except nut grass.
- To control nut grass in the field 50 kg neem cake is applied both at the time of ploughing and sowing.
- Dissolving 200 g salt in 1litre water and spraying to eradicate congress weed (*Parthenium hysterophorus*).
- Continuous submergence of field for some time controls the weeds.
- Keeping the irrigation channels free from weeds,

6.5. Pest and Disease Management

- Uninterrupted drizzling of rain leads to the occurrence of more number of pests and diseases.
- Winds blowing at the end of December bring lot of pests.
- Crops grown in alkaline soils are more prone to disease attack.
- Small lamps are placed on either side of the house entrance and light from the lamps acts asa light trap and the farmers are able to identify the pest outbreak.
- To prevent the attack of aphids and mite flies, sorghum or pearl millet is grown very closely in 4 rows around the fields to act as a shelter so that these pests can not enter the fields.

- Growing Thangarali' *(Tecoma stands)* and 'Sevvarali' *(Nerium oleander)* as border crops, which act as trap crops and control the insect pest.
- Thiruneeru' (Sacred ash) is dusted on the crops to reduce pest attack.
- Kitchen ash is applied to control aphids.
- To control the sucking pests, 5 kg tobacco powder is soaked in a mixture of 10 litres of cow urine and 5 litres of water for 5 days. Then it is filtered and diluted with 80 litres of water and sprayed.
- Mixing cow urine, neem oil and tobacco decoction together and spraying on crops controls sucking pests.
- Burning of rice stubble's and straw effectively reduces stem rot caused by *Sclerotium oryzae* and sheath blight caused by *Corticium sasakii.* Flaming was effective in controlling *Verticillium dahliae* in peppermint. The only disadvantage is that the beneficial organism may also be destroyed.
- Leaves of *Calotropis gigantea* and *Strychnos nux-vomica* and neem cake are soaked in water in a mud pot and fixed in the field. Moths are attracted towards the smell, fall inside and die.
- Leaves of notchi (*Vitex negundo)* and pungam *(Pongamia pinnata) are* also used to control moths.
- About 5 kg *Calotropis* leaves is soaked in a mixture of 10 liter cow urine and 5 litre water for 5 days. Then it is filtered and diluted with 80 li. of water and sprayed to control defoliation.
- Spraying sarvodaya soap solution to control mealy bugs.
- To control nematodes, pungam *(Pongamia pinnata)* or iluppai *(Bassia latifolia)* cakes are applied.
- Cow dung, Cow urine, *calotropi*s leaves and neem cake are put in a pit near the irrigation channel. After decomposing, it is mixed with irrigation water.
- Grinding the leaves of *Calotropis gigantea* with the fruits of *Datura metal,* soaking in water for 15 days, filtering and spraying to control all the pests
- Two handful each of leaves of thumb (*Leucas aspera),* kuppaimeni *(Acalypha indica),* thulasi*(Ocimum canum). Datura metal,* neem, nochi (Vitex *negundo), 5* fruits of *Datura metal* and handful each of neem cake and lluppai *(Bassia latifolia)* cake are pounded together and soaked in water in earthen pot for 10 days. Then it is filtered, diluted (100 ml./lit.) to which 100 ml sarvodaya Khadi soap solution and 100 ml neem oil added and sprayed to control all insect pests.

- Leaves of Tulsi (*Ocimumcanum),* seeds of *Nerium oleander and* fruits of *Datura metal* are taken in equal quantities, powdered and soaked in cow urine for 10 days. Then it is filtered and diluted (100 ml/lit.) to which 100 ml neem oil is added and sprayed to control all insect pests.
- Neem oil and neem seed kernel extract are the general organic pesticides used to control many pests.
- During the night time on full moon day of Tamil month 'Karthigai' (Nov.-Dec.) 'Chokkapanai' (Community firing) is performed as a part of celebrations in a common place in the village by which the pests get attracted and killed. Ash from this fire is dusted on the crops to control sucking pests.
- Crop wastes are burnt and its ash is dusted on the fields to control diseases.
- If neem cakes are applied as basal fertilizer there will not be any incidence of diseases.
- One kg leaves of seemaikaruvel *(Prosopis juliflora)* is pounded and diluted with water and sprayed to control yellow mosaic virus.
- Spraying cow urine to control many pests.
- A mixture of extracts of garlic and neem cake is *sprayed* to control aphids.
- Planting 'Pirandai' *(Cisssus quadrangularis)* vines around fields to protect against termites.
- Grow castor on the fields to control termites.
- Spread neem leaves over the nursery to control termite damage.
- Putting neem cake inside a gunny bag and placing it in the irrigation channel controls mites.
- Termites destroy the seedlings in nursery grown in dry land condition. To control these termites, apart from putting the neem leaves, sheep wool and human hairs are also put. Termites eating these hairs die.
- Pouring decoction of finger millet roots on the root zone of crops to control termites.
- Before planting tree seedlings, dried leaves and trashes are burnt in the pits to protect the seedlings against termite attack.
- Dusting ash in the pits before planting tree seedlings also helps to prevent termites.
- Sprinkling 5% common salt solution to reduce termite attack on the trees.

- After the harvest of tobacco leaves, their stems and roots are ploughed in-situ to control the termites
- Tobacco soaked water is poured on the ant mounds to control them.
- Any spraying is to be done in the early morning.
- Take 30gm. of gounded *Nerium* (Arali) seeds in 10 litre. of water for 1 hour and mix with khadi soap and spary to control thrips, aphids, whiteflies and leaf eating caterpillars.
- Clear polyethylene placed over moist soil, during summer days raises the temperature at the top 5 cm of soil to as high as 52^{0}C. The increased soil temperature from solar heat, known as solarization inactivates many soil borne pathogens and reduces the inoculum and the potential for disease.

6.6. Rat Control

- Rats do not live in fields where sheep penning is being practiced.
- Planting closely notchi (*Vitex negundo)* and erukku (*Calotropis gigantea)* around the fields as a fence helps to control rat problem.
- Putting the branches of Thangarali (*Tecoma stands*) around the fields to control rats.
- 'To control rats in paddy fields, Channampoo (*Cycas cercinalis*) flowers are cut in to pieces and placed in many places whose bad odour drives away the rats.
- Pieces of Palmyra (*Borassus flabellifer*) leaves are tied on the poles fixed on the field. The sound produced by the leaves scares away the rats
- Providing owl stands near the rat holes will help in reducing the rat damage.
- To reduce rat population, rat holes are dug and rats are killed after each harvest.
- To catch the rats, a trap made up of wire loops on bamboo pegs is being used.
- Big round shaped earthen pots are buried on the field at ground level. Half of the pot is filled with mud slurry on which baiting material is put on a coconut shell. Attracted rats fall inside the pot and they cannot climb up and get killed.
- Use of soaked rice as bait attracts more rats.
- Putting fresh cow dung on both the fields and bunds to reduce rat problem.
- Papaya (Latex/ milk yielding) 3 fruits/ acre

6.6.1. Eli viratti (Rat control)

- Palymyra frond is tied on the stick and kept in the rice field. It will create sound due to air movement, which will scare the rat away.
- Total numbers required: 15 numbers/ acre

6.6.2. Eli kattuppaduthuthal (Rat control)

- Cow dung balls are immersed in kerosene and kept in 8-10 places of the field randomly. Odours comes from from kerosene and cow dung will control rat damage
- By leaving space (nearly 1 ft) in all four sides of field, rat damage will be reduced.
- Pots are filled with water and buried in the bunds of the field. When rat enter into the field, it will be trapped and killed.

6.6.3. Neekalpoduthal (Rat control in rice field)

- Neekal poduthal is nothing but giving one feet space in 4-5 places of the field randomly when the crop is at milky stage in the rice.

6.6.4. Rat control method in coconut

Developed method of rat control includes an old bamboo basket, binding wire, plastic thread, snap trap and desiccated coconut pieces. An old bamboo basket is tied at four corners with binding wire. All these four binding wires are connected to a single plastic thread and put on a coconut frond which can be pulled up or down. Desiccated coconut piece is attached to a snap trap and placed inside the bamboo basket. Rats get attracted to desiccated coconut and they get locked inside the trap. Dead rats are removed and burried in the soil so as to prevent the spread of diseases as informed by Shri S.R. Arun Kumar, Karnataka, Mobile 08133-269564, 09900824420.

6.6.5. Extract of *Ipomoea fistulosa*

Leaves of *Ipomoea* are boiled in water andfiltered. Sorghum grains are boiled in this extract and placed near the rat burrows. It is believed that rats die after eating it.

6.6.6. Rat control in field

Soak the dead rats (4-5 nos) in field in cow's urine for a week time and sprinkle the decayed rat syrup in all the borders of the field.

6.1.7. Birds perching sticks

In the paddy field few places long sticks are planted and straw rolls are placed to over them, so as to attract to perch over on them. During night hours some predatory birds like owls will come and perch over the sticks. The field rats that are found in the vicinity are caught as prey and eaten.

6.6.8. Prevention of Rodents

- To prevent the damage by fields rats and squirrels in coconut trees, tar is heated and applied to the base of the palm all around upto a height of 0.5 meter from the ground level to a thickness of 15 cm.
- As the tar applied surface will be smooth, the field rats and squirrels cannot climb over it.

6.6.9. Control of Field Rats

- All around the paddy field, "Thalai" (*Pandanus fascicularis*) leaf sheath are fixed closely in the field hedges as fence to prevent entry of rats. The thorns found in the leaf margin will pierce into the abdomen of the rats.
- Keep mud pot filled with straw bits with a hole in the bottom. The pot is placed over the live burrow and by burning the straw pits, smoke will be formed which is directed into the burrow to kill rats. The pot is tightly closed with its lid.
- A mud pot is placed with its bottom hole. Either bamboo or banana bank or "Korai' (nut grass) stem is inserted up to the bottom and by frequent pulling up and down a typical sound is made, which will frighten the rats to leave off the place.

6.6.10. Using Mud pots

A small mud pot with a lid is used as a trap to catch field rats. Mud pot is erected over tripped of wooden sticks at a height of 9 inches from ground and set at certain angle (slanting position). The lid of the pot is put in the field ground, top side facing down and fixed with 3 small sticks.

Inside the pot small quantity of dried fish / fried kernels of ground nut kept as bait. When the rat attempts to get the bait by jumping the pot topple down over the lid and the rat gets trapped inside and subsequently removed and killed. Daily up to 8 rats trapped by this method.

6.7. Bird Scaring

- Stones are thrown with the help of 'Kavan' made by tying a long piece of gunny thread to both ends of a small piece of leather.

- Stones are thrown with the help of a leather rope called 'Kavattai' to scare the birds.
- Tying the carcass of a crow to a long pole and placing it in the center of fields.
- Tying a black cloth to a long pole and placing it in the center of fields to drive away the crows.
- Waste magnetic tapes taken from an audiocassette are tied in the fields in a criss cross manner which the birds will think as a net.

6.7.1. Paravaithangi (Bird perch)

- Coconut branches are kept in the field randomly.
- "T" shaped stick is used.
- Total numbers required : 15 numbers/acre.

6.7.2. Tying palmyra fronds

- Dried palmyra fronds are tied to the poles which makes sound when wind blows.
- Total numbers required: 5-10 numbers/acre to keep sparsely.
- Thus, by producing sound, birds are scared.

6.7.3. Tying unused recordable tapes

- Unused recordable tapes are tied across the field in a criss-cross manner and it looks like ant spread over the field.
- Birds perceive that net is bring spread to trap them and avoids to land on that field.

6.7.4. Tying polythene sheet and beating drums

- Polythene sheet/papers are tied to a pole as like that of dried palmyra fronds and kept in the field. When wind blows, sound is produced and this will scare the birds.
- To scare the birds, sound is produced by beating drums or by beating a plate with stick.

6.7.5. Use of small flags

- The use of ''small flags" made of plastic or paper sheets and strings attached to a long rope or plastic twine are examples of materials commonly used by local farmers to prevent birds from feeding on mature rice grains.
- These small flags are placed across the rice fields before harvest.

Fig. 23: Use of small flags from colored-synthetic materials

6.7.6. Inverted coconut fronds

- Coconut fronds are placed in strategic locations in the rice field in an upside down position resembling an owl or a cobra to prevent rodent infestation.

Fig. 24: Inverted coconut fronds

6.7.7. Owl perches

- 'T' shaped wooden sticks (3 feet height) are to be placed in the center / corner of the field at half kilometer distance.
- These sticks are surrounded by paddy straw on top portion act as a stand / resting places for owls in the night hours.
- During night time, owls would rest on this stick and catch rodent.
- About 20 "T" shaped wooden sticks per acre required.

6.7.8. Bird scaring in maize using tin and stick

- The labour or the farmer engaged in beating the tinbox moved around the field for effective scaring of birds to distance

6.8. Wild Boar Control

- Broadcast human hair waste (1-2 cm length) on the pathway of the wild boar route and all possible places of entry. In haling the hair bitswild boar will be driven away due to irritation caused by the hairs sticking on its nostrits (Inner mucous wet membrane).

7

Indigenous Technical Knowledge Techniques for Individual Crops

7.1. Rice

- Treatment of paddy seeds in diluted bio gas slurry for 12 hours increases resistance of seedlings to pests and diseases.
- During panicle formation in paddy, the flowers of *Cycas circinalis* are placed on sticks in paddy fields @ 4/ac. Its unpleasant odor repels ear head bugs.
- About 30 kg of tamarind seeds are applied for an acre of paddy field 1 day after transplanting to boost up the crop growth and yield.
- Soaking the paddy seeds in diluted cow's urine before sowing, considerably reduces the incidence of leaf spot and rice blast
- Presoaking of paddy seeds in milk increases its resistance against 'tungro' virus and 'stunt' virus
- For control of red leaf spot disease in paddy, the seeds are soaked in 'Pudina' leaf extract (*Mentha sativa*) for 24 hours
- 'T' shaped bamboo stands are placed in many places in the paddy fields so that birds can sit on them and feed on the larvae and adults of rice pests.
- Sowing on eighteenth day (Aadipperukku) of Tamil month Aadi (Jul-Aug.) ensures good harvest.
- Daincha (*Sesbania* spp.) seeds are sown on paddy main fields when paddy nursery is raised and the grown up daincha is ploughed in-situ during field preparation.
- Plough the main field for four to six times for better yield.
- Good harvest can be obtained from the crop transplanted during Aavani i.e. Aug. - Sep.

- The crop transplanted during October-November will give reduced yield.
- The rice crop will establish better if it is transplanted along the wind direction.
- Planting the 'samba (Aug), crop thickly and 'navarai' (Feb.) thinly.
- Practice sheep penning during summer to get more yield.
- Practice sheep penning for the first season and green leaf manure for the second season for better yield.
- Apply 100 kg of pig manure for one acre of rice at 10 days after planting to get higher yield.
- Apply the neem seeds @ 40 kg / ac as basal to get more yield as compared to the equal quantity of neem cake.
- Irrigate the fields, allow the weed seeds to germinate and then plough the fields to incorporate the weeds into the soil before sowing or transplanting of rice crop to control weed growth.
- Cultivation of sunnhemp or daincha helps to control the nut grass (*Cyperus rotundus*) weed.
- Application of *Calotropis gigantea* as green leaf manure will prevent thrips attack in the nursery.
- Neem (*Azadirachta indica*) oil cake extract is sprayed to control thrips in rice.
- Dragging the branches of country ber or Aloe sp. on the affected field to control the leaf roller.
- Neem oil is mixed with water @ 30ml / litre and sprayed to control stem borer in rice.
- Dusting chulah ash in the early morning to control stem borer and ear head bug.
- To control the ear head bugs, 10 kg cow dung ash is mixed with 2 kg lime powder and 1 kg powdered tobacco waste and dusted on the rice crop during morning hours.
- Hundred ml. of leaf extract of "Karuvel" (*Acacia nilotica*) and 10 kg cow dung are dissolved in 10 litre. of water and sprayed on the rice crop to control ear head bug.
- Growing or planting calotropis at 12 feet interval on all sides of paddy fields to control the hoppers.
- Applying neem cake before last plough to control root rot and nematode problem.

- A mixture of 5 kg common salt and 15 kg. of sand is applied for 1 acre to control brown spot disease.
- Soaking the paddy seeds in 20% mint leaves solution before sowing will control the brown leaf spot.
- Spraying the leaf extract of *Adathoda vasica* to control rice tungro.
- Palmyra (*Borassus flabellifer*) fronds are tied on to poles and kept on the corners of rice fields so that the noise produced by them scare away the birds like ducks, sparrows etc. and save the grains being damaged.
- When one ear head contains about 100 grains, the yield will be 20-22 quintals/ac.
- One hundred and twenty grains found in a rice ear head indicates the full yield.
- Use large mud pots called 'Kudhir' as high as six feet for storing paddy grains for longer periods.
- Putting the leaves of notchi (*Vitex negundo)* and pungam (*Pongamia pinnata)* inside the *Kulumai* to ward off storage pests.
- Mixing the paddy grains with the leaves of pungam (*Pongamia pinnata*) or notchi (*Vitex negundo*) or neem (*Azadirachta indica*) before storage to avoid storage pest attack.

7.2. Maize

- If the sheaths on maize cobs are not removed, they can be stored for more than three months.
- Maize seeds are soaked in cow urine for 12 hours before sowing
- Before sowing, maize seeds are soaked in warm water for 3-6 hours and shade dried to induce better germination and to control shoot borer.
- Maize is sown after tomato in the same ridges and furrows to reduce the cost of land preparation.
- When a sample of maize grains in chewed, metallic sound indicates their optimal dryness.
- Dried maize stalks are stacked as heap on stone slabs and covered with paddy straw. This cab be stored for more than a year and used as cattle feed.

7.3. Pearlmillet

- For bird scaring, the carcass of a crow is hung on a long pole in the center of the pearl millet field.
- Pearl millet seeds are mixed with the leaves of noichi (*Vitex negundo*) and then stored.
- Pearl millet is stored by building a structure called 'Pattarai' before threshing.
- Pearl millet ear heads are spread circularly to 1 foot height and cattle threshed
- When pearl millet seeds are immwersed in salt solution (1 kg. of common salt in 10 litre. of water) and the floaters and disease affected seeds are removed.
- Pearl millet seeds are presoaked in water overnight before sowing, for early germination.
- Pearl millet nursery is grown on raised beds for better seedling growth.
- Before transplanting, seedling tips are clipped to remove the eggs of pests and white flies and to control tip blight disease.
- Dusting Chula ash in pearl millet fields to control green leaf hoppers sitting on inner side of leaves.
- Grow cowpea as intercrop to control leaf miner in pearl millet.
- Storing pearl millet seeds by mixing with ash.

7.4. Sorghum

- Sorghum ear heads are sun dried for two days and stored without seed separation by building a storage structure called 'Pattarai'. This structure is opened and threshed when needed..
- Soak the sorghum seeds in common salt solution before sowing to secure good germination under adverse conditions.
- Soak the sorghum seeds in cow urine for half-an-hour and sun drying them before sowing to control head smut and to induce drought tolerance.
- To ensure quick germination of sorghum seeds and to avoid shoot fly attack, enough water is boiled and kept in an open place throughout the night for cooling. In the next day morning prior to sowing, the sorghum seeds are immersed in cold water for some time and sown in the field, which produces better seedlings.

- Country plough is run at the early stage of sorghum crop to ensure optimum plant population.
- Sow sorghum during the Tamil months *Vaikasi - Aani* (May-June) to avoid shoot fly and stem borer.
- Sow cowpea as an intercrop in sorghum to minimize stem borer attack due to its repellent smell.
- Sow lab-lab as an intercrop to reduce stem borer damage in sorghum.
- Pouring neem cake extract drop by drop on the sorghum shoot to control shoot borer.
- Dusting ash on the infected leaves of sorghum to prevent the pest incidence.
- Dusting ash at milking stage to control ear head bugs.
- Growing coriander as a mixed crop in sorghum to control the parasitic weed viz., witch weed (*Striga lutea*).
- A black cloth is tied to a long pole and fixed in the centre of the field to scare away the crows.
- Mixing sorghum seeds with ash to prevent storage pests

7.5. Fingermillet

- Long duration varieties are adopted in dry lands to avoid the rainy days coinciding with the harvesting stage.
- Finger millet seeds are treated with cow urine at 1:10 ratio to enhance germination
- When a sample of dried finger millet grain is chewed, metallic sound indicates its dryness

7.6. Redgram

- Spray the decoction of tobacco waste to control sucking pests and cater pillars.
- Red gram seeds are mixed with red earth slurry, dried and stored to avoid storage pests.
- Castor seeds are fried, powdered and mixed with red gram seeds to reduce pest attack during storage.
- Storing the red gram seeds after mixing them with ‘sweet flag’ (*Acorus calamus*) powder @1 kg per 50 kg seeds to preserve them for one year.

- Dry the red gram seeds well and store them in gunny bags after placing dried leaves of 'Naithulasi' (*Ocimum canum*) inside them to prevent pod borer attack. (Also for black gram)
- Putting the pods of dried chillies in the red gram container to control bruchids (beetle) attack.

7.7. Blackgram

- When a wooden plank is moved with pressing over the drying gram, splitting of gram indicates optimal drying.
- Bullocks pulling a heavy stone roller are allowed to trample over the harvested black gram crops spread out in the threshing yard so as to separate the grains.
- Yield will be higher in black gram crop, if it is sown in the second fortnight of September.
- Neem oil is sprayed @ 6 litre /ac. to control powdery mildew in black gram crop.
- Mixing the black gram seeds with ash and storing them in earthern pots for longer period (Also for cow pea and green gram).
- Coating the black gram with castor oil to increase the keeping quality.
- Mixing the black gram with sweet flag (*Acorus calamus*) powder for seed purpose.
- Black gram grains broken into halves will escape from weevil attack during storage.

7.8. Cowpea

- Putrefied buttermilk is sprayed on cowpea crop to control yellow mosaic disease (Also for green gram).
- Vegetable oil is mixed with cowpea before storage.
- For safe storage, cowpea seeds are filled in earthen pot to its 4/5th volume and the remaining volume is filled with ash (also for field bean)
- Mix cowpea seeds with red earth slurry, dry and store them in earthen pots for one year.

7.9. Peas

- Raising the peas during January (i.e.) during heavy mist time reduces the plant growth and leads to poor yield.
- Three harvestings with 10 days interval to ensure good harvest.

7.10. Groundnut

- Cowpea grown as an intercrop and broder crop in groundnut fields acts as a trap crop to control red hairy cater pillar.
- Adopt crop rotation to control red hairy caterpillar
- Kitchen ash is spread around the groundnut storage bags to prevent insect attack.
- Sandy soil is highly suitable for groundnut crop since it gives less number of ill-filled pods.
- During nights, burning the heaps of straw in several places near the fields along with a bucket full of water or castor cake dissolved water near the fire to attract and kill the insects.
- Spray lime solution and water to control leaf roller.
- Spray water to control leaf roller attack.
- Grow castor as a border crop (trap crop) to reduce the attack of tobacco cut worms.
- Summer ploughing is practiced to expose and destroy the pupae of red hairy caterpillar pest.
- Grinding well and dissolving 10 kg of the leaves of *Aloe vera* in water and spaying for an acre to control RHC (Red Hairy caterpillar) in ground nut.
- To control groundnut ring mosaic, dried sorghum or coconut leaves are powdered and boiled in water to 60% C for one hour, filtered, diluted and spray for two times at 10 and 20 days after sowing.
- Mix neem oil with irrigation water during second or third irrigations to prevent root rot in groundnut.
- Spray neem oil @ 6 litre / ac to control root rot.
- Neem oil solution 4% or neem kernel extract 6% is sprayed to control rust disease in groundnut (600 g in 10 litres of water).

7.11. Gingelly

- Gingelly seeds are to be mixed with dry sand and then sow to have uniform plant population. Gingelly needs more manures than other crops.
- Spray diluted cow urine to control leaf roller
- Store gingelly seeds after mixing them with ash.
- Mix gingelly seeds with activated clay for storage.

- Addition of palm sugar to gingelly during oil extraction gives higher oil recovery.
- Storing gingelly oil in mud pots increases its keeping quality for more than a year.
- Putting a small piece of palm sugar in to gingelly oil increases its keeping quality.

7.12. Cotton

- For delinting cotton seeds, the seed with wet cow dung and dried for 30 minutes and then slightly rub against a stone and clean.
- Coating the cotton seeds with red soil and sun drying them before sowing to ensure good *germination* and to make the dibbling easy.
- To minimize the problem of prevention of shedding of boll of cotton, pour castor oil (approximately 50 ml) near the stem in the soil.
- Farmers store rain water received during the *magha* (February) in the monsoon for future use. They believe that it works as growth promoter on the standing crop. According to them, the stored water does not get spoiled (Rajkot, Gujarat).
- Cut the top branch of the plant which checks the erect growth of the plant and imparts good flowering and retards vegetative growth.
- Practice fumigation and irrigation in the fields of cotton when they anticipate frost. Fumigation is done by cow dung cake, waste grass or makes tall hedges around the field to protect the crop from frost.
- Leaves of *Calotropis* plant are immersed in water channel during irrigation to minimize aphid infestation.
- *Talkidi* a soil pest, attacks fully grown cotton plant. Affected plant withers in a short time. To control *talkidi,* farmers take 20-25 kg onions in a jute bag, crush them using a wooden mallet, and this bag is kept in water channels during irrigation.
- Leaves of agave, vettikottai (*Strycchnos nuxvomica*), neem, and cakes of pungam *(Pongamia Pinnata)* and neem are pulverized together and soaked in cattle urine. The resultant solution is diluted in water in 1:8 and sprayed against pests and diseases of cotton.
- 20 to 25 kg of common salt is mixed with about 10 cart loads of natural manure and is applied in farm during May. It would be more effective if the mixture is applied uniformly in all the furrows.

- Mixing one kg of cotton seeds with 200 ml. of neem oil and pasting it with fresh cow dung and then drying this over night before sowing to avoid pests.
- Applying groundnut cake three times during the crop period to enhance plant growth and boll reduction.
- Growing castor as intercrop or border crop to attract tobacco cut worms.
- Maize and lady's fingers are also grown as intercrops to reduce boll worm attack.
- A mixture of fish meal and Jaggery at 2:1 ratio is broadcasted on cotton fields to attract cranes and bulbuls, which feed on tobacco cut worms
- Five kg. of neem kernels are powdered and soaked in 100 litre of water for one day and filtered in the next day, diluted, mixed with soap solution and sprayed to control white flies.
- One meter of white cloth is cut in to 10 pieces, soaked in turmeric solution and dried. They are again soaked in castor oil and tied in the fields in different locations under which chimney lamps are kept to control sucking pests like white flies.
- Powdering the seeds of *Neriurn oleander,* (Arali) soaking in water and spraying the extract solution to control white flies.
- About 600g of tobacco is soaked in water for 2-3 days, filtered and sprayed to control white flies.
- Sugar solution and neem oil are mixed with water and sprayed to control mealy bugs.
- Dusting ash on cotton leaves to control aphids/thrips.
- Cultivating sorghum or pearl millet around cotton fields to prevent whiteflies and thrips.
- Powdering the neem kernels with 'Vasambu' (Sweet flag- *Acorus calamus),* soaking in water over night and spraying the filtrate in the next day to control all pests.
- Kapas (lint + seed) are picked up only during morning hours to avoid dust.

7.13. Sugarcane

- Tank silt is applied before planting for higher yield.
- Sugar content of the cane is increased if sheep penning is practiced or sheep manure is applied.

  Cow urine 1 litre is mixed with a 10 litres of water and sprayed on the sugarcane crop once in a month to take care of consects pests

- Detrashing the canes to control scales and mealy bugs. It is locally called as *'Sogai Uriththal'* Thogai urithal.
- Providing frequent irrigation to control termites.
- Chulah ash is applied 2-3 months after planting to control early shoot borer.
- Neem kernels or neem cake are powdered, soaked in water overnight and the filtrate is diluted and sprayed to control early shoot borer.
- Dried neem fruits are powdered and applied @ 200 kg/ha to control stem borers and fungal diseases.
- Jaggery making: Country sugar is manufactured indigenously Sugarcane juice is poured in to a wide mouthed big iron utensil called 'Kopparai' and boiled for 1 hour. Then baking soda is put to remove the dusts and continuously boiled for another hour with addition of colouring agent. After reaching required consistency of sticky, colloidal nature (Pagu), juice is poured in to moulds or moulded by hands.

3.7.14. Banana

- Unripened banana bunches are piled in a vessel and keep incense sticks (Agarbati) are inside the vessel. Then if the lid of the vessel is closed, the bunches will ripe in about 12 hours.
- For quick ripening of banana fruits, lime solution is sprinkled over the bunches.
- For easy ripening of banana, neem leaves are inserted in between the bunches.
- About 25g mixture of neem oil cake and castor oil cake is applied around each banana sucker 60 days after planting to control nematodes (*Rodopholus similis*)
- Diluted tobacco leaf extract is sprayed on banana crop to control leaf spot diseases.
- Suckers, which are half foot in height and 2½ kg. in weight are used for banana planting.
- Banana suckers are immersed for a while in 1 litre of neem oil dissolved in 100 litre of water before planting in order to prevent rhizome rot.
- Ground nut cake is applied to banana crop for better yield.
- To control fruit rot in banana during storage, the fruit stalks are soaked in 10% thulasi (*Ocimum canum*) leaf extract or 1 % neem oil solution and stored.
- Avoid mattocking (Leaving the plant after bunch harvest for recycling of nutrients) is weevil endemic area.

- Intercropping of banana with *Crotalaria juncea* will reduce the infection of burrowing nematode.
- Banana crop raised after the marigold cultivation invites less nematode attack.
- Growing *Sesbania* spp. and Casurina (trees) as border crop around banana fields to act as a shelter crop in order to prevent the wind damage.
- Neem leaves are put inside a vessel containing banana hands for ripening of fruits. But ripening will take about four days.

7.15. Mango

- Neem oil is sprayed to control the hoppers (400 ml in 100 litres of water).
- Sunflower is cultivated in between the mango trees to attract honey bees which increase pollination and fruit production.
- To induce early ripening of mango fruits, they are spread on a layer of the branches of 'Aavaram' (*Cassia auriculata*) plant on the floor and again covered with its branches and finally.
- For sooty mould, 1 kg of cow dung is mixed with 10 litres of water and spary

7.16. Grapes

- Brown coloured cuttings, which are foot in length, having good and live buds are used for planting.
- Long trenches are dug at a spacing of 10 feet three months before planting. In these trenches green leaf manures like Kolingi (*Tephrosia purpurea*), *Agave* sp., Erukku (*Calotropis* spp.) are applied, covered with soil and allowed to decompose.
- Groundnut cake is powdered, soaked in water and poured to grapes @ 1 bucket/pit for better fruit quality and yield.
- Circular trenches are dug around vines in which green manures and FYM are applied in 3:1 ratio and covered with soil before the monsoon starts in order to get higher yields.
- Neem cake is applied to control the nematodes.

7.17. Guava

- Pounding 2 kg foliage of *Calotropis* spp. with 3 kg.of neem cake, soaking them in 20 litre. of water for four days and dissolving the extract in 200 lit water and spraying for one acre to control all the pests.
- Burrying the dog's carcass at the root zone for higher yields in acid lime.

- Applyig manure @ 1 basket/tree to increase the yield and to prevent flower shedding.
- Neem seed kernel extract or neem cake solution is sprayed to control leaf miner.
- Dried neem fruits are powdered and applied @ 500g per tree to control the nematode attack.

7.18. Mandarin Orange

- Orange trees attacked by stem borer are given lime wash, holes are cleaned and plugged with lime soaked cotton or wrapped with lemon grass.
- Greenish *Aloe vera* plants are cut into small pieces and spread to a radius of 2 feet around the tree during flowering to control powdery mildew.
- Collected orange seeds are mixed with ash to avoid ants' attack

7.19. Hill Banana

- Fruits of 'Karpoorvalli' are small sized with ash coated rind; higher water content and less taste and fetch low value.
- Farmers use fruits of Red banana variety for medicinal purpose
- Hill banana is grown in the system of 'Vazhaiyadi Vazhai' i.e. every year field is not prepared and new suckers are not planted. After harvesting from mother trees, side suckers are removed leaving only one of them to be allowed to grow as mother tree in the next year. It is repeated every year.
- Banana is planted either in 'Chithiraipattam' (Apr.-May) or in Aadipattam (July-August). Banana suckers are trimmed and roots are removed before planting in order to induce fresh roots.
- On sloppy lands, hill banana is grown
- Dried, drooping leaves are removed once in three months to avoid shade effect which may produce black spots on fruits and to reduce wind damage there by preventing lodging. It is called 'Kondai poduthal'.
- One kg each of powdered neem cake and tobacco wastes are soaked separately in 5 litre water each. In the next day, they are filtered and decanted solutions are mixed together in which suckers are immersed before planting to prevent nematode attack.
- To control banana wilt, affected plants are removed and burnt and 1 -2 kg of lime is applied in each pit.

- For quick ripening and for festive occasions, bunches are stacked in bigger earthen pots in to which fuming incense sticks are kept and mouth is covered with clothes.
- To induce quick ripening, the leaves of 'Kodipasalai' (*Basella alba*) are put in to the baskets or gunny bags and bunches are stacked inside and covered air tightly.

7.20. Plum

- One year old plum grafts are used for planting to control algal infestation, 1kg. of lime dissolved in 10 litre water is sprayed after pruning.
- Plum fruits are packed in bamboo baskets and leaves of a fern type called 'Idaivalai' (*Pteridium aquilinum*) are kept in between the fruits in order to slow down their ripening since these leaves are slow drying.

7.21. Tomato

- A decoction of Aloe vera, *Ocimum tenuiflorum* and *Aristolochia bracteolate* is prepared and sprayed on tomato crop to control pests and disease and to reduce flower shedding.
- Twenty ml. of leaf extract of *Bougainvillea glabra* is mixed with 1 litre water in which the tomato seeds are soaked for six hours to control damping off in nursery.
- About 25-30 days old tomato seedlings are preferred for planting.
- Spraying neem oil to reduce flower droppings in tomato.
- Dissolving Chula ash and cow dung in water and spraying it to reduce flower dropping in tomato.
- Growing marigold as border crop in tomato fields to control nematode attack.
- Raising garlic or onion as border crop in tomato fields to prevent fruit borer attack.
- About 1½ kg. of asafoetida for one acre is tied in a cloth and kept in irrigation channel to control fruit borer.
- Fumigation is done to tomato fields with benzoin in morning and evening during flowering to control fruit borers and aphids (Also for Lady's finger).
- Two kg of neem kernels are powdered and soaked, in water for 10 days after which it is filtered, mixed with water and sprayed for one acre of tomato crop to control fruit borer and thrips.
- Mixing neem cake with sheep droppings and dusting on the field to control thrips.

- Dusting ash on tomato crop in the morning hours to control aphids, thrips etc. To control most of the pests in tomato, 1½ kg ash is mixed with cow dung solution and sprayed to control aphids and other sucking pests in lab-lab.
- Dissolving sarvodoya soap in water and spraying to control mealy bugs by which a film is formed around its eggs which get died.

7.22. Brinjal

- In order to prevent fruit rotting in brinjal plants, a solution is made of 1 litre. of water and eight crushed leaves of *Aloe vera* and sprayed on the crops.
- *Chrysanthemum coronaries* is grown as a border crop in brinjal to control fruit borers.
- Poultry manure is applied for more yields in brinjal.
- Grinding and applying the neem seeds @40kg./ac. on 35th day after transplanting gives higher yield.
- Growing castor in brinjal fields as border crop to act as a trap crop for insects.
- Growing onion as intercrop in brinjal to control many pests including fruit borers.
- Mixing and grinding well neem cake with *Aloe vera* and soaking in water for 10 days, after which spraying the filtrate to control thrips.
- Ash and turmeric powder are mixed in equal proportion and sprinkled to control aphids.
- Sprinkling of lime powder to control mealy bugs.
- Cow urine, neem oil and tobacco decoction in 1:1:1 ratio in 1 litre of water are mixed and sprayed to control all sucking pests.
- Spraying neem cake extracts to control mites and the spotted beetle *(Epilachna octopunctata)* in brinjal.

7.23. Chillies

- Chilli seeds are immersed in biogas slurry for 1½ hours for vigorous growtj and disease resistance.
- Aadipattam (Jul. -Aug.) and Thaipattam (Jan.-Feb.) are suitable for sowing chillies. Own seeds are mostly used for chillies cultivation.
- Sheep penning is practiced for getting higher yield.
- Groundnut cake is applied to reduce the flower dropping and to increase the yield (100 kg /acre).

- To induce more flowering and to reduce flower dropping, asafoetida @ 1 kg/ac is powdered, tied in a cloth and placed in the irrigation channel and act as pest repellant.
- Growing castor as border crop to act as a trap crop for tobacco cut worms.
- Tow rows of maize or sorghum are grown for every five rows of chillies to control mosaic disease.
- Spraying the leaf extract of *Prosopis juliflora,* two months after planting to control leaf spot, powdery mildew and fruit rot in chillies.
- To control leaf curl and to encourage good growth in chillies the leaves of neem, oleander, noichi *(Vitex negundo),* nuna *(Morinda tinctoria),* thulasi *(Ocimum canum),* thumbai *(Leucas aspera)* and *Calotropis gigantea* are pounded and the solution is sprayed to control white flies.
- Leaf extract of "Vilvam" *(Aegle marmelos)* is sprayed to control fruit rot in chillies.

7.24. Drumstick

- Drumstick plants are pinched off ('Uchhi kolundukilludal') when they are at 4-5 feet height so as to facilitate more branching.
- Place finger shaped or a pinch of asafoetida just deep into the soil near the roots of drumstick trees to control hairy caterpillars.
- Crop wastes and other residues are burnt around the base of the drumstick tree to control hairy caterpillar.
- Three ratoon crops are obtained from annual drumstick

7.25. Cucumber

- About 10 kg of groundnut cake is soaked overnight and in the next morning this solution is poured @ 200 ml. /hill to get more yields. Ash is sprinkled on cucumber crop to control aphids and powdery mildew.
- Cucumber seeds are extracted from fully ripened fruits, washed well with water, mixed with ash, dried and stored up to one year.

7.26. Snake gourd

- Soaking the seeds of snake gourd in cow dung solution for 6 hours before sowing helps for early germination and withstanding drought conditions.
- Asafoetida is dissolved in water @ 25g per litre and sprayed to control flower dropping.

- After the harvest of snake gourd, cabbage foliage is ploughed *in-situ* to serve as a manure.

7.27. Ribbed gourd

Ribbed gourd having even number of raised ridges on its skin is likely to be sweeter and the one with odd number of ridges is likely to be bitter.

7.28. Bitter gourd

- Bitter gourd germinates faster and grows well when its seeds are soaked in milk for a day prior to sowing.

7.29. Cauliflower

- To control diamond back moth in cauliflower neem oil is mixed with water @ 30 ml. /litre of water and sprayed.

7.30. Beans

- Sow seeds behind the country plough for rainfed beans cultivation.
- Soak bean seeds in water for few minutes, then mixing with wet red soil and wood ash and sowing seeds after drying under sun to prevent pest infestation during crop growth.
- Dust of the fresh ash made immediately after burning the weeds over the pest attacked plants.

7.31. Pumpkin

- Storing of pumpkin seeds in dried fruit storage containers. After completely drying the fresh fruits, the inner pump and seeds are removed. Then the seeds selected for storage are mixed with ash and placed inside the fruit container. The entrance is covered with moist red earth and a small hole is made for aeration.

7.32. Bottle gourd

- Store the seeds by sun drying them with fruit itself, as the dried fruits act as a storage container.
- Soak the seeds in water and wrapping in a moist cotton cloth to enhance the germination of bottle gourd.
- Practice of stripping the lateral branches induces formation of more number of such lateral branches and ensures good quality gourds.

7.33. Cabbage

- Frequent weeding in the cabbage nursery for better establishment of cabbage plants in nursery and main field for getting matured heads. Grow garlic as an center crop against diamond back moth attack.
- Spray of the spray solution made of the mixture of ash and cattle urine on the cabbage plants to control leaf hoppers.
- Application of 1.5 kg lime mixed in 2 – 3 litres of buttermilk and weekly intervals by using broomstick to control cabbage caterpillar.
- Hoist five-balled sticks in the cabbage fields at an interval of 20 to control insect pest attack on cabbage.

7.34. Radish

- Raise the radish crop after potato, needs little expense because of the loose soil after digging the potato tubers.
- Raise radish as a rainfed crop, comes up well by utilizing the available residual moisture in the field.
- Raise radish after the potato requires no manures or fertilizers because of the residual fertilizers and manures available in the field.
- Wash of the harvested radish tubers with water for better shipping.

7.35. Beet root

- Raise Beet root as an intercrop with carrot requires little individual care, as the operations done for carrot itself is enough for beet root.
- Selection of big seeds having better striving for seed purpose.
- The stripping of beet root tubers indicates good quality.

7.36. Carrot

- Crop rotation of carrot with potato and peas to get more income because same next crop will not give more yield.
- Fencing with the thin branches of Seegai tree from the local forest to arrest soil erosion.
- Construction of stone wall across the slope in the field to arrest soil erosion.
- Forking and clod breaking in the main field before taking up cultivation to make the bottom soils come up and top soils go down.
- Spray the solution made from the mixture of ash and cattle urine on the carrot crop to kill leafhoppers.

- Assess the harvesting time by digging carrot plants sample. If the carrot is of orange colour and big size that indicates the maturity.
- Wash the carrot tubers after harvesting in the running water of small streams for better shining of carrot tubers.
- Use the stem and leaves after harvesting either as cattle feed or for making compost to use as manure in the next crop.

7.37. Potato

- Crop rotation of potato with crops like marigold and onion is practiced for golden nematode control.
- Grow potato increases soil texture.
- Mixed cropping of potato with Marigold (*Tagetes* spp.) reduces the risk of root nematode.
- Lime (100 kg) applied is potato fields at planting to reduce acidity inorder to control brown rot.
- Crop rotation of potato crop followed by cabbage or radish or peas gives good returns to the farmers.
- Crop rotation of potato with carrot or beet root or turnip utilizes the time and moisture available.
- Dip the potato seeds in cowdung slurry for 30 mins to control tuber rot.
- Fallowing of land during the months of November, December and January to make the field hardened.
- Heavy application of farm yard manure to potato every year before ploughing to increase the number of tubers formation.
- Soil amendment with FYM for the reduction of cyst nematode in potato.
- Land stirring by using hand fork to expose the bottom soils to sun.
- Gathering more soil near the stem of the potato crop helpful for the roots to spread over unobstructed and the potatoes were also believed to grow in bigger size.
- Harvest when the leaves of the potato turn yellowish brown colour or starts drying after 90–100 days of planting.
- Cut the plant stem above the ground level before harvesting of tubers to facilitate the harvesting operation.
- Non-washing of the harvested tubers with water because it may reduce the original colour of the tubers.

- The formation of less than five big sized tubers per plant indicates the failure of potato crop in that particular season.
- The formation of more than 10 big sized tubers per plant indicates the good yield of that particular season.
- Plant of potato eyes towards the slope to prevent water logging and provide good drainage during rainy season.
- Construction of shallow well with 8' depth and 4' width at the lower end of the field to store the draining water for irrigation purposes.

7.38. Jasmine

- Naattu malli' is a local variety producing small sized, but highly fragrant flowers with short stalks and hence suitable for preparing perfumes.
- Every year, FYM is applied or sheep penning is practiced to replenish soil fertility.
- One kg. of oleander fruits (Sevvarali) is soaked in water for one day after which, they are crushed and extract is taken. Then it is diluted with water for 10 times and sprayed to control all the pests in jasmine.
- Five kg of leaves and small branches of 'Oduvanthalai' *(Cleistanthus collinus) are* boiled with 2 litre of water until the volume gets reduced to 1 litre. This decoction is mixed with water @ 50 ml. in 15 litre. of water and sprayed with hand sprayer in jasmine fields to control most of insect pests and diseases.
- Neem cake powder is applied @ 50g./plant to control nematodes.
- While tying and pruning the bushes in Nov.–Dec. sheep penning is practiced and sheeps are allowed to feed on the old foliage to induce fresh growth.

7.39. Crossandra

- To control the root knot nematodes in 'crossandra dried flowers of red oleander (*Nerium oleander*) are mixed with neem cake and applied at last ploughing.
- Neem cake is applied to control nematodes (Also for Chrysanthemum).

7.40. Chrysanthemum

- Chrysanthemum planted during Chithirai (April – May) exhibits slow and dense growth and prodcues small flowers but the yield is higher and hence it is profitable.

- Groundnut cake is powdered and applied to get higher yields and big sized flowers (Also for tuberose)
- Pinching of terminal shoots is done two months after planting to have more branches.
- In case of one day delay for sending to market, harvested flowers are spread on gunny bags, sprinkled with water and covered with white cloth in order to preneut.

7.41. Red Oleander

- FYM is abundantly applied during January and August for increased flower yield.
- Oleander plants are severely pruned to 1-2 feet height at 5-6 years after planting for getting fresh shoots and higher yields. This pruning is repeated for another 2 times once in 4 years after which they are uprooted and fresh planting is done.

7.42. Pepper

- 'Karimundan' is another local variety suitable for higher elevations with short spikes.
- 'Kattumilagu' is a type of wild variety, yielding once in 2-3 years with more pungent berries and it is found in reserve forests only.
- When a sample of pepper is chewed, metallic sound indicates its optimal dryness.
- For producing white pepper, the fruits are allowed to ripen in the climber itself. Then they are collected, put in tanks, foot pressed to remove the skins and they are washed with water and dried to produce white pepper which is having medicinal value.

3.7.43. Cardamom

- 'Valukkai' is suitable for high rainfall areas producing semi prostrate panicles with smooth surfaced pods.
- Cardamom seeds are sown immediately after harvest to get better germination.
- Seeds, after extraction, are washed well with water for 2 or 3 times to remove the mucilaginous substances, mixed with ash and dried for 2 or 3 days.

- A mixture of neem cake powder and sheep manure is applied @ 200g/plant.
- A mixture of extracts of neem cake and tobacco waste is sprayed to control stem borer and capsule thrips in cardamom.
- Cardamom capsules after the harvest are cured thrrough fumigation. Fumigation is done in a room, locally called as 'store'. There will be 11 trays fitted with sand sieves in a stand and four such stands will be in a store. Capsules are spread on these trays and fumigated. During cardamom curing in the first day morning, the capsules are first preheated at 40°C for 1 hour after which windows are kept open for 1 hour. Then second heating is done at 60° C for 5 hours followed by the opening of windows for 2-3 hours. Then a mild third heating is given for another 10 hours and kept in the trays up to the second day evening. Total curing period takes 30-36 hours. Purpose of curing is to dry the pods without losing their green color to fetch good market value with better keeping quality.

7.44. Garlic

- 'Singapore red' is a local variety called as 'periumpoodu' with duration of 4 1/2 months produce bulbs having reddish skin, big sized cloves with less water content and high keeping quality.
- 'Malaippoodu, is a local variety with a duration of 4 months producing big sized bulbs with more pungency and more water content.
- Sheep penning is better for getting higher yields.
- Neem cake is applied 4kg /ac to reduce the infestation for root grub.
- Time of harvest is indicated by yellowing and withering of leaves which turn to pale green colour and start drying from the top. Over maturity causes damage to the bulbs.

7.45. Coconut

- To prevent rats from climbing coconut trees, a large palm leaf is split along its mid rib; one set of leaflets is wrapped around the trunk below the crown and the other set is wrapped in the opposite direction.
- To control flower shedding in coconut, salt is poured on the apical portion of the flower buds and also spread at the root zone ad given plenty of water.
- Seed nuts are collected from high yielding mother palms having dense and longer leaves and bigger nuts.

- Seed nuts, which are round in shape and produce metallic sound on tapping are selected for raising nursery.
- Seed nuts are planted in sand bed nursery and kept for six months with irrigation in alternate days.
- About 6-8 months old coconut seedlings having 5-6 leaves are selected for planting.
- For coconut planting, pits are dug and filled with Kolingi *(Tephrosia purpurea)* allowed to decompose for six months.
- Suitable seasons for Coconut planting are 'Aadi' (July-Aug.) and Karthigai (Dec–Jan) months.
- Before planting coconut seedling, roots are removed in order to induce fresh roots.
- Application of 10 -15kg of FYM per tree every year.
- Application of Kolingi *(Tephrosia purpurea)* @ 10 kg/tree every year.
- Applying *Calotropis gigantea,* (1kg.), Kolingi (1kg.), Pothakalli *(Poeciloneuron pauciflorum)* (1 kg.) fishmeal (1Kg.), salt (1kg.) and sand in a semi-circular basin around the tree to get higher yield.
- Mulching by burying of coconut husks around the tree to conserve moisture and to control weeds.
- Inter space in the coconut garden is ploughed twice in a year during June-July and Dec–Jan to facilitate aeration to the roots and to control weeds.
- Spraying neem oil to reduce flower shedding.
- To prevent button shedding, common salt is applied around the growing tip @ 2 kg / tree during rainy season.
- Application of ash to control button shedding.
- Kolingi *(Tephrosia purpurea)* and *Calotropis gigantea* are applied in circular basin just before flowering to control button shedding.
- Application of neem cake in the pits before planting coconut, to avoid the attack of insect pests and ants.
- Earthen pots are placed in small pits in coconut gardens and 50% the of the pot is filled with water and ½ kg castor cake. After three days due to the smell, rhinoceros beetles get attracted, fall in to the pot and die.
- Pouring neem cake extract on the growing tip and adjoining fronds to control rhinoceros beetle.

- To control stem weevil in coconut, the hole bored by it, is cleaned and plugged after putting common salt.
- Putting 1-2 kg of common salt in the pit, while planting coconut, to control termites and to conserve moisture.
- While planting coconut seedling, one leaf of *Agave spp.* is planted in the pit to retain soil moisture and to control termites. Flooding the coconut garden to wash off termites
- Lime washing for 2 feet height at the base of coconut trees to control termite attack.
- To control termites, 500 g. of common salt is dissolved in 5 litre. of water and poured on the trunk.
- Grow poultry birds in coconut gardens to feed on termites.
- To control Thanjavur wilt of coconut, green manures like kolingi *(Tephrosia purpurea),* daincha (*Sesbania* spp.) etc. are raised and ploughed in situ or well decomposed FYM is applied followed by the application of neem cake.
- To control stem bleeding, the bleeding mouth on the trunk is cut to certain extent, cleaned and poured with lime solution.
- Branches of Seemai karuvel *(Prosopis juliflora)* or barbed wires are tied around the mid trunk to a height of 2-3 feet to prevent climbing of rats and squirrels.
- Greenish yellow coconuts are harvested.
- Adding a piece of jaggery (country sugar) in coconut oil to separate the dusts and make the oil more clear.

7.46. Ginger

- Use rhizomes collected near the bunds of the field for seed purpose.
- Application of lime to the field before planting of rhizomes to avoid pests.
- Spreading the available fresh or dried green leaves over the planted rhizomes to avoid weeds and give shading.

7.47. Arecanut

- Mixing the selected big nuts with sand and soil mixture prepared in 2:1 ratio keeping in gunny bag in shade and sprinkling water for 2 – 3 times/ week over it for 30 days for sprouting.

- Protecting the young seedlings against heavy rain, hot sun and mist by covering it with the branches of local trees.

3.48. Tobacco

- For eradicating broom rape weed (*Orabanche cernua*) in tobacco, a drop of ground nut or gingelly oil is placed above the growth during its emergence.
- A solution is made of 5 litre of milk in 100 litre of water and sprayed after a month of planting for 1 acre of tobacco crop to prevent tobacco mosaic virus.
- Wider spacing is provided to get more yield in tobacco.
- Sheep penning is practiced for tobacco fields for getting higher yield.
- Topping of tobacco plants along with flowers is done after 8-10 leaves appearance. This is locally termed as 'Nuni Killudhal'
- Neem seed kernel extract is sprayed to control tobacco cut worm.
- Sorghum or gingelly is grown as a trap crop to control broom rape weed *(Orabanche cernua).*
- Spraying buttermilk @ 50ml. dissolved in 1 litre of water to control tobacco mosaic.

7.49. Coffee

- Half a kg of fresh cow dung is dissolved in 15 litre. of water and sprinkled on coffee seedlings at nursery stage for better growth.
- Coffee is grown in any one of the following multi-tier systems of cultivation. (a) Banana - Coffee (b) Jack-Mandarin orange – Coffee (c) Mandarin orange - Coffee - Pepper
- Forming semi circular basins around the coffee plants to conserve moisture.
- Stems of coffee plants are rubbed with cloves made up of coir rope during Apr-May to prevent stem borer.
- Coffee berry borer mostly spreads from one plantation to another through the gunny bags used for transport of coffee beans for curing factories. Hence, these gunny bags are impregnated with neem cake extract and dried before packing.
- To control berry borers in coffee, 3 kg of neem cake and 6 kg of forest marigold are soaked in water separately for 3 days after which they are dissolved together in 200 litre of water and sprayed for an acre twice with an interval of 15 days.

- Lime is applied to control wilt disease in coffee.
- Good quality coffee is obtained at third or fourth picking.
- Disease free coffee plants having good viguor are selected for seed collection.
- Stems of coffee plants are rubbed with cloves made up of coir rope during April- May to prevent stem borer.
- *Robusta coffee* is resistant to nematode attack.

7.50. Tea

- Raise Vettiver/napier grass grass in bunds across the slope to arrest the soil erosion in the tea plantations.
- Grow tea at higher elevation gives good quality tea.
- Grow tea at higher elevation gives less yield.
- Intercrop of coffee variety Kaveri in tea plantation comes up well.
- Place the seeds above one layer of sand, covering with one layer of sand and water spraying for three weeks for getting sprouting.

8

Moisture Conservation Techniques

8.1. Documented ITKs on Moisture Conservation

The ITKs were documented under following specific categories

- Agronomic Measures
- Tillage
- Bunding & Terracing (Mech. & Vegetative barrier)
- Land Configuration
- Soil Amendment / Mulching
- Erosion Control & Runoff Diversion Structures
- Water Harvesting, Seepage Control & Ground Water Recharge

8.2. Agronomic Measures

8.2.1. Intercropping (Coriander with Bengal gram)

- The practice consists of sowing of coriander and bengalgram in 7:1 ratio is done during mid November.
- Initially coriander is sown behind the country plough and after 7 days of coriander, one row of Bengal gram is sown
- The interspace between row is 30 cm and between plants is 10 to 20 cm
- Seed rate of coriander 18 kg/ha & bengalgram 6-8 kg/ha

8.2.2. Intercropping of Cotton + Blackgram

- The practice consists of sowing of Cotton + Blackgram in 1:1 ratio manually through seed hoppers attached to the cultivator.
- Row spacing of 30 cm and plant to plant spacing of 15 to 20 cm is maiantained.

8.2.3. Mixed intercropping as vegetative barrier

- Groundnut is intercropped with pigeonpea mixed with pulses like cowpea, field bean, greengram and horsegram in 20:1 or 30:1 ratio.
- Groundnut @ 100 kg/ha and mixed pulses are sown simultaneously.

8.2.4. Cultivation and sowing across the slope

- Cultivation and sowing of groundnut is done across the slope to reduce runoff losses. It is an alternate method to contour cultivation.
- In this practice, depth of ploughing is up to 15 – 30 cm and depth of sowing is 5 cm.
- The crops are sown in July or August.

8.2.5. Wider row spacing and deep interculturing

- In this practice, sunflower and pearlmillet of 135 cm row spacing are adopted in *kharif* season. Due to frequent deep intercultivation by closing the soil cracks.
- The intercultivation id done through blade harrow.
- For sowing of sunflower and peralmillet with a row spacing of 135 cm, the seed drill of 45 cm row spacing is used; the seeds are sown only in middle coulter by closing the two end coulters. This automatically maintains a row spacing of 135 cm.
- The cost of operation is Rs. 250 to 300/ha

8.2.6. Strip cropping

- Sowing of pulses and maize is done with 10:10 row ratio in broad strips.

8.2.7. Raising green manure crop in Mango

- He raises wild indigo (Kolinchi) as an inter crop in slopy land of young mango garden, to prevent silt erosion and for improving soil fertility.

8.3. Bunding & Terracing (Mech. & Vegetative barrier)

8.3.1. Vegetative fencing with Kiluvai (*Blasmo dendron verii*)

- Cuttings of kiluvai are planted manually at field boundaries to a depth of 15 to 20 cm at a spacing of 30 cm during the month of April-May.
- The cost adoption is Rs. 50/m.

8.3.2. Vegetative barrier with Agave

- Agave suckers to be palnted in the month of Oct/Nov manually at a spacing of 60 cm on field boundaries and in the fields adjacent to gullies to reduce the run off.
- The cost planting is Rs. 3-5/m.

8.3.3. Loose stone waste weir (Clear overfall type)

- Boulders are arranged manually in layers. The length of the check ranges from 1 to 2 m and the width from 0.5 to 1.0 m.
- The gaps between the boulders are filled with small stones, no sealing material is used.

8.3.4. Nala check with soil filled in cement bags

- Embankments are constructed across the nala with cement bags filled with soil for strengthening the embankment. The run off water is stored without much seepage loss and is utilized for irrigation purpose.

8.3.5. Peripheral bunding with Agave spp. for gully control

- Agave suckers to be planted during rainy season manually at a spacing of 30 cm × 30 cm in two rows across and along the field boundaries.
- In a due course, a thick vegetative bund is formed that restricts cattle trespassing besides restricting soil erosion.
- The cost of planting approximately is Rs. 20/m.

8.3.6. Peripheral stone bunding

- Stones are piled upto 1 m height.
- The cost involved may be Rs. 30/m.

8.3.7. Ipomea as vegetative barrier

- *Ipomea* cuttings are planted thickly in 2 rows across gullies costing approximately Rs. 1000/ha

8.3.8. Stone bunding

- Bunds are constructed during summer season with stones which are locally available.
- The height of the bund is 1.3 m and bottom width is 1 m.
- The cost of adoption may be Rs. 6000-7000/m.

8.3.9. Clear over fall waste weir

- The size of the waste weir with 15 m length and 2.5 m height is constructed across the valley with locally available stones to dispose runoff water without causing damage to the downstream land.
- The cost of construction may be Rs. 48000.

8.3.10. Loose stone surplus bund

- The available stones of different sizes (15-30 cm) are alid in layers to some height at the downward side of gradient in order to serve as an outlet for safe disposal of runoff.
- The cost may be Rs. 10/m

8.3.11. Stone and soil bunding with an opening

- A bund is constructed by stone and soil in layers of 15 cm at the lower end of the field.
- The size of bund is 5 m length × 2.5 m width × 1.5 m height.
- A small opening of 1 m length × 2.5 m width and 1.5 m height is made between to allow runoff through the opening
- This has to be maintained seasonally before starting of monsoon.
- The cost of construction may be Rs. 2000.

8.3.12. Stabilization of field boundary bund with Vitex negundo-Notchi

- In this method *Vitex negundo* is planted on the filed bund. The plant to palnt spacing is 8-10 cm.
- The method of planting is vegetative propagation by sticks.
- The sot of planting may be Rs. 1 / m length.

8.3.13. Grass Plantation on field boundaries (filter strip)

- Grass specis like *Marvel, Madras anjan, Khus, Stylosanthes hamata* (Pencil flower)are planted on the field bund in rainy season.
- Method of planting is by suckers or seed.

8.3.14. Bund farming of pulse crops in kharif under rainfed situation

- Spread puddle soil on the bunds and sow pigeonpea or blackgram seeds by dibbling after the soil attains optimum soil moisture for germination.
- Pulse crops are grown on the 2 sides of the bund leaving a small space in between for movement of the farmers.

8.3.15. Earthen bunds

- Construction of bunds across the slope with a length of 12 m, base width 2 m, top width 1 m and 1 m height. The bund is covered with local grasses.

8.3.16. Stone-cum-earthen bunding

- On the lower side of the land, a bund is prepared by using locally available stone at the base and then soil is put over this to make a stable bund.
- The height of the bund varies from 1.5 to 2.5 m and the width is up to 1.5 m.

8.3.17. Live bunding by raising Cactus

- A trench of size 20 × 10 cm approximately is made along farm boundary with the help of hand tools. Cactus clumps are put vertically in the trench and the soil around the clump is compacted.
- The palnting time is summer season before the onset of monsoon.

8.3.18. Contour cultivation

- Line joining the points of equal elevation is called contour. All the cultural practices such as ploughing, sowing, intercultivation etc. done across the slope reduce soil and water loss.
- By ploughing and sowing across the slope, each ridge of plough furrow and each row of the crop act as obstruction to the runoff and provide more time for water to enter into the soil leading to reduced soil and water loss.

Fig. 25: Contour cultivation

8.4. Tillage and Land configuration

8.4.1. Conservation furrows with traditional plough

- Furrows of a depth ranging from 15 to 20 cm are opnend in between the crop rows (when the crop is 20 days old) with animal drawn country/iron plough.
- The cost of ploughing is Rs.400 to 500/ha
- The area coverage is about 0.5 ha/day

8.4.2. Broad bed & furrows (*Aala saal agala paathi*)

- Broad bed & furrows are formed by deepening the furrow at every 1.25 to 1.5 m while ploughing the land during the off season i.e. during july-August with bullock drawn iron plough. The width and depth of the furrow are 30 cm.
- The cost of ploughing is Rs.500/ha
- The area coverage about 0.5 ha/day

8.4.3. Traditional compartmental bunding (*Paguthi paathi*)

- The field is prepared and divides into small comparments during off-season.
- The length of the compartments is 25-30 m and width is 10-15 m.
- In many places, compartmental bunds are covered with legume and vegetables.
- The cost adoption may be Rs. 750-1000/ha.

8.4.4. Deep ploughing (*Azha uzhuthal*)

- Deep ploughing is done once in 2 or 3 years with tractor drawn mould board plough to a depth of 30 cm during summer.
- The cost adoption may be Rs. 2000/ha.

8.4.5. Ploughing across the slope

- Ploughing is done across the slope using bullock drawn wooden plough.
- The depth of ploughing is to 10 to 20 cm. This is an alternate to contour farming.
- The cost adoption may be Rs. 1000/ha.

8.4.6. Criss cross ploughing

- Ploughing is done by country plough; shallow ploughing is done.
- If the first ploughing is done along the slope then the second ploughing is across the slope and vice versa.
- After ploughing, a square or rectangular pattern is formed to help in seed sowing.

8.4.7. Tied ridging

- It is a modification of the ridges and furrows.
- Wherein the ridges are connected or tied by a small bund at 2-3 m interval along the furrows to allow the rain water collection in the furrows which slowly percolated in to the soil profile.

8.4.8. Scooping

- Scooping the soil surface to form small depressions or basins help in retaining rain water on the surface for longer periods.
- They also reduce erosion by trapping eroding sediment

8.4.9. Furrows for In-situ moisture conservation *(Oodu uzhavu)*

- In this practice make furrows of depth 20 cm in between the crop rows
- Furrows served both as drainage channel and storage structure.

8.4.10. Summer ploughing *(Kodai uzhavu/Malaikkumun uzhuthal)*

- Plough the land up to 45 cm (tractor drawn plough) repeatedly for 4-5 times or upto 30 cm (bullock drawn plough) during summer season.

8.5. Soil Amendment / Mulching

8.5.1. Application of Tank bed silt *(Karambai Adithal)*

- Tank silt is removed from the village tank and spread uniformly by tractor over the fields at the rate of 150 to 180 t/ha before ploughing and then mixed with soil thoroughly during summer season.
- After 10-15 days of mixing tank silt in soil, ploughing is done.

8.5.2. Application of ground nut shells

- Available ground nut shells are incorporated in the field after preparation and before sowing of groundnut.
- The groundnut shells are transported to the field and kept in the form of small heaps.

- These are then uniformly spread throughout the field.
- The cost of adoption approximately Rs. 200/ha.
- The recommended rate of application is 5t/ha

8.5.3. Sand mulching

- In alfisol, sand mulching can be done for any crop under dryland conditions.
- It is done before sowing in the field.
- The thickness is less than 1.0 cm.
- The recommended rate of application is 40 t/ha

8.5.4. Gravel sand mulching

- This practice is adopted in alkaline soil.
- Before spreading gravel sand, perennial weeds are removed.
- During summer season about 250-300 tractor loads/ha of gravel sand is spread on the soil surface with a thickness of 7.5 – 10 cm.
- Only harrowing is done in the sand layer.
- Approximate cost involved is Rs. 40,000/ha

8.5.5. Retention of pebbles on the soil surface

- Pebbles have to be retained on the soil surface.

8.5.6. Retention of sunflower stalks

- After harvest of sunflower crop, the stalks are retained on the soil surface up to sowing of *rabi* crops.
- The stalks are removed before 15 days of sowing of main crops.

8.5.7. Sand mulching *(Mannai Kothividuthal)*

- In between the crop rows disturb the soil with hand hoe by breaking away the crust formation/compactness.
- Approximate cost adoption for this technology is Rs.200/acre.

8.5.8. Bark clippings

- These are good mulch materials as they are long lasting and allow proper aeration to the soil underneath.
- Hardwood bark clippings contain more nutrients than softwood but bark clippings are not easily and abundantly available, and some bark products may cause phytotoxicity.

8.5.9. Coir mulching

- Coir fibre comes from inside the coconut shell and is a natural byproduct of the processing of coconuts.
- The coarse, strong fiber is used to make rope, floor mats, brushes, mattresses, and, for the past 20 years, landscape mulch.

8.5.9. Compost mulching

- The compost is one of the best mulch materials. It increases microbial population, improves the soil structure and provides nutrients.
- It is the excellent material for improving the health of soil.

8.5.10. Dry leaves

- Leaves, an easily available material, are good for mulching. Though leaves are good for protecting dormant plants during winter by keeping them warm and dry but due to lightweight they may be blown away even by light wind.
- To counter this problem, it requires anchoring which can be done with stones, chipped bark and covering with net or some form of sheet.

8.5.11. Grass clippings

- This is one of the most abundantly and easily available mulch materials across the country. It provides nitrogen to the soil, if incorporated fresh.
- However, application of green grass in rainy season may result into the development of its own root system which will be detrimental to plant growth.
- Therefore, use of dry grass as mulch material is suggested.

8.6. Water Harvesting, Seepage Control & Ground Water Recharge

8.6.1. Seepage control by lining farm ponds with white soil

- Locally available white soil (Kaoline clay) is used for liming farm ponds.
- The plastering of tank surface with white soil (5 cm thick) is done manually once during initial stage to control seepage loss.
- Cost involved is Rs. 10,000 is spent for a pond.

8.6.2. Harvesting of seepage water

- Earthen embankments are constructed across the gully carrying seepage water from hills for storage.

- The land slope of catchment area is more than 6 %.
- The cost may range from Rs. 500 to 10,000 depending on the size of earthen checks.

8.6.3. Wells as runoff storage structures

- Existing water ways are used to divert the *runoff* water to dried open wells.
- In this practice, an additional cost of Rs. 10,000 is needed for construction of diversion channels, besides the cost of existing open well.

8.6.4. Rain water management using indigenous rain gauge (Role)

- A hole of 20 cm depth × 18 cm diameter is made in the granite stone size of 1× 0.5 m to collect rain water, whenever there is rain in the area; this is called ROLE (indigenous rain gauge).
- When the role if filled with rain water, ploughing has to be done. After ploughing, the land sowing of seed is done in the field.

8.6.5. Farm pond (Pannai kuttai)

- The farm pond is constructed at lower gradient of the filed in order to catch the runoff water from the higher gradient.
- The average size of the pond is 7m × 4m × 2m for 1-2 ha land and is constructed manually.
- Approximate cost involved is Rs. 2500/-.

Fig. 26: Farm pond in a corner of a field

8.6.6. Percolation pond / tank

- Size about 40 m length × 25 m width × 3 to 4 m depth is constructed at lower level in the farmer field.
- Removal of silt from pond in summer season is to be done every year.

8.6.7. Ground water recharging through ditches and percolation pits

- V shape ditch shuld be made at the lower level part of the land to divert runoff water to the ditch.
- At the end of ditch, one pit has to be made and filled with stones, sand and stone grits to work as filter.
- PVC pipes of diameter 5 cm were used to connect the pit and wells. The runoof water from upper land flows into ditch and from the ditch it goes to pit. From the pit, the water flows to well for recharging the well.

8.6.8. Small check dams

- A check dam having one side a bund of cement concrete (17 m length, 4 m height and base and top width 3 m and 1 m, respectively) to be constructed.
- Earthen bunds with vegetative covers to be made on the other two sides.

Fig. 27: Concrete check dam for gully control

8.6.9. Well recharging through runoff collection pits

- Water is diverted to the collecting pit of size 1.5 m × 1.5 m whose vottom is filled with pebbles upto a height of 0.5 m. the depth of pit depends on the soil topographic condition.
- Pebbles are used to trap sediment.
- The outlet pipe is kept above the pebbles to put water into the well and its maintained individual.

8.7. Erosion Control & Runoff Diversion Structures

8.7.1. Sand bags as gully check

- This is done in off-season before onset of monsoon.
- Empty fertilizer and cement bags are filled with sand and piled one above the other in 5 rows, generally filling the gully width.
- Whenever the sand bags are damaged they ar replaced.
- Approximate cost involved is Rs. 100/ structure

8.7.2. Surplus waste weir at the outlet of the field

- In this practice, surplus weir of size 5.5 × 0.45 × 0.35 m of length, width and depth respectively constructed at the outlet of the field.
- The cost of construction is Rs. 3000 to 9000/m length.
- The structure is usually constructed using stones/boulders as rubble masonry.
- For longer life, can be construted with stone masonry using cement as binding agent.

8.7.3. Loose boulder checks

- This is practiced in hilly and slopy terrains where the rocks or stones are easily available in surrounding areas.
- Stones of size 15-30 cm are used for construction of checks across the rills.
- By visuall observation the eroded patches at the lower gradient of the field where runoff concentration takes place (rills) are identified.
- Stones are laid in 2 layers throughout the length of the lower gradient of the field.
- These are constructed using spade.
- These require to be maintain every season.

8.7.4. Stone waste weir

- Stones of size 20-30 cm in diameter are laid in layers at the lower most part of (outlet) the filed
- The dimension of the stone waste weir varies with site situation and size of the field.
- The height of the crest and base width is 90-100 cm.

8.7.5. Field bunding

- This practice is adopted during off season
- There is a variation in size of the bund and it varies according to soil type, slope, accordingly the cost varies.
- Earthen bund of 1-1.5 m height and 1.5 -2.0 m width at bottom is constructed around the field. The length varies with field boundary.
- A suitable outlet is required for removal of excess runoff water.

8.7.6. Waste weir (Stones/Sorghum stubbles) at the outlet of field

- Collection of stones & sorghum stubbles and making a bund at lower most spot of the field boundary. If the slope is higher than to 2 -3 % bunds may be required at suitable vertical intervals.
- Initially ground surface, wherever bund is constructed, is leveled with a help of spade.
- The size of the bund is 0.5 to 1.0 m width, 0.5 to 0.75 m height and length varies depending on the plot situation.
- The stones are arranged in such a manner that the bigger stones are put at the bottom and smaller at top, then stubbles are put in the gaps in between th stones.
- Some bunds are prepared using stubbles and earth.

8.7.7. Brushwood structure across the bund

- It is followed in the hill base end with abrupt fall.
- Constructing low cost brush wood structures in series along the drainage lines or in their field bunds to remove excess water.
- The structure consists of locally available wood logs.
- The loose boulders are packed in between 2 rows of wooden barrier and runoff water flows through the structure with non erosive velocity.

8.7.8. Grassed waterways

- The width of the waterway is about 1 m and depth is 0.5 to 0.7 m.
- A bund on one side of the waterway is high about 1-2 m and on the other up to 1 m high.

8.7.9. Spur structure

- After transplanting rice, spur (temporary barrier like structure) is prepared with the help of locally available materials in the way of runoff water in order to divert the water flow.
- Some grasses and weeds are also placed on these structures in order to make them semi-permanent.

8.7.10. Nala plugging

- Drop structures are constructed during dry season by excavating the bed to a depth of 0.55' placing a brick and stone pieces in the bed raising it to the original bed level.
- Afterwards soil layers are placed and raised to the desired height lower than the exixting surface layer, then facilitating flow of water downstream.
- In a place of soil, bricks and sometimes plank or local materials like bamboo, boulders etc. are used.
- Grassed water ways are adopted or vetiver (*Vetiveria zizanoides*) is planted along the bunds in red soils to check the soil erosion (Farmers in Karnataka, India)
- Terracing is a traditional conservation method employed to prevent soil erosion particularly in steep areas (Farmers in Philippines)
- Leaves of *Butea monsperma* (flame of forest) are spread in the field and burnt along with the dried wheat stalks to improve the soil fertility (Farmers in Gujarat)

8.7.11. Vetiver to control soil erosion (Vetiver Naduthal)

- Plant the Vetiver grass (khus) with 1 feet distant along the bunds and across the slope.

Table 29: List of some documented ITKs on soil and water conservation measures

No	Categories	Name of ITK
1.	AgronomicMeasures	1. Cover cropping 2. Criss –cross ploughing 3. Hoeing with local hoes 4. Set furrow cultivation 5. Application of manure (FYM) 6. Green manuring
2.	Tillage	1. Summer ploughing/ Off-season tillage 2. Repeated tillage during monsoon season
3.	Bunding & Terracing (Mech. & Vegetativebarrier)	1. Vegetative barrier 2. Compartmental bunding 3. Peripheral bunding/ Field bunding 4. Conservation bench terrace 5. Strengthening bunds by growing grasses 6. Growing of *Saccharum* sp.
4.	Land Configuration	1. Use of indigenous plough for formation of broad bed & furrows 2. Levelling the plots by local leveler 3. Opening up set furrow 4. Conservation furrow : Gurr
5.	Soil Amendment / Mulching	1. Retention of sunflower stalks 2. Mulching of Sal leaf in turmeric 3. Crop residue application in the field
6.	Water Harvesting, Seepage Control & Ground WaterRecharge	1. Dug wells 2. Haveli / Bharel system 3. Bandh system of cultivation 4. Earthen check dams 5. Field water harvesting 6. Nadi farming system 7. Collection of sub-surface runoff water and recycling in Diara land 8. Rain water harvesting in Kund / Tanka
7.	Furrow opening in standing crops	After intercultivation, a small furrow is opened with the help of blade hoe after every 2 crops, in standing crops *viz.,* cotton sorghum having row spacing of atleast 45 cm. it is done for rainwater conservation
8.	Nadi farmingsystem	To collect runoff during kharif for lifesaving irrigation during drought spell or presowing irrigation (Palewa) for rabi crops
9.	Application of white soil aslining material in farm pond	To work as a sealant material for lining dugout farm pond
10.	Wider row spacing in pearlmillet	Rainwater conservation and weed control

Contd.

No	Categories	Name of ITK
11.	Rainwater harvesting inkund/tanka	The harvested water in kund / tanka is used for drinking and establishment of tree
12.	Mulching in turmeric	To conserve rainwater
13.	*Agave* sp. As vegetative barrier	To reduce runoff velocity and to increase infiltration opportunity time
14.	Broad bed and furrow practice	To harvest rain water and dispose of runoff
15.	Set-row cultivation	For harvesting rain water and maintaining soil structure
16.	Summer / pre monsoon tillage	Conservation tillage-to harvest early showers, facilitate timely seeding and weed control
17.	Ridge & furrow planting for modulation of overland flow	Conservation of rain water, modulating excess water, control soil loss and boosting productivity
18.	Application of tank silt	To increase the fertility and water holding capacity of soil

9

Techniques for Seeds and Sowing

9.1. ITKS for Seed Treatment and Sowing

- Yield of almost all the crops depends on seed quality.
- Even if it is a quality seed, it should be a dried one because well dried seeds will have higher longevity and keeping quality.
- It better to change the seeds at least once in two years.
- To maintain the seed viability and prevent it from outside damage, the outer shell is not separated.
- The seeds are generally stored along with the leaves of neem *(Azadirachta indica), pungam (Pongamia pinnata),* notchi *(Vitex negundo)* and thumbai *(Leucas aspera).*
- Six weeks nursery period is enough for *six* months duration of the crop.
- Shallow sowing is followed in dry lands.
- Severing the broadcasted seeds in dry lands with soil.
- Dragging, over the fields, the thorny branches with weight over them for covering seeds the sown on dry lands.
- It is better to start planting from 'Sani moolai' (north east) of the field to get higher yields.
- It is better to perform sowing and planting operations during evening hours.
- Even if the seeds are of good quality, they must be sown in the right season only.
- Sowing in the right season is better even on a poor land.
- The best seasons for crop cultivation/sowing are the Tamil months viz. 'Chittirai' (Apr.), Aadi (Jul.), Aavani (Aug.) and Thai (Jan.).
- Crops sown on eighteenth day *(Aadipperukku)* and new moon day of the Tamil month 'Aadi' (Jul./Aug.) yield better.

- On new moon day or up to 48 hours before the new moon day, it is better to do sowing/planting crop and cutting a tree.
- During 'Keelnokku' days (i.e. when moon moves from north east to south east direction) it is better to do sowing / planting / harvesting of crops that bear under ground, starting preparation of compost and application of compost to field, pruning and cutting of trees, planting tree seedlings and ploughing the fields.
- During 'Melnokku' days (*i.e.* when moon moves from south east to north east direction) it is better to do budding and layering operations, sowing/ planting and harvesting of crops that bear above the ground.
- Seeds sown 48 hours before full moon day germinate quicker and grow faster.
- It is better to avoid sowing on Ashtami (eighth) and Navami (ninth) days from full moon and new / moon days.
- Sowing is done on Tuesday and Saturday will germinate better.
- The crop sown on new moon day escapes from pest and diseases.
- Higher yield of tamarind is considered as an indication for good agricultural season and higher yield of mango for poor season.

9.2. Seed Treatment

9.2.1. Mint extract

- 200 g of puthina (mint) leaves crushed with water and extract the solution
- Soak the paddy seeds nearly for 1-2 hrs in the extracted solution
- Then it can be taken for sowing after drying.

9.2.2. Sweet flag rhizome extract

Pound 10 gm of sweet flag rhizome to a coarse powder and add 50 ml of water. Leave the solution undisturbed for one hour and filter the sweet flag rhizome extract.

For seed treatment, boil one litre of water and add 50 ml each of cow urine and sweet flag rhizome extract the following day. Soak the seeds in water for six hours and then in the above solution for about 30 minutes. Filter the seeds, shade dry and sow. This procedure protects the seed against a number of bacterial and fungal diseases

9.2.3. Pre soaking of cotton seeds (*Oora vaithal*)

- Soak the cotton seeds in water for an overnight.
- After soaking, the seeds have to be separated and dried for an hour.
- Sowing can be done after drying of seeds.

9.2.4. Cow dung slurry seed treatment

- Dissolve 1/2 kg of cowdung in a litre of water in an earthern pot.
- Then put the vegetable and cotton seeds in cowdung slurry for overnight.
- In the next day morning, the seeds have to be separated from the slurry and sowing can be done.

9.2.5. Neem oil seed treatment for cotton

- About 200 ml Neem oil and 20 g of cow dung have to be mixed per kg of cotton seeds and keep it for overnight.
- Then this treated cotton seeds can be sown in the early morning hours.

9.2.6. Salt water seed treatment for sorghum

- About 5 Kg of table salt was mixed in 15 liters of water.
- The required quantity of seeds for an acre (8 Kg) is taken in a vessel or bucket and salt water is used for soaking for a period of 3–4 hrs.
- The soaking of seeds in salt water facilitated the floating of ill-filled grains leaving good grains at the bottom of the vessel.
- By discarding ill filled and diseased grains, good ones suited for sowing are to be collected and used after shade drying them.

9.2.7. Nursery bed heating

- Dried leafs of plants and wood dusts were burnt in the nursery bed for 10-15 minutes so that heat generated destroyed disease causing micro organisms.
- Then cool the soil and use it for sowing seeds.

9.2.8. Water soaking for sorghum (*Oora vaithal*)

- For water soaking, 1 kg of sorghum seed needs about 3 litres of water (1:3 ratio).
- The required quantity of seeds is taken in a vessel or bucket and cold water is used for soaking for a period of 3–4 hrs.

- 12.5 kg of water soaked seeds were enough for sowing 1 ha of the field.
- Paddy -12 hrs, maize – 24 hrs soaking time

9.2.9. Chilli seed treatment (*Millakayil vithainerthi*)

- Cowdung solution is prepared by dissolving 1 kg of cowdung in 2 litres of water.
- The chilli seeds (required for sowing) are tied in a cotton cloth and soak in cowdung solution for 2-3 days.
- Then the seeds to be dried in shade (2 hrs) and then use for sowing.

9.2.10. Cow's urine for ragi seed (*Komiyathil Oora Vaithal*)

- Ragi seeds (6 kg/ha) are soaked with 8 litres of 10% of fresh cow's urine for 16-18 hrs.
- 10% of cow's urine was prepared by mixing 100 ml cow's urine in one litre of water (8 litres of cows urine solution/6 kg/ha).
- Then the seeds are dried in shade for about 24 hrs and used for sowing.

9.2.11. Hot water seed treatment for sorghum (*Venthanni Vithai nerthi*)

- For treating 15 kg of seeds about 20 litres of water is required.
- Well-graded sorghum seeds are soaked in hot water (lukewarm temperature) and then the water is drained out immediately (within 2 – 3 minutes) with the help of bamboo basket.
- After this hot water treatment vasambu (*Acorus calamus*) is powdered and mixed with the grains.
- For 15 kg of seeds about 100 g of powdered 'vasambu' is required.

9.2.12. Broadcasting gingelly seeds with sand

- Sand with gingelly seeds in the ratio of (4:1) and to be broadcasted them in the field.
- By this way, gingelly seeds will evenly spread in the soils which in turn help in covering of seeds and its good germination.

9.2.13. Seed treatment in coriander (*Pothumai Vaithal*)

- A day before the sowing, the coriander seeds are to be wetted in the late evening and keep heaped for overnight.
- Then the next day sowing can be done (8 kg/ha) without breaking the seeds.

9.2.14. Seed treatment for paddy

- Take 15 litres of water in a 40 litre capacity bucket.
- Immerse egg in the water to test its purity. The egg is said to be pure if it sinks in water but if it floats vertically, it should be discarded and replaced by a good one.
- Go on adding the crystal salt to the water by - stirring it continuously till 25% of the egg is visible. One kilo of salt will be usually sufficient for one bucket of water.
- Add 15 kilos of paddy to the water and remove the chaffy seeds that float on water.
- Soak these seeds in salt water for fifteen minutes and collect the treated seeds of paddy from the bottom of the bucket. Wash it with clean water two times and dry it in shade

9.2.15. Seed treatment for groundnut

- Pound the neem seeds well and extract juice by addition of water.
- Heat four or five agaves strips at low flame and extract the juice by wringing.
- Mix the neem juice and agaves juice in one litre of cow urine and soak it overnight.
- Add 10 litres of water to this extract next morning.
- Sprinkle this mixture on the groundnut seeds used for sowing and mix thoroughly with hands before sowing.

9.2.16. Millets and sorghum treatment

- Add 100 g of asafoetida to 10 litres of water and prepare a solution.
- Add 6 to 8 kgs of maize seeds to this solution and soak it for 15 minutes.
- Similarly add 3 kg of millets to this solution and soak it for 15 minutes.
- Remove the half filled immature seeds and dry them under shade.
- Now the seeds are ready for cultivation.

9.2.17. Biofertilizer with jaggery Seed treatment

- Add 100 gm Jaggery to 1 litre water in a container and prepare a jaggery paste by heating it on low flame for 5-10 minutes.
- Cool the solution and add the biofertilizer into this jaggery paste @ 4gm of *Trichoderma* and 4 gm of *Pseudomonas* to 1 kg of seeds.

- Add the seeds to be treated into this container and mix well so that the biofertilizer is smeared properly over the seeds.
- These seeds are dried under shade and used for sowing. All type of vegetable seeds, fruit crops and food crops can be treated using biofertilizers.

9.2.18. Seed treatment using CPP manure (Cow Pat Pit)

- Water has to be sprinkled over the seeds and CPP manure (@ 5-10 grams of CPP manure is applied to 1 kg seeds) has to be applied on the seeds.
- It is mixed well and dried under shade before sowing. This can be used for all the types of crops.

9.2.19. Milk seed treatment

- 200 ml milk is added to one litre water in a container.
- Pour approximately 2 kg of seeds into this mixture and soak it for about 15-30 minutes. Remove the seeds and dry them under shade before sowing.

9.2.20. To induce sprouting of seed nuts of coconut

- Selected seed nuts of coconut are allowed to float in well water for 20 days. Thereafter the sprouted nuts are planted in the nursery bed.
- The sprout emerges out quickly in the nursery bed. This will subsequently develop into vigorous, healthy seedlings.

9.2.21. Seed treatment for improving germination in soybean

- When soya seeds are sown in salt affected soil, its germination is as low as 30 percent. To improve its viability, treat the seeds with 'Usilai' (*Albizia amara*) leaf powder.
- For 1 kg of seeds take 150gm of 'usilai' (*Albizia amara*) leaf powder by mixing them with rice gruel.
- By this seed treatment, the germination will improve up to 90 percent.

9.2.22. Seed treatment by soaking in cow urine and *Acorus* powder

- Previous night water is boiled and cooled. In the next day morning, both cow's urine and powdered ' vasambu' (*Acorus calamus*) are mixed with that water.
- Then the sorghum seeds are soaked in that solution for a day. All the floating seeds are removed.

- The next day the solution is decanted and the soaked seeds are sown in the field. It is believed that cow's urine and Acorus powder had anti pathogen effect to kill disease causing organisms.

9.2.23. Wood ash

- Prepare a solution by mixing 10 grams of ash (approximately 2 table spoons) in one litre of water.
- Dip the vegetable seeds in the above solution for 15-30 minutes.
- Seeds are dried in shade and sown immediately.

9.2.24. Seed treatment with cow urine

Cow urine – 2 litre (preferably buffalo's urine)

Cow dung: 1 kg

Mond or live soil: 1 kg

- Mix all the above mentioned materials with seeds and allow them for drying up to one hour.
- This method is suitable for the crops with seed rate of 30-60 Kgs, e.g., Ground nut

9.2.25. Sugarcane seed treatment

- Keezhanelli *(Phyllanthus niruri),* Poovarasu (*Thespesia populnea*) and Pungam *(Pongamia pinnata)* (Dry and boil 1 kg of leaves of each type with enough water and filter after 2 days and use).
- Dip the sugarcane setts in an extract and cover it with a wet gunny bag for a day and plant on the next day. This will control the seed borne diseases.

9.3. Sowing

9.3.1. Heating of nursery bed soil (*Soodu paduthuthal*)

- Dried leafs of plants and wood dusts are to be burnt in the nursery bed for 10-15 minutes so that heat generated will destroy disease causing micro organisms.
- Then cool the soil and used it for sowing seeds.

9.3.2. Covering Nursery beds with coconut leaves (*Thennai olai iduthal*)

- After sowing vegetables seeds (like tomato, brinjal, chillies) in the nursery beds cover the bed using coconut leaves.
- Irrigation has to be done over this coconut leaves.

9.3.3. Paddy straw mulching in tomato nursery (*Nilaporvai amaithal*)

- Paddy straw to be applied in nursery bed with a height of 2-3cm as mulch

9.4. Seed Treatment for Individual Crops

Seed treatments

Paddy

i) Seed Treatment for Improved Germination

Dry seeds in bright sun light (between 12.00 p.m. to 1.00 p.m.) for half an hour before sowing to improve the germination and seedling vigour.

- Soak the paddy seeds along with a gunny bag in water for 12 hours and then soak in biogas slurry for 12 hours before sowing.
- Soak paddy seeds tied in khada cloth in sweet flag extract (500 gms of sweet flag rhizome powder in 2.5 litres of water) for 30 minutes and shade dry before sowing.
- Fill the paddy seeds in a closely-knit bamboo basket lined with *Salvadora persica* (Meswak tree) leaves at the bottom and pour about 10 to 12 litres of water over the basket. Cover the basket with *Salvadora* leaves and place a weight over it. Leave the setup undisturbed for 24 hours before sowing. This will help in early and vigorous germination.

ii) Seed Treatment for Healthy Seedlings

- Take the paddy seeds in a tightly closed gunny bag and soak it in the biogas slurry for 24 hours before sowing to get green and healthy seedlings with well developed root system. These seedlings will get established well soon after the transplantation.
- Collect cow's urine in a mud pot and keep it for 48 hours. Soak paddy seeds in 10% of this cow's urine (100 ml cow's urine in 1 litre of water) before sowing for healthy crop. Seeds should be shade dried for half an hour before sowing.
- Mix Vitex, Tulsi and Pungam leaves extract (pound 3 kgs of each leaves and extract) with fresh cow dung solution and soak 25 kg of paddy seeds tied in a gunny bag in this solution for 12 hours. Seeds should be shade dried for half an hour before sowing. This will produce healthy and disease resistant seedlings.

Note

3 kgs of each of the leaves should be collected and pound. This should be added to 1 litre of water and the extract is filtered. This should be added to fresh cow dung extract (5 kg cow dung in 15 litres of water).

iii) Seed treatment for the prevention of pest and disease attack

- Soak seeds in water for 12 hours and then mix it with 10% cow's urine (10 ml cow's urine + 90 ml water) or 5% *prosophis kashaayam* (5 ml *kashaayam* + 95 ml water) and dry it for 30 minutes. Use the seeds for sowing within 24 hours. This will enhance the resistance of the paddy against bacterial leaf blight disease.
- Soak paddy seeds tied into small bundles using kada cloth in cow's urine solution (500 ml of cow's urine with 2.5 litres of water) for 30 minutes and shade dry before sowing. This method of seed treatment prevents the crop from seed borne fungal and bacterial diseases.
- Soak paddy seeds in 20% mint (*Mentha sativa*) leaf extract (200 ml of leaf extract mixed with 800 ml of water) for 12 hours before sowing. This will increase the germination rate and vigour of seedlings. This will also help in the control of *Helminthosporium* leaf spot disease in paddy.
- Soak the sprouted seeds of paddy tied in small bags in sweet flag solution (500 gms of sweet flag rhizome powder in 2.5 litres of water) for 30 minutes and shade dry before sowing. This will improve the resistance of the seedlings against seed borne bacterial and fungal diseases.

Seed treatment techniques for pulses

Chickpea

- Soak seeds in water before sowing to enhance the germination percentage of the seeds.
- Smear seeds (1 kg) with a mixture of turmeric and sweet flag powder (50 gms turmeric powder and 15 gm sweet flag powder with 10 ml of water) and sow after 10 minutes. This will enhance the disease resistance of the crop.
- Smear seeds with mustard oil @ 100 ml /40 kg of seeds before sowing to prevent wilt disease.
- Soak seeds of pigeon pea / chickpea in curd for 24 – 48 hours before sowing to control wilt disease.

Bengalgram

Mix seeds with well fermented (sour) butter milk and shade dry before sowing. The acidic nature of the butter milk reduces the incidence of wilt and dry root rot diseases.

Seed Pelleting in Greengram

Take the seeds in a plastic tray and add a small quantity of adhesive (10% maida solution) to the seeds. Shake this gently to enable the seeds to spread evenly on all parts of each of the seed. Add *Arappu* powder (*Albizia amara*) as filler material evenly over the seeds and continue shaking until the uniform coating is ensured. Remove the seed clumps manually and also the excess filler material by sieving. Shades dry this before sowing. This process helps to handle small and irregular shape seeds. It also enables precision sowing of seeds and physiological characters of seedsare strengthened.

Seed treatment techniques for millets

Sorghum (*Jowar*)

- Treat the seeds with asafoetida solution (75–100 gms in 1 litre of water) and shade dry before sowing. This seed treatment method prevents ergot disease in sorghum.
- Mix the seeds with the extract of *Ashwagandha* and *Datura* (for 1 kg seeds, pound 250 gms of *Ashwagandha* / *Amukura (Withania somnifera)* roots and 50 gms of *Datura* /Oomathai *(Datura metel)* leaves by adding water) and shade dry before sowing. This will help in the production of healthy and disease free seedlings.
- Treat the seeds with dried cow dung powder and cow's urine (100 gms cow dung powder and 250 ml cow's urine per kilogram of seeds). This will break the dormancy and improve germination
- Soak the seeds in lime water (1 kg lime in 10 litres of water kept for 10 days and the superficial water is collected and used) for overnight. Dry the seeds before sowing.

Pearl Millet and Finger Millet

Mix the seeds with the extract of *Ashwagandha* and *Datura* (for 1 kg seeds, pound 250 gms of *Ashwagandha* / *Amukura (Withania somnifera)* roots and 50 gms of *Datura* / Oomathai *(Datura metel)* leaves by adding water) and shade dry before sowing. This will help in the production of healthy and disease free seedlings.

Seed treatment techniques for oilseeds

Groundnut

- Soak the seeds tied in a gunny bag in water for 4-6 hours. Then untie the gunny bag and cover it with another wet gunny bag for 12-14 hours.

- Shade dry the germinated seeds for 3–4 hours and treat with *Rhizobium* (@ 600 gms / 110–120 kg of seeds) and sow within 1 or 2 days.
- Smear the seeds with *kallipal* (milky latex from leafy spurge or milk hedge) before sowing @ 100 gms of *kallipal*/10 kgs of seeds. This will protect the crop from pest and diseases.
- Soak the seeds in asafoetida solution (250 gms in 2 litres water) for 12 hours before sowing to prevent blight disease.

Coconut

Allow the seed coconuts to float on the surface of irrigation well or water. This will enhance the germination and seed coconuts will germinate within a month.

Seed treatment techniques for vegetables

Soak all kinds of vegetable seeds in biogas slurry for 30 minutes before sowing.

Bhendi

- Treat seeds with 15% or 25% raw cow's milk (150 ml of milk in 850 ml of water or 250 ml of milk in 750 ml of water) for 6 hours and then sow. This will increase the germination percentage and seedling vigour.
- Soak seeds in cow's urine at 5% or 10% concentration (50 ml of cow's urine in 950 ml of water or 100 ml of cow's urine in 900 ml of water) for 12 hours before sowing for good germination percentage.
- For summer crop, soak the seeds in water for 12 hours before sowing.
- Soak the seeds in sweet flag *rhizome* extract or cow's urine solution (dilution 1:5 ratio – 1 part of extract or cow's urine in 5 parts of water) for 30 minutes before sowing.

Brinjal

- Soak the seeds in 12% raw cow's milk (120 ml of raw cow's milk in 880 ml of water) for good germination percentage and seedling vigour.
- Seeds should be soaked in a solution of cow's urine (1 part cow's urine + 5 parts of water) for 30 minutes prior to the sowing.
- Seeds should be bundled using a thin cotton cloth and soaked in the biogas slurry for 12 hours prior to the sowing.

Bitter Gourd

- Soak the seeds in diluted cow's urine for 12 hours and in diluted cow's milk for 6 hours before sowing. The dilution should be at the ratio of 1:1 (1 part of cow's urine or cow's milk with 1 part of water). Soak the seeds in raw cow's milk for 24 hours before sowing.

Tomato

- Fumigate the seeds with *Vasambu (Acorus calamus)* and *Vaividanga (Embelia ribes)* powder. Take seeds in a metal sieve. Take hot coal in a metal plate and sprinkle *Vasambu* or *Vaividanga* powder over the hot coal and hold the sieve with seeds against the fumes in a standing position for 2–3 minutes. This will enhance the germination rate and protect the seedlings from fungal pathogens. For treating 100 gms of seeds 5gms of Sweet flag or *Vasambu* and 5 gms of *Vaividanga* is required.
- Soak the seeds tied in a khada cloth in diluted milk solution (75 ml milk and 425 ml water) for 6 hours and then sow.
- Soak the seeds in a mixture of fermented buttermilk (3 days old) and water in 1:4 ratio for 6 hours and shade dry before sowing. The practice is applicable only for 6 to 12 months old seeds. Buttermilk can be replaced by Coconut or Palmyra toddy.
- Soak the seeds in sweet flag rhizome extract (dilution 1:5 ratio – 1 part of extract in 5 parts of water) for 30 minutes before sowing.

Chillies

- Soak seeds in sweet flag extract or cow's urine at 1:5 ratio (1 part of extract or cow's urine with 5 parts of water) for 30 minutes before sowing. This will inhibit the seed borne diseases like fruit rot and die back.
- Soak seeds tied in a cotton cloth in biogas slurry for 12 hours before sowing. This will kill the disease causing microbes and enhance the seed vigour.

Bottle gourd

- Soak seeds in water for 24 hours before sowing to break the dormancy and to quicken the germination.
- Soak seeds in warm water for 30 minutes before sowing.
- Soak seeds in cow's urine solution (1 part cow's urine + 5 parts of water) for 30 minutes prior to the sowing.

Snake gourd

- Treat the seeds with cow dung @ 1 kg per kg of seeds for 30 minutes.
- This will increase the drought resistance and make the seeds germinate quickly.

Beans

- Soak the seeds in raw cow's milk for 24 hours before sowing for good germination and yield.

Root Vegetables

- Soak the seeds of beetroot and radish tied in a cotton cloth in water overnight or in warm water for 30 minutes before sowing.
- Soak seeds in a solution of cow's urine (1 part cow's urine + 5 parts of water) for 30 minutes prior to the sowing.

10

Regenerating the Soil

Soil regeneration is about bulding or making topsoil. It is a particular form of ecological regeneration within the field of restoration ecology is the act/idea of replenishing the top soil with beneficial nutrients casing natural methods. Time required from inorganic to organic would be about 60 to 200 days depending upon the situation. Once the inorganic fertilization is stopped, following actions facilitate quick conversion:

- Apply Farm yard/Poultry manure available in the farm as a basal dressing.
- Sow 25 kg of multi varietal seeds (MVS) in one acre.
- Allow it to grow up to 60 days.
- Inoculate the soil with the bio-agents like *Azospirillum, Phosphobacterium, Pseudomonas flouresens, Trichoderma viride* on 30 days after sowing (DAS).
- Incorporate in situ 60 DAS (30 t ha^{-1} biomass is generated on 60 DAS) green manures.

Assumptions

Regeneration capacity of the soil is enhanced with the sowing of MVS which provides ideal condition for regeneration. It creates green cover / dry leaf cover in the top soil (35-40 cm height) which will facilitate the multiplication of inoculated beneficial microbes in the soil.

10.1. Multi Varietal Seeds Sowing Techniques (MVS)

MVS contains seeds of three to four crops belonging to each category viz., cereals, pulses, oilseeds, nitrogen fixing green manures, spices and condiments. The seeds of above crops may be mixed in different proportion based on their growth habit and seasonal requirement. For example in the summer the following mixture can be practiced.

Table 29: Crop mixture for sowing

Cereals	Sorghum, Pearlmilet; Foxtailmillet, Kodomillet
Pulses	Blackgram, Greengram, Cowpea, Redgram
Oilseeds	Soybean, Castor, Sesame, Sunflower
N giving GM (Green manure)	Sunnhemp, Daincha, Wild indigo, Sesbania spp.
Spices and condiments	Chillies, Coriander and Aniseed

Take up sowing using appropriate crop mixtures at the rate of 20-25 kg/acre.

Under normal soil conditions

At the time of MVS sowing, 100kg of well decomposed cowdung manure + palm sugar (1-3kg) slurry spray to create more friable soil texture which conducive for the microbes. The above preparation may be split into equal four parts and inoculate each one with the bio-agents separately and allow it for 30 days to multiply the microbes.

After that, take a part of digested manure in to a gunny bag and place it in the entry point of irrigation water for slow dissolution and uniform spread into the field. Another part may be applied as basal dressing to soil at an interval of 30 days from the previous application. All manures get converted into available nutrients within 60 days after the incorporation of MVS.

Under inert / dead soil conditions

Due to the intensive cultivation using modern varieties with indiscriminate use of chemicals heavily depleted the organic matter content of the soil. Low organic matter content may not support the growth of microbes. Soil which having poor physical conditions and biological activity with low buffering capacity may be called a sick or dead soil. These soils can be revived for cultivation within a year through MVS technique as detailed be low.

- Apply FYM / Poultry manure as basal
- First MVS sowing
- Allow it to grow for 30 DAS
- Inoculate microbial cultures and incorporate on 30 DAS
- Second MVS sowing, inoculate microbial cultures and incorporate on 60 DAS
- Third MVS sowing and keep it for 100-110 DAS, collect seeds and incorporate the entire biomass.

Methodology

- Applying bio-digested gas slurry (BDGS) (25 litres) + Cow's urine (5 litres) + 10 lit of water as slurry mixture applied @ 1050 litres / acre along with irrigation water.

- Bio-digested gas slurry (BDGS) + RK (Decayed fruits fermented in BDGS for a week) and applied as *foliar spray* @ 10%.

10.2. Soil Management

- For soil improvement in 'theri' lands of Tuticorin district, 200 tonnes to tank silt are applied per acre followed by 50 tonnes per year for the next few years (Farmers of Erode and Tuticorin District in Tamil Nadu).
- Crop residues and tree branches are burnt on the soil surface at the end of summer to improve the structure of clayey soil and to make ploughing easy (Farmers of Konkan Region in Maharashtra).

10.2.1. Different methods for soil conservation

Coir pith waste, farm waste, dried leaves, dried grasses, sugarcane trashes and groundnut shell can be used as soil conserving agents.

After the harvest of sugarcane crop, the trashes should be spread over on the ridges for conserving the moisture.

After planting banana, well decomposed sugarcane trashes can be spread over.

10.2.2. Different conservation systems for soil

No-tillage (Slot planting)

It is a specialized type of conservation tillage consisting of a one-pass planting and manuring operation in which the soil and surface residues are minimally disturbed. A narrow (2-3 cm wide) strip or slot on the ground is made for sowing. Compare to other tillage system, no-till will be more useful for effective control of soil erosion, increased water storage, lower energy costs per unit of production and higher grain yields.

Mulch tillage

It is based on the principle that least disturbance and leaving the soil with maximum of crop residues (6 t ha^{-1}) for obtaining quick seed germination adequate stand and for satisfactory yield. The use of live mulching is also accommodated in this system. Small and marginal farmers can also adopt this system.

Strip or zonal tillage

Here the seedbed is divided into a seeding zone and a soil management zone (5 to 10 cm wide) is mechanically tilled to optimize the soil and micro-climate environment for better seed germination and stand. The interrow zone is left undisturbed and protected by mulches.

Reduced or minimal tillage

This covers minimum number of ploughings ie. tyne cultivation before puddling the soil or directly seeding and covering. Possibly 30% residue over the surface maintained under dryland farming. It also means non-turning, superficial tillage.

Stubble cutting

Moisture conservation can be enhanced by trapping more overwinter snow with "tall" or "sculptured" stubble. Tall stubble refers to stubble which is cut 12 or more inches high, usually when straight combining. Sculptured stubble refers to when a swather-mounted clipper or deflector cuts strips of taller stubble with each swath, or when alternate swaths are cut at normal height and taller.

Direct seeding

Zero-till, direct seeding is not usually associated with organic crop production because herbicides cannot be used. Some Saskatchewan organic producers, however, do practice direct seeding. Thus, producers may wish to consider this conservation practice when fall and spring weed pressure is low, and previous crop straw and chaff have been adequately spread.

Extended crop rotations

Summer fallowing is destructive to the soil because no new organic matter is returned to the soil during the fallow year. Tillage also speeds the breakdown of soil organic matter and predisposes the soil to erosion. Extending crop rotations is a conservation practice because it reduces the incidence of summer fallow.

The benefits of an extended crop rotation are numerous and include improved fertility, tilth, aggregate stability, moisture storage, and resistance to soil erosion and degradation, as well as reductions in insect, weed and disease problems. All of these factors contribute to increased productivity and most have a significant positive effect on soil sustainability. Decisions regarding cropping strategies should consider not only short-term benefits, but also their long term effects on soil and environmental quality.

A diverse rotation should include cereals, oilseeds, pulses, fall-seeded crops and forages. The level of crop diversity determines the significance and degree of the rotational benefits. Selection and management of legume species (pulses and forage legumes) within a rotation is a vital aspect of achieving diversity and supplying nitrogen through symbiotic nitrogen fixation.

Forage cropping

Forages contribute significant amounts of organic material to the soil. They also offer an alternative commodity in the form of hay, silage or seed. Semi-permanent forage production should be considered on soils which are inherently low in productivity and/or vulnerable to erosion. Forage production for two to four years should also be considered as part of a normal crop rotation. Selection of forage species and management practices can be tailored to specific problems such as drought, excessive soil moisture, salinity, poor soil structure, low pH and other problems.

Complementary rotational crops

An ideal rotation should be as diverse as possible including cereals, oilseeds, pulses and forages. A diverse crop rotation, in addition to many other benefits, can help soil nutrient availability, because different crops have a different demand for and ability to remove particular nutrients. All crops require 16 essential nutrients, and in our country, most soils are deficient in nitrogen and phosphorus besides micronutrients.

Growing legumes in the rotation provides both nitrogen and non-nitrogen benefits to subsequent crops. Properly inoculated/nodulated legumes fix 50 to 90 per cent of their nitrogen requirement from the air. The remainder is obtained from the soil. However, nitrogen is also exuded from legume roots during the growing season and the legume residue decomposes and recycles the nutrients faster than non-legume residues, thus more nitrogen is usually available to the subsequent crop than if a non-legume had been grown. Furthermore, research has shown that the non-nitrogen benefits (such as disease suppression, tilth improvement, etc.) of growing legumes in the rotation may result in increased yields.

10.2.3. Wind barriers

Annual crop barriers on summer fallow

Annual crop barriers can be used to prevent wind erosion on summerfallow fields. Annual crop barriers are two to five rows of plants seeded every cultivator width or two in July during the summer fallow year. The greater the potential for erosion, the closer the barrier strips should be placed. Crops such as flax and oriental or yellow mustard have good "standability", but cereals and sunflower have also been used.

A number of commercial strip seeders are available for purchase and mounting on tillage equipment. Following seeding, these barriers are left intact, to protect the soil until planting the following year. This may require adjustments of summer fallow tillage widths or removal of two or more cultivator shanks to compensate for the barriers. The following crop may look poor where the barriers were located the previous year. However, research has found yield losses amount to less than two per cent over the entire field.

Annual crop barriers in crop

Barriers of "taller" annual crops have been used to a limited degree in low residue-producing crops. A divider is placed in the seedbox so that two rows of wheat or flax are seeded every seeder width or two of lentil. At harvest, the lentil is combined and the barrier strip left standing to trap snow and prevent wind erosion during the upcoming winter.

Perennial grass barriers

Perennial grass barriers are two rows of grass planted perpendicular to prevailing winds to reduce wind erosion, trap snow and reduce evaporative losses. Barriers should be placed 30 to 60 feet apart, depending on soil type; closest on sands, moderately spaced on clays and furthest apart on loams. Barriers may be placed further apart if other soil conservation practices are also being used.

Species such as tall wheat grass (*Thinopyrum intermedium*) work well, as it is usually a weak competitor with most field crops and will not spread beyond the seeded rows. It also grows high enough without lodging to trap snow, helping in soil moisture recharge.

Shelterbelts

Shelterbelts can effectively reduce wind velocity for a distance approximately 20 or more times their height. This is usually sufficient to control wind erosion for a distance of approximately 10 or more times their height when planted perpendicular to prevailing winds. Their effectiveness, however, does depend upon shape, porosity and maintenance as well as height. Shelterbelts have the added benefit of increasing crop yields, particularly of less drought-tolerant crops such as canola and alfalfa.

Strip cropping

Strip cropping consists of alternative strips of crop and summer fallow at an angle perpendicular to the prevailing winds. The strip width varies depending on soil texture. Sandy soils are the most prone to wind erosion, followed by clays, then loams. Strip cropping works well for loams to clays where eight to 10 strips per quarter section will significantly reduce the potential for wind erosion. With sandy textured soils, however, too many strips are required to be manageable. Keep field equipment sizes in mind when establishing strip widths.

Strip cropping is a more common practice in the drier areas of the prairies where often too little crop residue is present to prevent wind erosion. It can be used in wetter areas, however, provided the strips also run perpendicular to the slope, so that water erosion does not become a problem.

Cover crops

Rotations should also include the use of cover crops to protect the soil from wind and water erosion during vulnerable periods such as summerfallow or partial fallow where normal standing stubble is not available. Cereal crops should be seeded at a rate of one-third to one-half bushels (1 bushel = 27.2 kg) per acre.

Green manures

Green manuring involves the incorporation of any green, fresh vegetative material into the soil. Green manure crops add organic matter to the soil, improve soil tilth, and if a legume, contribute nitrogen fixed from the atmosphere. Even weeds can be regarded as green manure. The extent of soil improvement depends on the type and quantity of plant material returned to the soil. The greatest benefits usually occur by using biennial (sweetclover) or perennial legumes (alfalfa or true clovers) as green manure on poor structured soils with low organic matter levels. If sweetclover is used as a green manure crop, it should be incorporated into the soil by mid-June (approximately 10 per cent bloom) to allow soil moisture recharge for the following year's crop. Crop quantity and quality are good by this time and little is gained by delaying incorporation.

Grain legumes (pulses) can be used effectively as green manure. Due to their annual growth habit these crops will not contribute nitrogen or crop residue to the same degree as a biennial or perennial legumes. They may, however, be more adaptable to an existing crop rotation. Pulses such as Indianhead lentil have been specially developed for this use.

Non-legume crops such as buckwheat have also been used as a green manure crop.

To protect against soil erosion, do not over-incorporate green manure crops.

Animal manure

The spreading of livestock and poultry manure provides not only nutrients required for plant growth but has a major beneficial effect on soil tilth and particle aggregation. The organic materials contained in manure act as binding agents in stabilizing soil structure. This positive change in soil structure caused by the addition of manure is equally, if not more important, than the nutrient contribution provided. Changes in structure of this nature positively affect water infiltration, water holding capacity and aeration, as well as resistance to wind and water erosion.

The nutrient value of manure is highly variable depending on numerous factors such as animal type and age, type of feed, amount of straw, and method and time of storage. Typically, barnyard cow manure contains approximately three to five lbs. (2.30kg) of crop available nitrogen, four to 11 lbs (5 kg) of phosphate, nine to 16 lbs (7.25 kg) of potassium and about three lbs. of sulphur, per ton of manure. Manure usually has sufficient micronutrients present to prevent plant deficiency symptoms from occurring. Soil testing laboratories test animal manure for nutrient content and make recommendations on manure application rates.

Application rates of manure will vary depending on availability, soil type, location, slope, crop rotation and production practices. To prevent leaching losses and potential environmental contamination, rates of manure application should not exceed what a crop can use in one growing season. Following the application of manure, it should be incorporated as quickly as possible into the soil to prevent nitrogen loss. In addition, changes in soil nutrient levels resulting from manure should be monitored by soil testing on a regular basis.

10.3. Saline Sodic Soil Reclamation

- Addition of plant residues, FYM and compost improve the hydraulic properties, thereby facilitating leaching of soluble salts.
- Helping to mobilize calcium through decomposition products.
- Serving as source of plant nutrient.
- Invigorating microbial activity that is at a low in sodic soils.
- Under favourable temperature and moisture condition, microbial decomposition of organic matter initiates natural reclamation through production of organic acids, which in turn reacts with calcium carbonate and release calcium for exchange of sodium.
- This is a slow process can be enhanced by adding easily decomposable organic materials like green manure and FYM.
- Organic materials of narrow C:N ratio are very efficient ameliorants.
- Green manuring with *Sesbania* has been considered a very useful practice in reclaiming sodic soils.
- *Sesbania* green manuring has three special attributes owing to its very narrow C:N ratio it decomposes rapidly and produces organic acids under high soil moisture during monsoon.

- Green manuring with *Sesbania rostrata* also improves the reclaiming effect Gypsum.
- FYM application is one of the first treatments introduced for the reclamation of sodic soils in Uttar Pradesh.
- Its use in conjunction with Gypsum has been very beneficial.
- *Argemone Mexicana* is used as a substitute for chemical amendments
- *Sesbania* and *Ipomoea* proved to be more effective.
- *Cassia auriculata* , *Calotropis* have also been tried and found promising for reclaiming the soil.
- Molasses alone or in combination with pressmud is very effective soil ameliorant.
- Spreading rice straw together with impounding water on alkali soils also fastens their reclamation.
- Palmyra leaf (*Borassus flabellifer*), Pungan leaf (*Pongamia glabra*), Tamarind fruit shell (Tamarindus indica), Pirandai (*Cissus quadrangalaris*) vines are cut into small pieces and applied as organic amendment and allowed to decay in the wet land (after puddling the soil).
- Good quality water / rain water is let in to the field and the soluble salts are leached out by draining the water.
- Daincha (*Sesbania acculeata*) raised with high seed rate as green manure and ploughed *in situ* during its flowering stage.
- Organic manures like farm yard manure, compost and tank silt applied liberally and incorporated in to soil.

10.2.1. Muthalai poondu/Agasathaamarai – *Pistia stratiotes*

- It is a common weed found in all water bodies. In the month of January, whole plants are removed from the water bodies and allowed for drying.
- The fully dried plant materials are incorporated to the saline and sodic soil for reclamation.
- It can be used as green leaf manure in wetland condition.
- After incorporation, discoloration of water i.e. red colour indicates that salt is dissolved and removed from soil. No specific dosage in agasathamari incorporation.

10.2.2. pH maintenance in soil

- The two locally available plants like Kolinji (*Tephrosia purpurea*) and Avaram poo (*Cassia auriculata*) were cut in to pieces and leaves were placed in irrigation channels. Water in the channels flow through these leaves extracting its essence and carry over it to the cropped field. These leaves believed to have a property in maintaining the soil pH.

10.3. Organic Amendments to Reclaim Sodic Soil

Farmers in Tamil Nadu follow traditional practices for reclamation of saline / alkaline soil. The following practices separately or in combination have been followed by farmers and they are given below:

Palmyra leaf (*Borassus flabellifer*), Pungan leaf (*Pongamia glabra*), Tamarind fruit shell (*Tamarindus indica*), Pirandai (*Cissus quadrangalaris*) vines are cut into small pieces and applied as organic amendment and allowed to decay in the wet land (after puddling the soil).

- Good quality water/rain water is let in to the field and the soluble salts are leached out by draining the water.
- Daincha (*Sesbania acculeata*) raised with high seed rate as green manure and ploughed in situ during its flowering stage.
- Organic manures like Farmyard Manure, compost and tank silt applied liberally and incorporated in to soil.

11

Microclimate Modification Techniques

Change in weather affects growth and development of plants. The extent of these effects depends upon the range or severity of change of weather. It is very much possible to change the weather in relatively small area to reap better harvest. Micro weather modification is defined as, "The manipulation of weather elements at increasing or decreasing the duration, intensity, quality etc., for desired effects at microlevel of crop under investigation".

The climate near the ground can be managed partly by influencing the moisture status and temperature of the upper soil layers, e.g. by mulching and partly by action above the ground e.g. by shading or use of wind breaks.

11.1. Mulching

Mulching is the application, or creation of some soil cover which reduces vertical transfer of heat and water vapour. Mulches vary in surface albedo, in thermal conductivity and capacity and in porosity to vapour and all of these are important. Mulches may consist of;

i. A loose layer of top soil (harrowing produces a mulch, apart from other effects);

ii. Cut or gathered vegetable material, such as grass, weeds, straw, tree leaves;

iii. Redeployed surface material, such as stubble, litter and stones:

iv. Manufactured materials, such as paper, plastics and reinforced aluminium foil

11.2. Surface Geometry

Slope and aspect (direction) affect the soil temperature and soil heat flux since capture of radiation is determined in part by these factors. Slope and aspect are more important in summer than winter because the low sun angles in winter season increase the proportion of diffuse radiation. Whatever the season, the slope and aspect are of lesser importance in cloudy than in clear conditions for the same reason.

It is possible to design methods that beneficially alter the temperature regimes and ridging and shaping is one such method. In a north-south running furrow the temperature will be lowest in the furrow. Generally the west facing slope was warmer than the opposite slope. This is because the east facing slope is illuminated at a time of a day when the soil is most cool. Evaporation of dew at this time also tends to keep temperatures low.

11.3. Wind Breaks and Shelter Belts

The use of barriers to provide shelter from the wind is an old and well-developed procedure. The barrier may be a line of trees (shelter belt) or any other arrangement of trees, bushes, hedges, soil embankments, stone walls or fences. The principal aim is to reduce the horizontal wind speed near the ground in areas open to damaging or otherwise undesirably strong winds.

Such areas include coastlines and open prairie landscapes where frictional retardation is weak because of small terrain roughness, or locations open to topographic or other local wind systems possessing undesirable features (e.g. strong and / or cold katabatic winds). The objects of protection could be sensitive agricultural crops, domestic animals, houses and farm buildings, transportation routes, or the conservation of such properties as the snow cover, soil moisture and the top soil. The climatic effects of shelter are not restricted to the simple reduction of wind speed.

12

Indigenous Technical Knowledge (ITK) Based Inputs/Techniques for Storage

12.1. General Practices for Storage

- Lime juice is mixed with grains and then sun dried before storage to prevent insect pests (Farmers in Nigeria).
- Seeds are safely stored in earthern pots after mixing with the leaves of neem and Vitex *negundo* (Farmers in Gujarat, India).
- Drying seeds and grains of all the crops on the new moon day before storing to avoid pest attack.
- Store seed materials in earthen pots which are kept on the 'Paran' directly above the furnace in kitchen. The smoke acts as a repellent.
- Grains, trees etc. harvested on full moon day are raqre prone to storage pest attack.
- Mixing the dried leaves of neem with seeds while storing them.
- Mixing the dried leaves of notchi (*Vitex negundo)* with seeds while storing them.
- Storing the seeds after mixing with pungam *(Pongamia pinnata)* leaves.
- Mixing 1 kg. of Vasambu *(Acorus calamus)* powder with 50kg. of grains for storage even for one year.
- While storing the seeds of food crops, they are filled in a container to its 3/4" height, covered with a rough cloth on which leaves of neem, pungam and notchi are placed to the remaining volume and finally covered with sand up to its mouth.
- Grains are filled in earthen pot to its 3/4" height and the remaining volume with dried cow dung. Then the cow dung is set on fire and mouth of the pot is tightly covered so that fire puts off and carbon monoxide formed inside kills the pests.

- Pulses and food grains are stored in gunny bags, which are previously wet with 10% salt solution and dried, in order to avoid storage pest attack.
- Generally seeds stored with their outer coat/shell escape from storage pests.

12.2. Use of Mustard oil for Storage of Pulses

- Farmers reported that 10-20 ml of mustard oil is sufficient to be applied on one *Nali* of the pulse (one *Nali* is approximately 2 kg. *Nali* is a container which is made of wood)

12.3. Use of Dried Leaves for Storage of Food Grains

- Leaves of Persian walnut (*Juglans regia*) and red cedar (*Toona ciliata*) have to be kept under sun for 1 day
- Then the dried leaves must be crushed and mixed wih food grains.
- When the grain needs to be consumed the crushed leaves should be separated from the grains.
- Young leaves of red cedar used as natural insecticide

12.4. Use of Table Salt

- To store rice grains table salt is mixed with these grains.
- About 50 g of salt should be mixed with one kg of the grains.
- 200 g of salt should be mixed with 1 kg of redgram and stored in jute gunny bags and the bags must be stitched.

12.5. Use of Ash for Storage

- Ash obtained from the burning of fuel must be mixed with food grains.
- The main reason for its usage may be its ability to absorb moisture, which otherwise will be absorbed by the grains and make prone to pest attack.

12.6. For the Control of Rice Grain Stored Pests

- Keeping a pot of water over the stored rice grain will control the rice moth
- Keeping neem leaf or ber branch over the stored rice will control rice weevil
- Placing of curry leaves in grain storage will control the stored grain pests.
- Dry leaf powder of Indian ash tree (*Begunia*), Poduthalai (*Lippia sp*), Vilvam (*Aegle marmelos*), Thulasi (*Ocimum americanum)* @ 1: 100 part

- Fresh pungam leaves must be placed as layers in between the gunny bags arranged above other against grain moth and rice weevils.

12.7. For the Control of Storage Pests of Cereals and Pulses

- Mix dried leaves with stored grains

12.8. For the Control of Storage Pest in Green Gram

- Dipping gunny bag in the oil of white batch vasambu (*Acorus calamus*) 0.1% with water for 5 mintues then dry the bag and use

12.9. Storage of Grans Using Camphor

- 1 g of camphor piece per 5 kg grains must placed in a jute gunny bags.
- 3 months possible with this traditional method

12.10. Ash Seed Treatment in Sorghum

- Ash should be mixed with sorghum at the ratio of 1:4
- After that, must placed in a jute gunny bags

12.11. Ragi Storage with Neem and Thumbai Leaves

- Mix dried leaves of neem and thumbai with ragi

12.12. Storage of Seeds with Lime

- 10 g of lime per kg of grains, must placed in a jute gunny bags

12.13. Use of Sweet Flag

- 1 g of sweet flag powder must be used to treat 1 kg of grains
- Then it must be stored effectively

12.14. Pasu Komiam Vithai Nerthi (Seed Treatment with Cow's Urine)

- 1 litre of cow urine is reuwired to treat 25 kg of seeds. This prevents moth damage during storage

12.15. Manjal and Mor Vithai Nerthi (Seed Treatment with Turmeric and Butter Milk)

- 1 kg of turmeric is taken and it is soaked in 2 litre of skimmed butter or Neeragaram. It should be pounded well and the extract is used for seed treatemtn.
- This can be done 2 days after harvesting of rice grains.
- Treated seeds are shade dreid for 8-10 hours

12.16. Water Extract of Bulbs

The water extract (0.02%) of bulbs mixed with the grains at the rate of 2 litres/ 100 kg grains can give good protection to paddy grains against insect attack

12.17. Paddy Husk in Managing Storage Pests

- Store the paddy grains in earthen pots and placed paddy husk in top layer (5 cm) above it.

12.18. Cucumber Seed Storage with Soil

- Mix wet soil (of any type) 200 g for treating 1 kg of cucumber seeds. After the seeds were shade dried for 2-3 days and then packed.

12.19. Naithulasi and Milakai for Pulses

- Only dried naithulasi leaves and millakai fruits were used. Farmers approximately applied about 100 gram of dried naithulasi leaves and 4 - 5 number of dried millakai for storing 10 kg of pulse grains n the jute gunny bags.

12.20. Safe Seed Storage in Mud Pot

- The pot is filled with seed with 1 foot space at the bottom of the pot. The grain is covered with cloth / paper.
- Then over this a layer, the leaves of neem, notchi (*Vitex negundo*), Pungam (*Pongamia pinnata*) are spread as a layer of 3" height and above that sand is spread to 6 "height.
- The stored pests that pass through the sand layer will be affected in the neck portion and subsequently by the leaf layer consisting of the above 3 types of leaves having insecticidal properties.

12.21. Stepping Method or Salt and Chilli Powder-Application Method

- Apply ½ kg powdered salt and 250 gms dry chilly powder to 25 kgs horse gram seeds and fill it into a 30 kg capacity plastic bag. Tie it with a thread tightly and keep the bag near the door where people walk often.
- Repeated stamping of this bag would not only avoid pest attack but also destroys the eggs and maggots that are already present in the seeds. This method also known as stamping method prevents the horse gram from pest attacks and could be stored upto one year.

12.22. Sand Mixture Method

- Take a 5 kg capacity mud pot. Any variety of seeds to be stored are collected, cleaned and dried under- sun and shade.
- Into this pot, add a thick layer of sand at the base and spread the seeds to be stored over this sand. Again add sand over the seeds.
- Continue the same process of filling sand seed mixture layer by layer till it reaches upto the brim of the pot. See that the upper layer is again a thick layer of sand. Close the container with a lid and air tight it with cow dung paste. By doing so, seeds can be stored for more than a year.

12.23. Moode Method

- Moode is a container made of ropes of bamboo stripes, paddy straw, kodo millet straw and some fibrous weeds (hanchi hullu). Arrange the weeds to form a basket.
- The structure must be air tight with 3 ½ feet height and 30 cms radius. The top and base of the container are lined with ragi husk and the grain/ seed to be stored are filled in.
- Red chilli powder is mixed with the seeds before storage. Close the container with bamboo lid and tighten it with bamboo straw.
- This method is usually used to store seeds and grains of pulses, cereals, oilseeds etc for nearly 6-10 months

12.24. Sweet Flag Paste Coating in Earthen Pots (Vasambu Thadavuthal)

- Sweet flag (10 g /pot) was powdered and mixed with ½ litre of water to form paste and coated over earthen pots had significantly reduced infestation and damage to grain during a 3 month storage period.
- Coat the inner sides/walls of earthen pots and shade dried them. After drying the pots, the seeds were stored inside the pot, which helps in repelling most of the storage pests (upto 60%).

12.25. Gingelly Seeds Storage

- In gingelly seeds storage, mixing a handful of (nearly 100 g) paddy in storage container significantly reduced the infestation of Indian meal moth and prevented the damage of seeds for the next 3-month storage period.

12.26. Storage of Vegetable Seeds Using Cotton Cloth

- First the seeds separated from vegetables were thoroughly washed and dried in shade for 2-3 days. A clean cotton cloth of required size (30x30cm) was first spread in the floor.
- The vegetable seeds to be stored were heaped in the centre of the cotton cloth. Then using the edges of the same cloth a tight knot was made. This tied cloth bag was placed in the wooden shelf and preserved for even a year.

12.27. Cucumber Seed Storage with Soil

- Mix wet soil (of any type) 200 g for treating 1 kg of cucumber seeds.
- After the seeds were shade dried for 2-3 days and then packed

12.28. Storage of Lettuce

- Extend the storage period to a short period of 1 day by preserving them in a wet cloth.

12.29. Storage of Sugar Crystals with Camphor

- About 5-6 pieces of camphor were packed in a white paper and kept inside the sugar stored container.

12.30. Coating Bamboo Baskets with Sheep Dung

- The sheep dung coated over the baskets filled the gap / holes between inter woven basket sticks thus help in avoiding the spillage of grains. After coating, the baskets were kept in sundry for a day and then used.
- In this practice, farmers dissolved the sheep dung with little water and applied that dung in both in sides/walls (1kg in ½ litre of water) of the basket.

12.31. Thulasi Katti Vaithal

- Only dried naithulasi leaves and Millakai fruits were used. Farmers approximately applied about 100 gram of dried naithulasi leaves and 4 - 5 number of dried millakai for storing 10 Kg of pulse grains n the jute gunny bags .

12.32. Neem Leaves Against Storage Pests

- The healthy neem leaves of approximately 50 in numbers were collected and kept along with the grains stored in gunny bags (60 kg capacity). This could be kept for nearly 6 months provided good ventilation was available.

12.33. Post Harvest Storage in Cotton (*Ambaram poduthal*)

- Ambaram was an arrangement made by spreading a thin layer of dry sand on the ground and the kapas were kept over it. The dry sand was spread to a height of 10 cm and length and breadth according to farmer's necessity. Above that, jute gunny bags were placed covering the sand and over it the kapas were arranged like a mat in an air tight condition to a height of 6 feet.
- The dry sand and jute gunny bags would absorb the moisture and prevent the kapas from coming into contact with moisture which would otherwise stain the kapas and lower its value. The top of the "Ambaram" was also covered with a gunny bag to avoid dust and direct sunlight. When the kapas were subjected to direct sunlight and dust, its fibre strength and luster would be lost. This practice "Ambaram Poduthal" was an age-old practice followed by the farmers for more than a decade.

12.34. Storage of Pulses with Sand (*Mannkoodaiyil Vaithal*)

- The pulse grains after cleaning and grading were tied in a clear cotton cloth. Bamboo baskets were taken and quarter of its capacity was filled with dry river sand (eg: basket of 10 kg capacity means 2½ kg of sand) and were placed inside the bamboo basket. Above cotton cloth, another quarter of river sand (2 ½ kg) was poured as a covering.

12.35. Post Harvest Storage in Pulses (*Ennai katti vaithal*)

- About 50g of oil was enough for treating 1 kg of pulse grains. Oil treatment also increased the keeping quality of pulses and prevented the damage caused by storage pest up to 80%.
- The oil treatment was more effective when the grains were exposed to sun once in a 3-4 months interval.

12.36. Storage of Groundnut oil with Tamarind

- For storing 5 litres of groundnut oil, about ¼ kg of tamarind was placed inside the oil container. The mouth of the container/vessel was then tightly closed with cotton cloth.

12.37. Storage of Groundnut oil Using Coriander Seeds and Salt

- For a litre of oil, farmers placed 100g of coriander seeds and a spoon of salt inside the oil stored tin container/earthen pot. The oil was exposed to sun for few hours and kept closed in airtight condition.

12.38. Storage of Oil with Tamarind and Salt

- In this practice to avoid the spoilage of oil about 2g of salt and 10g of tamarind was added per litre of any edible oil.

12.39. Storage of Gingelly Oil with Palm Jaggery

- For oil tins of 15 litres capacity about ½ kg of palm jaggery pieces was used. Oil could be stored by this method even up to 1½ years.

12.40. Storage of Gingelly Oil by Dipping Hot Iron Bar

- In this practice, a long iron rod of 8 cm width and 2 feet length was exposed in fire (using an earthen stove) for 30 minutes.
- When the iron rod turned to reddish tinge, it was dipped in the stored oil for just 5 minutes. Then the mouth of the oil container was closed with the help of cotton cloth tightly.

12.41. Oil Storage by Heating

- Heat the oil containing mud pots with the help of local earthen stove for 15 minutes.
- Warm heating of oil help in removing the fresh odour of the oil and also preserved it without spoilage.
- Some farmers had also kept this oil containing vessel under direct sunrays for 2 days. After, the mouth of the container was closed with a cotton cloth to avoid dusts. It can be stored oil by this way even upto1 year.

Table 30: Traditional post-harvest pest management methods of the ethnic tribes

S.No	Plant used	Procedure
1.	*Artemisia vulgaris* (L).	For repelling insects and Rats, put fresh or dried branches/ leaves in and around granaries
2.	*Azadirachta indica* *Elsholtzia blanda* *Cannabis sativa* (L). *Cymbopogon sp* *Melia composita*	Crush and put fresh or dried leaves in and around granaries to repel insects
3.	*Cyathula tomentosa*	Branches with inflorescences are put in and around gr anaries to repel rats
4.	*Dendrocnide sinuata*	Place fresh leaves in between two layers of boiled rice; after 24 hrs, put treated rice in possible rat runways as rodenticide
5.	*Entada pursaetha*	Mix equal proportion of ground seed kernel and ground rice and put in possible rat runways as rodenticide
6.	*Scorulla parasitica*	Prepare gum from the fruits, fix in sticks and put in possible rat runways for trapping

Table 31: Traditional post-harvest pest management methods

Method	Effect	Remarks
Fresh or dried leaves of *Annona spp.* added to the commodity in layers	Strong repulsive and insecticide effect during 3 to 4 months on bruchids and sorghum and millet pests.	Very widespread in Africa
Mint *(Mentha spp.)* leaves. added to the grain / seed at 0.5 to 2 weight %	Insecticide effect supposed acts on pests of cereals.	Quick effect on *Sitophilus oryzae,* which is a pest rather difficult to control
Crushed *Lantana* parts -sandwich technique or as a top layer.	Repulsive effect on bruchids of grain legumes acting up to 6 months.	Extremely widespread in Africa and thus readily available.
Dry or powdered Neem or *Melia* leaves mixed to the grain /seedor applied in layers.	Insecticide and repulsive inhibition of development. Acts mainly on stored product beetles up to one year.	Well-known multiple use plant originating from India. Seed powder oil or extracts have a better effect.
Ocimum canum (hoary basil) leaves entire or as powders applied in sandwich technique.	Insecticide effect on beetles	Very good immediate effect. but insufficient persistence for long-term storage
Entire or powdered fruits of Red pepper (*Capsicum spp.)*	Insecticide and repulsive effect	Attention: irritation eyes possible during application!
Entire or powdered fruits of Black pepper (*Piper spp.)*	Comparable to Red pepper effect lasts for 3 months.	irritation
Neem kernel powder added at a rate of 0.5 to 4 vol. %.	Strong repulsive and insecticide effect, but stronger.	Neem kernels have the highest content in active ingredients.
Annona grain powder added at a rate of 0.5 to 2 weight %	Strong repulsive and insecticide effect	Attention: the powder has an irritant effect on the eyes!
Powder of dried rhizomes of Acorus calamus (added at a rate of 0.2 to 1 weight %).	Insecticide, repulsive effect and inhibition of development against many pests for more than 6 months.	The powder can be stored for 2 months without any loss of effect. There are some doubts concerning adverse effects on humans in high doses.

13

Other Forms of Organic Farming

13 1. Rishi Krishi

Drawn from Vedas, the Rishi Krishi method of natural farming has been mastered by farmers of Maharashtra and Madhya Pradesh. In this method, all on-farm sources of nutrients including composts, cattle dung manure, green leaf manure and crop biomass for mulching are exploited to their best potential with continuous soil enrichment through the use of Rishi Krishi formulation known as "*Amritpani*" and virgin soil. 15 kg of virgin rhizosperic soil collected from beneath of Banyan tree (*Ficus bengalensis*) is spread over one acre and the soil is enriched with 200 lit Amritpani.

It is prepared by mixing 250 g ghee into 10 kg of cow dung followed by 500 g honey and diluted with 200 lit of water. This formulation is utilized for seed treatment (*beej sanskar*), enrichment of soil (*bhumi sanskar*) and foliar spray on plants (*padap sanskar*). For soil treatment it needs to be applied through irrigation water as fertigation. The system has been demonstrated on a wide range of crops i.e. fruits, vegetables, cereals, pulses, oilseeds, sugarcane and cotton.

13.2. Panchagavya Krishi

Panchagavya is a special bioenhancer prepared from five products obtained from cow; dung, uine, milk, curd and ghee. Dr Natrajan, a Medical practitioner and scientists from Tamil Nadu Agricultural University, has further refined the formulation suiting to the requirement of various horticultural and agricultural crops. Ingredients and methods of preparation of Panchagavya and enriched Panchagavya (Dasagavya) have already been described in Chapter 3.

The cost of production of Pancahgavya is about RS. 25-35 per lit. Panchagavya contains many useful microorganisms such as fungi, bacteria, actinomycetes and various micronutrients. The formulation act as tonic to enrich the soil, induce plant vigour with quality production. Strength of various microorganisms detected in panchagavya are as follows:

i. Total fungi 38,800/ml
ii. Total bacteria 1,880,000/ml
iii. Lactobacillus 2,260,000/ml
iv. Total anaerobes 10,000/ml
v. Acid formers 360/ml
vi. Methanogens 250/ml

Physico-chemical studies have revealed that panchagavya possess almost all macro and micronutrients and growth hormones (IAA, GA) required for plant growth. Predominance of fermentative microorganisms such as yeasts and Lactobacillus helps improve the soil biological activity and promote the growth of other microorganisms.

13.2.1. Dose

- For foliar spray 3-4% panchagavya solution is quite effective. Four to five sprays ensure optimum growth and productivity: (a) two sprays before flowering at 15 days interval, (b) two sprays during flowering and pod setting at 10 days interval and (c) one spray during fruit/pod maturation.
- Application of panchagavya has been found to be very effective in many horticultural crops such as mango, guava, acid lime, banana, spice turmeric, flower-jasmine, medicinal plants like Coleus, Ashwagandha, vegetable like cucumber, spinach, okra, radish and grain crops such as maize, green gram and sunflower.
- Panchagavya has also been found to be reducing nematode problem in terms of gall index and soil nematode population. As due to application of panchagavya a thin oily film is formed on the leaves and stem, it reduces evaporation losses and ensures better utilization of applied water.

13.3. Natueco Farming

The Natueco farming system follows the principles of eco-system networking of nature. It is beyond the broader concepts of organic or natural farming in both philosophy and practice. It offers an alternative to the commercial and heavily chemical techniques of modern farming. Instead, the emphasis is on the simple harvest of sunlight through the critical application of scientific examination, experiments and methods that are rooted in the neighborhood resources. It depends on developing a thorough understanding of plant physiology, geometry of growth, fertility, and biochemistry.

Natueco Farming Step by Step

Natueco Farming emphasizes 'Neighborhood Resource Enrichment' by 'Additive Regeneration' rather than through dependence on external, commercial inputs. The three relevant aspects of Natueco Farming are:

a) **Soil** - Enrichment of soil by recycling of the biomass by establishing a proper energy chain.

b) **Roots** - Development and maintenance of white feeder root zones for efficient absorption of nutrients.

c) **Canopy** - Harvesting the sun through proper canopy management for efficient photosynthesis.

13.4. Homa Farming

Homa farming has its origin from Vedas and is based on the principle that "you heal the atmosphere and the healed atmosphere will heal you" The practitioners and propagators of homa farming call it a "revealed science". It is an entirely spiritual practice that dates from the Vedic period. The basic aspect of homa farming is the chanting of Sanskrit mantras (Agnihotra puja) at specific times in the day before a holy fire. The timing is extremely important. While there is no specific agricultural practice associated with homa farming, the farm and household it is practiced in, is energised and "awakened". The ash that results from the puja is used to energise composts, plants, animals, etc. Homa Organic Farming is holistic healing for agriculture and can be used in conjunction with any good organic farming system. It is obviously extremely inexpensive and simple to undertake but requires discipline and regularity.

Agnihotra is the basic Homa fire technique, based on the bio-rhythm of sunrise and sunset, and can be found in the ancient sciences of the Vedas. Agnihotra has been simplified and adapted to modern times, so anybody can perform it. During Agnihotra, dried cow dung, ghee (clarified butter) and brown rice are burned in an inverted, pyramid-shaped copper vessel, along with which a special mantra (word-tone combination) is sung. It is widely believed that through burning organic substances in a pyramid-formed copper vessel, valuable purifying and harmonizing energies arise. These are directed into the atmosphere and are also contained in the remaining ash. This highly energized ash can successfully be used as organic fertilizer in organic farming.

Besides the practice of Agnihotra and the frugal distribution of Homa ash on beds and fields, a variety of further applications have also been recommended. Here are some examples:

Impregnation of Seeds and Bulbs

Before planting/sowing, seeds and bulbs are treated i.e., impregnated with a mixture of Agnihotra ash and cow urine. It is recommended to prepare a mixture of cow urine and water in a ratio of 50:50, to which up to 4 tablespoons of Agnihotra ash per 5 liters of solution are added and stirred. Seeds and bulbs should soak in this solution for 30-40 minutes. This strengthens the germinating plant and makes it more resistant to pests.

Like cow dung, cow urine has antibacterial effects and provides a protective coating around the seeds and bulbs. After this time of treatment, seeds are spread on filter paper, or other absorbent paper, to dry. They should be dry enough to spread, but moist enough so that the core of the seed doesn't dry out. Through the impregnation, germination is started-which would be ended if the seeds completely dried out. Bulbs may be planted immediately after being treated with the solution.

Fertilizers

In addition, plants can be fertilized with a mixture of Agnihotra ash, stinging nettles, and water. This special liquid fertilizer strengthens plants. The stinging nettles are fermented i.e. decomposed in the water for 7-14 days, depending on weather conditions and the amount of nettles needed. This mixture should then be diluted to a solution with a ratio of 1:9. In other words, 1 part stinging nettle solution is mixed with 9 parts water and filtered with a fine screen (sieve) into a spraying container or watering can.

Plant Nutrient Solution

To make an Agnihotra plant nutrient solution, up to 4 tablespoons of Agnihotra ash and up to 4 tablespoons of pulverized, dried cow dung are stirred in approximately 5 liters of water and then applied to plants. This may be repeated every 14 days, depending on how much it is needed.

Spray Solution

A nutrient solution to be sprayed can be made by mixing up to 4 tablespoons of Agnihotra ash with 5 liters of water. This spray solution is left standing for 3 days and then filtered through a fine screen before it is used to protect plants against pests and diseases. A spray solution can also be made from certain fern blossoms, in which approximately 150 grams of the blossoms, mixed with 2 liters of water and 2 tablespoons of Agnihotra ash, are left standing to ferment for 7-10 days. Filtered through a narrow-meshed screen and then finally distributed on the plants with a sprayer, this helps keep away pests such as snails.

13.5. Biointensive Farming

The biointesive method is an organic agricultural system which focuses on maximum yields from the minimum area of land, while simulataneously improving the soil. In order to achieve the greater productivity, the biointesive method uses are following.

13.5.1. Companion planting

Companion planting refers to the practice of planting more than one crop in the row or planting a different crop between two rows of a like crop. It is theorized that or aroma of certain plants are inhibitory to some diseases or insects. Companion cropping can effectively use garden space and is particularly useful with small gardens. If companion cropping is used, select companions with care to avoid shading or excessive, competition for water and nutrients.

Table 32: Different companion crops for planting

Vegetable	Companion crop	Antagonist
Beans	Potato, carrot, cucumber,cauliflower, cabbage, sum	Onion, garlic, gladiolus
Bush beans	Potato, cucumber, corn,strawberry, celery, summer	Onion
BeetCrucifers	Onion, Kohl rabiAromatic plants, potato,celery, sesame, chamomile,sage, peppermint, rosemary,	Pole beansStrawberry, tomato
Carrot	Peas, leaf lettuce, chive,onion, leek, rosemary, sage	sSesame
Celery	Leek, tomato, bush beans	
Chive	Carrot	Peas, beans
Com	Potato, peas, beans, cucumber	
Cucumber	Beans, com, peas, radish	Potato, aromatic plants
Onion and Garlic	Beet, strawberry, tomato,lettuce, summersavory,	Beans, peas
Peas	Carrot, turnip, radish, cucumber	Onion, garlic, gladiolus
Pole beans	Com, summer savory, sun	Onion, beet, kohl rabi
Potato	Beans, com, cabbage, horse	Pumpkin, cucumber, sunflower
Strawberry	Bush beans, spinach, forage	Cabbage
Sunflower	Cucumber	Potato
Tomato	Onion, parsley, asparagus, marigold	potato, fennel, cabbage

13.5.2. Double dug raised beds

- 12 inch deep trecn is dug across the width of the bed with a flat spade, and the soil from that first trench is set aside.
- The 12 inches below the trench are loosened with a spading fork.
- When the next trench is dug, that soil is dropped into the empty space of the firt trench, and the lower layer is again loosened with a spading fork.

- This process is repeated along the full length of the bed.
- The final trench is filled with the soil that was removed from the first trench. The result is a bed that has been tilled to a depth of 24 inches.

13.5.3. Intensive planting

- In order to plant intensively, beds are 4 to 6 feet wide, and at least 5 feet long, forming a bed of 100 square feet.
- Planted in a hexagonal or triangular pattern in the bed so that no space is left unnecessarily unused.

14

Water and Rainfall Prediction

14.1. Water Source Identification

- In places where *Ficus religiosa* tree is found, underground water will be more.
- Underground water will be more in places where the neem tree with more knots is found.
- In places where termite hills are found, underground water will be more.
- Farmers are practicing traditional method of using neem stick, metals etc. for finding the water resource in underground for digging wells. While they are walking with this stick on hand it will be rottate automatically at the place where underground water is surplus.
- Some farmers are using the magnetic stones for identifying water sources. Magnet is tied with a thread and held like a pendulum while walking over the fields. The magnet will rotate automatically at the place where underground water is more and this is taken as the indication for digging wells.

14.2. Predicting Rainfall

- The rain birds lays egg at the ground level then, there will be less rain however; if the eggs are laid at higher elevation then it is the indication of more rains. The local people assume that eggs of rain birds are laid on such a height that in case of more or less rains, the egg will not be submerged in rain water. Similarly, if the narrow ends of all the four eggs of rain bird are downwards, then if is the indication of good rain fall through out the season.
- When the adventitious roots of the banyan tree [*Ficus bengalensis*] start sprouting [tillering] and then the local people assume that the rains will appear with in 2-4 days.

- In castor (*Ricinous* spp.) and ber (*Ziziphus nummularia*) when the buds start sprouting, then it is predicted that rains will appear within 10-15 days.
- The rains will appear after 10-15 days of flowering in babul tree (*Acacia nilotica*).
- As soon ass the neem kernels ripen and start falling, it is expected that there will be rains after 10-15 days.
- Rain may come if damsel fly flies at ground surface, frogs make noise and ants move in line from one place to another.
- The farmer also forecasting rains by observing the direction of wind/ clouds. According to them Westerly winds/clouds meant good rainfall. Similarly Northwesterly clouds will bring hailstorm.
- Thunder in summer and lightening in rainy season bring heavy rain.
- If rain occurs in the last week of 'Aavani' (Aug.-Sep.) month, crop yield will be more in the succeeding month.
- Rainfall occurring during and after the Tamil month 'Marghazhi' (Dec.-Jan.) does not benefit the crop and soil but affects the grains.
- If a rainbow appears in the east in the evening or west in the morning, it will rain.
- Streaks of lightening in the east with rainbow in the west would lead to rain.
- If the sky becomes dark near the horizon, there will be instant rain.
- If the North wind blows, it will surely rain.
- Instant rain follows the winds blowing from many directions.
- Ants raising mounds is an indication of rains.
- Rain can be forecasted by looking at the sight of burrow of ants.
- If ants move to higher ground with eggs, it will rain.
- If white ants take wing in the evening, it will rain excessively.
- If termites fly in the evening time, there will be rainfall.
- Big sized white ant hills (termite mounds) existence in an area, also indicates good spring of underground source of water. Farmers tell that "the depth of the hill of termite extends up to the ground water level".
- Dragonflies flying low indicate the rainfall.
- Flight of butterflies from North to South and death of more insects indicates the arrival of rains.

- Rain follows the frogs crocking. Flocking (crowding together) or fast grazing of sheep's indicates occurrence of rainfall.
- It will rain if the cattle look at the sky frequently.
- If the sky acquires a faint yellow colour, there is less hope of rain.
- If crow-coloured clouds are observed throughout the day while the night sky remains clear, a drought is indicated.
- If the velocity of wind is not high during the Mrighirsh constellation and high. Heat is not experienced during the *Rohini* constellation; a drought can be expected to follow.
- Movement of clouds in one direction is a sign of rainy weather while the opposite direction is meant dry weather.
- A red sky at sunrise and sunset considered a warning for rainless days ahead.
- Cold nights with mist and dew considered signs of impending dry weather while hot and warm feelings during the day signaled rainy evenings.
- Thunder in summer and lightening inrainy season brings heavy rain.
- Drizzling occurring in 'Aani' (June-July), assures rain in 'Avani' (Aug-Sep).
- Rainfall during 'Marghali' (Dec-Jan) does not benefit the crop but affects the grains

Table 33: Flowering and foliage of tree species as indicators of rain

Name of Species	Indicator	Expected outcome
Mahuda, *Madhuca latifolia*	Good foliage	Good monsoon
Bamboo species	Good foliage	Drought, rat attack
Ber, *Zyzyphus mauritiana*	Heavy flush of fruit	Average monsoon
Darbha grass	Appearance of good foliage	Good monsoon
Bilva, *Aegle marmelos*	Good foliage	Subnormal monsoon
Pipal, *Ficus religiosa*	Good foliage	Adequate rain
Khejro, *Prosopis cinerea*	Heavy foliage	Drought
Kothi, *Limonia acidissima*	Good growth	Stormy rain
Neem, *Azadirachta indica*	Heavy flush	Drought

Table 34: Behaviour of birds and animals as predictors of rain

Indicator	Expected outcome
Sparrow bathing in dust	Good rain
Kachinda (chameleon) climbs the tree and assumesblack-white-red colours	Immediate rain
Frogs start singing in the initial days of *Jayestha* (May)	Early rain
Batairs (a bird) sing in pairs	Certainty of rain
Peacocks cry frequently	Rain within a day or two
Crows cry during the night and foxes during the day	Severe drought
Titodi or lapwing bird lays eggs during the night,especially on river banks	Heavy rains
Klheu/Bapaiya (a bird) sings songs early in the morning	Rains within a day or two
Snake climbs up on trees	Drought
Camel keeps facing the north-east direction, goat does not browse, crow scratches its nest	Immediate rains
Birds take bath in the dust on the full moon day of*Jayestha* (May)	Plenty of rain

15

Integrated Farming System — A Pathway to Organic Farming

15.1. IFS Components Tested for Wetlands

Area Allocation;

In 1 ha of land

75 % - Cropping

10 % - Fodder cultivation

3 % - Goat shed

12 % - Fish pond

Possible Integrations:

Crop + Poultry (20 pairs) / Pigeon (40 pairs) / Goat (Telicherry 20+1) + Fish (400 fingerlings)

Crop + Fish + Poultry (20 pairs)

Crop + Fish (A.F.) + Pigeon (40 pairs)

Crop + Goat (Telicherry 20+1) + Fish

Rice + Fish + Azolla

Rice + Fish + Azolla + Vegetables

Benefits: More productivity, employment and higher net returns

Fig. 28: Integrated farming system for wetland

Table 35: Nutrient value of recycled manures

Particulars	Poultry	Pigeon	Goat
Birds /animals used to satisfy the feed requirement of 400 fingerlings	20 layers	40 pairs	3 animals
Quantum of dropping received in an year	700 kg	700 kg	810 kg
Silt cleared after one year from 0.04 ha pond	4.5 t	4.5 t	4.5 t

Nutrient	Raw poultry dropping		Pond manure		Additional nutrient gained (kg)
	%	kg / 700kg	%	kg/ 4500 kg	
N	3.22	22.5	1.96	88.2	65.7
P_2O_5	2.50	17.5	1.02	45.9	28.4
K_2O	1.05	7.4	0.72	32.4	25.0

Nutrient	Raw Pigeon dropping		Pond manure		Additional nutrient gained (kg)
	%	kg / 700kg	%	kg/ 4500 kg	
N	1.82	12.7	0.84	37.8	25.1
P_2O_5	0.56	3.9	0.30	13.5	9.6
K_2O	0.98	6.9	0.56	25.2	18.3

Nutrient	Raw Goat dropping		Pond manure		Additional nutrient gained (kg)
	%	kg / 810kg	%	kg/ 4500 kg	
N	1.40	11.3	0.70	31.5	20.2
P_2O_5	0.85	6.9	0.62	27.9	21.0
K_2O	0.70	5.7	0.48	21.6	15.9

Table 36: Economic analysis of integrated farming system

Farming Systems	Production Cost (Rs./ha)	Gross Returns (Rs.)	Net Returns (Rs.)	B:C Ratio	Per day Return (Rs.)
Cropping alone	27822	64975	37153	2.43	178
Cropping + Fish + Poultry	4833	146035	97731	3.02	400
Cropping + Fish + Pigeon	47090	145868	98778	3.06	400
Cropping + Fish + Goat	55549	186667	131118	3.36	511

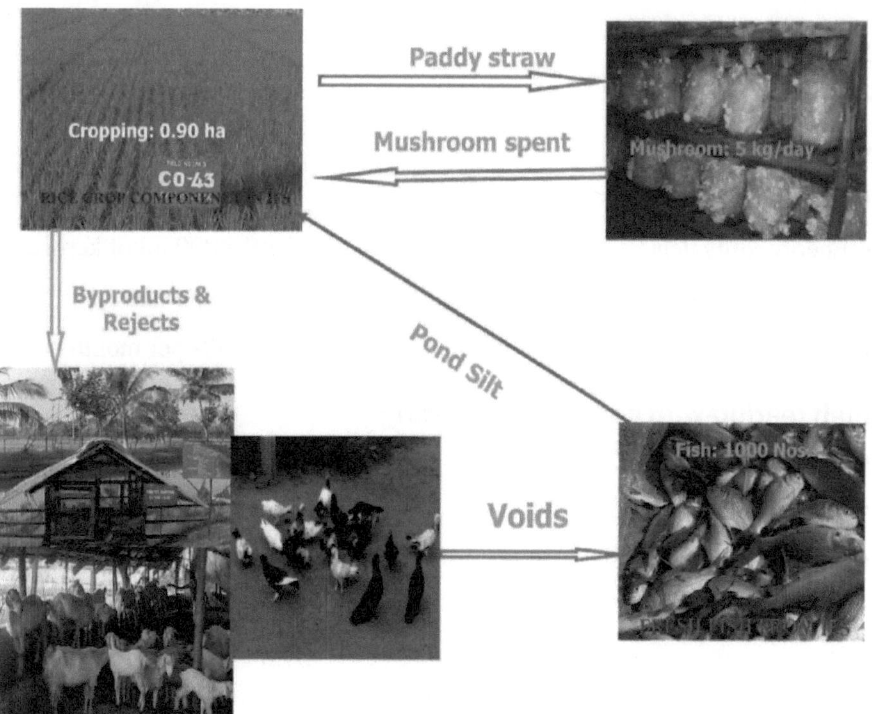

Fig. 29: Component integration and resource recycling in lowlands

15.2. Suitable Components for Integration in Irrigated Dry (Garden lands)

Crop + dairy + Biogas + Trees

Dairy

- Breed : Jersey crossbreed
- No. of cows : 3 (2+1)
- No. of calves : 2
- Average milk yield anticipated : 5475 lt./year
- Green fodder requirement : 25 to 30 kg day^{-1} $animal^{-1}$
- Production cost per liter of milk through IFS : Rs. 5/-
- Procurement price of milk : Rs.8/-
- Net income from 3 cows : Rs. 18,000 /- per year

Biogas

- Size of family : 4
- Required capacity : 2 m^3 day
- Quantum of dung required : 60 kg per day
- Energy equivalent : 1 m^3 = 800 ml of kerosene
- Quantum of slurry ejected : 56.5 kg per day
- Value of gas : Rs. 300/- per month

Fish rearing with cattleshed washing

- Pond area : 150 m^2
- No. of fingerlings : 150
- No. of cows : 3 (2+1)
- No. of calves : 2
- Quantum of washing required : 1501 /day
- Fish yield : 150 kg
- Net income : Rs.4000/-

Crop + Goat (Telicherry 10+1) + Horticultural crops

Table 37: Nutrient gain from 20 + 1 productive Tellichery goats - Deep litter system

No. of kidding, year^{-1}	1.5		
No. of kids, kidding^{-1}	1.5		
No. of kids year^{-1}, adult^{-1}	2.25		
No. of kids from 20 adults year^{-1}	45		
Mean wt. of dropping recorded day^{-1}(Excludes the pellets dropped during open grazing)	Adult900 g	Kid110 g	
Dropping received in an year kg	6800 + 1800 = 8600		
Wt. of coir waste used for the stall/year^{-1} (50 kg week^{-1})	2600 kg		
Deep litter waste obtained year^{-1} kg	8600 + 2600 = 11200		
Analytical value of deep litter waste obtained from the unit	N 1.79	P 0.95	K 0.82
Total nutrient available from the goat unit, year^{-1} kg	200.5	106.4	91.8

Horticultural crops: Amla, guava, sapota etc., (0.10 ha)

15.3. IFS for Rainfed lands

Crop+silvipasture + goat rearing

The organic manures like litter from the goat unit can readily be used for soil application, and thus will help in enriching the soil. Goat droppings are found to be a good source, which can also be linked with biogas unit before it is utilized as manure. This will generate good volume of gas (22 kg of goat dropping will generate one cubic meter of gas against 30 kg of cattle dung as well as enhance nutrient availability. This quantum is sufficient to enrich the soil fertility under rainfed situation as farmers hardly apply inorganic fertilizers. Thus, through recycling of organic in the farming systems approach, the potential of each produce can be exploited to a greater extent.

Deep litter waste obtained per year was 11200 kg along with bedding material. The nutrient content of deep litter waste was 1.79, 0.95 and 0.82 % of NPK, which supplied 200.5, 106.4 and 91.8 kg of NPK.

16

Organic Paddy Farming

Paddy is cultivated about 45 m ha in India, less than 1% is under organic cultivation.

16.1. Varieties

Good quality fine varieties like Ponni, Super ponni, insect and disease resistant varities must be selected for getting higher yield and returns.

16.2. Seeds

Organically produced seeds only must be used.

16.3. Seed Treatment

For 1 kg of seeds 20 g *Azospirillum*, *Pseudomonas* and 10 g *Phospobacteria* must be used with rice gruel.

Panchagavya spray @ 1% for treating the seeds.

16.3.1. Soaking the seeds in water

Tie the seeds in a small gunny bag or cloth bag and soak it in water for 12 hours. Later, remove the bag from the water and cover it with a moist gunny bag. The following day, soak the seeds in water for eight hours again. Later, remove the seeds from the water and sow them in the nursery. This method helps to improve the germination capacity of the seeds.

16.3.2. Using cow dung solution

Treating paddy seeds in a cow dung solution enhances their germination. Take 0.5 kg of fresh cow dung and two litres of cow urine and dilute them with five litres of water. Soak 10–15 kg seeds first in water for 10–12 hours and then in the cow dung solution for 5–6 hours. Dry the seeds in the shade before sowing them in the nursery.

16.3.3. Using goat dung solution

Treating 30-day old seeds for one day in a goat dung solution increases their germination.

Using cow's urine solution Dilute 500 ml of cow's urine in 2.5 litres of water. Tie the seeds in small bags and soak them in the urine solution for half an hour. Dry the seeds in the shade before sowing them.

16.3.4. Using sweet flag extract

Dissolve 1.25 kg of sweet flag rhizome powder in six litres of water. Tie the seeds in small bags and soak them in the extract for half an hour. Dry the seeds in the shade before sowing. (This is the quantity required for treating seeds to be sown in one hectare.)

16.3.5. Using *Salvadora persica*

Spread the leaves of *Salvadora persica* at the bottom of a closely knit bamboo basket, then fill it with seed and pour about 10 to 12 litres of water over the basket. Cover the basket with the Salvadora leaves and place a weight over it. Leave the seeds undisturbed for 24 hours. The seeds are then ready to be used for sowing in the nursery. This procedure helps in early and vigorous germination.

16.4. Nursery

For 20 cents of nursery area

Well decomposed farmyard manure	–	1 ton
Basal – Vermicompost	-	200 kg
Groundnut cake	-	10 kg
Neem oil cake	-	10 kg

16.5. Transplanting

- At the time of last ploughing, 12.5 tonnes of farmyard manure must be applied nad incorporated.
- Before transplanting of paddy, green manures must like kolinchi, sunhemp; sesbania should be grown upto 45 days and incorporated into the field.
- If green manure is not possible to grown, apply 5 tonnes of vermicompost.
- Rockphospate and bonemeal can be used for enrichment of phosphorus nutrient.

16.5.1. Seedling treatment

- The paddy seedlings can be treated with ash and neem seed mixture before transplanting. For this, the seedling bundles are kept in small plots of standing water mixed with ash and pulverized neem seeds from 30 minutes to an hour.
- One kilo of ash and 500 gm of neem seed are sufficient for treating 50 bundles of seedlings. The treated seedlings produce a crop free from pests and disease.
- Soak groundnut cake and neem cake in water overnight and filter. Treat the seedlings in this solution before transplantation. The treated seedlings are less vulnerable to pest attack.
- The paddy seedlings can also be dipped in a solution of amrut pani/ panchagavya/jeevamrut.

16.6. Biofertilizers

- Blue green algae @ 10 kg should be applied 10 days after transplanting
- Azolla @ 250 kg/ha should be applied 3-5 days after transplanting and well grown azolla must be incorporated.
- 10 pockets of *Azospirillum* and 10 pockets of *Phospobacteria* must be mixed with 25 kg of fertile soil and 50 kg of farmyard manure and apply this into the field
- 2.5 kg *Pseudomonas* must be mixed with 25 kg of fertile soil and 50 kg of farmyard manure and apply this into the field before transplanting

16.7. Weed Management

- Summer ploughing and incorporation of soil.
- Roatary weeder and cono weeder 10 days once @ 15 DAT.
- Intercropping with green manures and azolla.
- Bunds should be cleared off weeds.
- Best suited method of irrigation and drainage facilities.
- Not much application of Nitrogen rich nutrients.
- Keep light traps and pheromone traps.
- Use of calotropis (*Calotropis gigantea*) as green manure checks the growth of the weed *Marsilea quadrifolia*. The fibrous pericarp of coconut applied @ 25 baskets/ha also controls this weed to some extent. It releases a tannin-like substance that inhibits the growth of the weed.

- While preparing the land, apply leaves and small twigs of *Strychnos nux-vomica* (poison nut) and incorporate them into the soil. This helps to suppress the weeds.

Table 38: Remedial measures for various pests and diseases

Pests and diseases	Management techniques
Control of in breeding of insects	Before transplanting of rice, both sides of the bunds are cleared with hoe and pasted with mud to check the inbreeding of insect pests.
Stem borer Leaf folders	• Neem seed kernel extract @ 5 % spray • *Trichoderma japonicum* cards @ 5 cc/ha • *Bacillus thurinjiansis* spray @ 1 kg/ha • Erection of bamboo branches or top, or other stick in the rice field • Removal of grassy weeds from bunds around the paddy field • A type of citrus fruit (*Citrus grandis (L)*) cut into pices and applied in paddy fields.
Leaf hoppers	• Correct application of nitrogen fertilizers • Alternate wetting and drying • Neem seed kernel extract @ 5 % spray or neem seed oil @ 1%
Rice brown leaf spot	• Seed treatment with 20% mint leaf extract for 24 hours. • Spread the mature leaves of *Cleistanthus collinus* all over the field (25 quintals/hectare) and allow them to decay. Irrigate after three days. • Dusting of ash checks the spread of this disease.
Bacterial leaf blight	• Slurry is prepared by mixing 20 kg of cow dung with 200 litres of water. The mixture is strained through a gunny bag. The filtrate is further diluted with 50 litres of water and allowed to stand. The water is then decanted, strained and sprayed.
Rice thrips	• Raw cow dung is mixed with water to prepare a suspended solution, which is sprayed in the rice field.
Rice case worm	• Rope, dipped in kerosene stirred in the water of rice field. • Spreading of raw cow dung in the field water • Siam weed (*Chromolaena odorata)* crush the branches of this herb and throw uniformily in fields when there is water·
General management of diseases	• Take cow's urine in a mud pot and allow it to ferment for one week. Spraying this over the crops controls bacterial and fungal diseases.· Mix one litre of cow's urine with one litre of buttermilk and eight litres of water. Spraying this extract over the crop also controls bacterial and fungal disease. • Mix 300 ml of sweet flag extract with one litre of cow's urine and 8.7 litres of water. Spraying this extract controls the spread of disease.
Rice blast	• Remove collateral weed hosts from bunds and channels • Use only disease free seedlings • Avoid excess nitrogen • Apply N in three split doses (50% basal, 25% in tillering phase and 25% N in panicle initiation stage)

Contd.

Pests and diseases	Management techniques
	• Use resistant variety CO 47. • Seed Treatment with TNAU Pf 1liquid formulation @ 10 ml/kg of seeds • Seedling root dipping with TNAU Pf 1liquid formulation (500 ml for one hectare seedlings) • Soil application with TNAU Pf 1liquid formulation (500ml/ha) • Foliar spray with TNAU Pf 1liquid formulation @ 5ml/lit
Sheath blight	• Apply Neem cake at 150 kg/ha • Apply gypsum @ 500 kg/ha • Foliar spray with Neem oil at 3% (15 lit /ha) starting from disease appearance
Sheath rot	• Apply Gypsum @ 500 kg/ha at two equal splits once basally and another at active tillering stage. • Neem oil 3% • Ipomoea leaf powder extract (25 kg/ha) • Prosopis leaf powder extract (25 kg/ha). First spray at boot leaf stage and second 15 days later. • Seed Treatment with TNAU Pf 1liquid formulation @ 10 ml/kg of seeds· Seedling root dipping with TNAU Pf 1liquid formulation (500 ml for one hectare seedlings) • Soil application with TNAU Pf 1liquid formulation (500ml/ha)· Foliar spray with TNAU Pf 1liquid formulation @ 5ml/lit
Storage pests	
Rice weevil Rice moth Angoumois grain moth Lesser grain borer	• While filling grain in storage bags, place 200 gm of salt for every 50 kg of grain. This helps to control moth and weevil infestation during storage. • Place red chilies in storage bags (@10–15 fruits per every one quintal bag). The pests stay away due to the pungent odour. • Mix turmeric powder with paddy to protect the grain from weevils. • Mix leaves of Ipomoea carnea along with the grain to prevent common storage pests. • Place about 1–2 garlic bulbs in 5–10 kg of rice in the bin for storage. • *Vitex negundo* leaf powder 4 kg/100 kg seeds • Neem seed kernel powder 2 kg/100 kg seeds • Tobacco leaf powder 2 kg/100 kg seeds • Neem leaf powder – 4 kg/100 kg seeds·Common ash – 2kg/100 kg seeds

16.8. Other ITKS

16.8.1. Against yellowing

- Sprinkle fresh cowdung solution in rice fields (particularly in nursery bed).

16.8.2. Bird perches

- Putting upper portion of bamboo, in paddy fields is a common practice to facilitate the migratory birds to stay on this support so that they can pick up insects and larva eating the growing plants.

16.8.3. Trap crops

- At the time of first transplanting of rice, put a banana sucker, turmeric plant as a trap crops.

Table 39: Traditional tools used in agronomic package

Country plough	Weeder (Aruguvetti)
Dry land weeder (*cycle gundu*)	Spade (Mammutty)
Sickle (Kathir arivaal)	Knife (Kambar kathi)
Pulikokki	Ezumichai karandi
Kodun kol	Salladai
Chekku (stone grinder)	Muram (Bamboo winnower)
Uruttu kal	Thattupalagai
Ulakkai	Pukka (Padi)
Marakkal	Naali
Moonghil thattu (Bamboo pan)	Sakkaipiratti (floor cleaner)

17

Organic Methods for the Management of Phytoparasitic Nematodes

Realizing that nematodes cannot be eliminated, and that we must live with them, the overall goal is to keep the population density as low as possible. The economic threshold is the density of nematodes where the losses incurred exceed the cost of nematode management. It is not advisable to depend on a single method to control nematodes. Efficient management requires judicious and careful integration of several which will result in reduction of nematode populations. Important components of non chemical methods of nematode management program are considered in this chapter.

17.1. Organic Amendments

Manure and fertilizer applications are essential part of growing any crop. Besides supplying required essential nutrients they help in many ways directly or indirectly by suppressing harmful organisms including phytonematodes. The use of organic amendments against nematodes was demonstrated by Linford *et al.,* in as early as 1938. FYM and cattle urine have found to reduce nematode population. Biogas plant waste slurry was found to be effective for nematode management. Compost of water hyacinth, oilcakes, cellulose amendments like sawdust, wheat straw, rice husk, rice hull ash, green manuring with legumes, green leaf incorporation are useful to keep nematode populations low. Non - edible oilcakes like those of neem, karanj, mahua, caster, cotton as well as edible oilcakes groundnut, mustard sesame, coconut have been recommended at 2 - 2.5 t/ha for reduction of root knot and other nematode population in soil. Miscellaneous substances like decaffeinated tea waste, rice hull ash, chicken manure, tannery and food and fruit processing waste, market and domestic organic refuse, dead vegetation and crop stalks, wood shaving, etc. have been recommended for root knot nematode management. Castor leaves (40g/kg) along with NPK at 75:25:37.5 kg/ha achieved reduction of *M. javanica* population on tomato.

Horn meal and bone meal were found to be effective against *M. incognita* and *Tylenchorhynchus brassicae in* a wide variety of vegetables. Forest litter at

6t/ha could enhance predatory nematodes and reduce the mandarin orange nematode, *Tylenchulus semipenetrans* as well as *Helicotylenchus dihystera* associated with its decline in Darjeeling. Application of 50 per cent N as neemcake was found to be very effective for the management of root - knot and reniform nematode in betelvine. Addition of shade dried leaves of calotropis was found to reduce the nematode infestation in betelvine. Pressmud @ 15t/ha or neemcake @ 2t/ha was found to be very effective for the management of banana nematodes.

Various interrelated mechanisms have been suggested to explain the mode of action of organic amendments on plant parasitic nematodes.

- Changes in soil physico-chemical properties not condusive to nematodes
- Release of plant nutrients to accelerate root and overall plant growth helps plants to escape from nematode attack
- Induction of resistance / tolerance in plants
- Nematotoxic substances and other harmful products are released after dissolution of organic amendments in water. Ammonia, other volatile substances and phenols may be released from oilcakes, which accumulate as decomposition end products and microbial metabolites may kill nematodes.
- Increased microbial activity in amended soil increased enzymatic activities.

17.2. Cropping Sequences

Planting crops in special sequences (rotation) is an old but effective and widely used land management practice for reducing nematode populations in the soil. This practice was used by early civilizations in Peru to control the potato cyst nematode, although the natives were not aware of the presence of a specific pest. To be effective, however, crops that are resistant, or preferably, non-hosts for the nematode to be controlled, must be rotated with susceptible crops. Growing a resistant or non-host crop for only one year is helpful, but generally inadequate. Two resistant or non-host crops between susceptible crops may give fair to good control; however, three or four years of resistant or non-host crops between susceptible crops may be necessary for effective control of some nematodes.

Certain cyst and root – knot nematodes attack only a limited number of crops. This situation provides a wider choice of non-host crops available for use in the rotation. For example, the principal host for the soybean cyst nematode, the most important cyst nematode in the United States, is soybean. Planting corn, cotton, peanut, tobacco and many other crops in the rotation rapidly decreases the population of this nematode each year. To illustrate, yield of soybean from plots following two years of corn were three times 1615kg/ha versus 538kg / ha the yield of soybeans from plots planted to soybeans for three continuous years. In the fourth year of this same experiment, plots of soybeans following three years of corn yielded 2018 kg / ha, compared with 269 kg / ha for the continuous soybean plots.

Similarly, the northern root – knot nematode *Meloidogyne hapla,* a serious pest of peanut, soybean and many other crops, is effectively controlled by rotation with corn or cotton, which are non-host crops for this species. Planting cotton, a non-host crop, for one year prior to planting peanuts, resulted in a yield increase of 1060kg/ha and an increase in value of $ 242 over that of peanut following soybeans, a favorable host for *M. hapla.*

Growing banana after rice in wetland system of cultivation was found to be very effective in reducing the infestation of banana and rice nematodes. Growing intercrops *viz.*, marigold and daincha were found to be effective for the management of sugarcane nematodes. The intercrops can be sown at the time of planting cane setts and can be incorporated on 50 days after sowing the intercrops. Similarly growing marigold in banana as intercrop was found to reduce the infestation of lesion nematode *Pratylenchus coffeae* in cv. Nendran. Crop rotation, in spite of its effectiveness, has important limitations. The amount of damage inflicted on a crop depends on the level of resistance in the crops used in the rotation, and on the number of years of planting these crops between susceptible crops. Furthermore, the non-host or resistant crop, which effectively controls a specific nematode, may be a host for another plant – parasitic nematode species existing in the same field. For example, *Pratylenchus* species will build up on corn, a non-host for the soybean cyst nematode, if corn is planted too often in the rotation. Another disadvantage is the low commercial value of some non-host of resistant crops. Moreover, some farmers have inadequate land to rotate crops.

17.3. Resistant Cultivars

The uses of resistant and tolerant cultivars are economically attractive management tactics because most infested fields have a dominant or target nematode. Furthermore, nematodes spread slowly, persist in the soil for long periods of time, and management by physical and chemical means is relatively expensive. Available information indicates that plant resistance to nematodes is genetically controlled by one gene in some cases, and by more than one gene in other plant nematode combinations. Practical results have been obtained in breeding for resistance to a number of major nematodes. A prime example is the development of a soybean cultivar highly resistant to the soybean cyst nematode. Potato cultivars resistant to the potato cyst nematode have been developed and clover and alfalfa cultivars resistant to the stem nematode are available, commercial varieties of cotton, cowpea, grape, lespedeza, lima beans, peach, soybean, sweet potato, tomato and tobacco resistant to the root - knot nematode and cultivars of alfalfa, oats and barley resistant to the stem nematode are also grown extensively.

Availability of resistance is more common in case of endo or semi endoparasitic nematodes. Generally screening for resistance is being done by the scientists and genotype for resistant reaction is being identified. Efforts should be made for breeding nematode resistant varieties, location wise, and nematode wise. Once a resistant plant is developed, then it will be able to maintain its characters of antibiosis against the target nematode pest. However, the success in resistance breeding against nematodes is limited to a few crops only *i.e.* potato (Kufri Swarna) tomato (Hisar Lalit, PNR - 7 etc.) chillies (Pusa Jawala) wheat (Raj MR - 1) barley(C - 164, Rajkiran, RD 2035 and RD 2052 etc) and cowpea (Barsati Mutant). The development of new races and biotypes among different species of plant parasitic nematodes due to selection pressure exerted will have impact on resistant cultivars. Hence advances in biotechnology, tissue culture, protoplast fusion technology etc. can help in development of resistant cultivars in crops against key nematode pests.

17.4. Trap Crops/ Antagonistic Crops

The basic concept of trap crops is when a susceptible crop is planted in a nematode infested field; it (the crop) is ploughed up or destroyed before the egg laying by the female of the nematode. If the uprooting or ploughing up of the crop is delayed and nematode completed its life cycle it can result in higher population. So, it is imperative to know the exact life cycle of the nematode. Commercially this is rarely used as it is expensive and also dangerous to employ it successfully.

A more practical approach is to plant the crops which are highly susceptible to invasion by nematodes to be managed but do not allow further development of juveniles into adults. The crops can be harvested as a fodder or green manure. Many species of crotolaria have been used in other countries population of *Meloidogyne* spp. Other plants *e.g. Tagetes erecta T. patula, Chrysanthamum* and *Ricinus communis* are also very effective in reducing the sedentary endo-parasites. Such crops can be included in the farming schedules.

Root exudates of certain plant like onion, garlic, tagetes etc., contain nematoxic compounds. Such crops known as antagonistic can be used in vegetable cropping system or as intercropping. Mixed or strip cropping in orchards to suppress phytonematodes particularly in the early years of orchard planting.

Intercropping of marigold with mulberry gave a good control of *M. incognita* but yields were not improved over control. However, the reduction in number of galls and egg masses was 73-60 per cent. This decrease was attributed to the toxic root exudates terthienyl by marigold which also acts as a trap crop. Marigold intercropping in wider spacing and mulching them in soil improved the yield and reduced the nematode infestation.

17.5. Fallowing and Soil Solarization

This is a common practice of keeping a land free of any vegetation, including weds, for a specified period by occasional ploughing during hot and dry weather. By this method nematodes will not have any plants (host) to feed and reproduce. But under the present day scenario of multi cropping system wherein land holdings are getting smaller and smaller, it is not affordable proposition to keep a land without any crop for a complete season. Alternatively the fields after harvesting of Rabi season can be ploughed up and left during hot summer season when the temperature is high. This practice exposes eggs, and other stages to desiccation. Under field conditions the root knot nematodes get killed by drying of soil for two days in 2.5cm depth, 4 days in 5cm depth and 32 days in 15 cm depth. The nematodes can also be killed by giving frequent turnings of soil. After 133 days of last summer ploughing there was 2.7, 5.6 and 11.2% increase in yield over check with 1, 2 and 3 summer ploughings respectively at 10 days interval. One to five deep summer ploughings during hot summer months can cause reduction *H. avenae* population between 9.3 and 42.4% with a corresponding yield increase of 4.4 to 97.5%. Soil solarization during the month of May was found to be very effective for the management of nematodes in nursery beds of tomato and brinjal in order to obtain nematode free healthy seedlings.

17.6. Green Manuring

This is one of the conventional practices whereby green plants are ploughed into the soil to decompose and provide nutrients for the succeeding crop. The organic matter thus applied during its decomposition changes the physical, chemical and biological properties of the soil. The practice also changes the status of microflora and microfauna in the soil qualitatively and quantitatively. Initially the phytonematode population may increase due to more number of weeds in the field owing to higher moisture level during growing of crop for green manuring. However, at alter stage nematodes may be killed due to biological/biochemical actions of decomposed matter. Rice root nematode population increased during the months of July and August where FYM and *Sesbania aculeate* were grown and incorporated as green manure. But, in the same plots population declined during September and October due to decomposition of green manure and FYM and release of chemicals. However, endoparasites which have penetrated may not be much affected by decomposing matter, rather the individual that are present as free living forms are much affected due to action of decomposing matter.

Incorporation of green manuring reduced phytonematode population due to increase in microbial population and the nematophagous fungi which reduced nematode population in soil. Addition of green manures also adds nitrogen in the soil upto 56.3, 77.1, 84 and 63.2 kg/ha for cowpea, sesbania, sunnhemp and clusterbean respectively. The accumulation of nitrates and ammonia in high concentration is highly injurious to several nematodes. Accumulation of CO_2 in high amounts within first week of decomposition of green manuring which is injurious to nematodes.

17.7. Selection of Healthy Propagating Material

With plants that are propagated vegetatively it is often possible to eliminate nematodes selecting uninfected plants or parts of plants for propagation e.g. by rooting stem cuttings of shrubs in nematode free soil infestation of *Meloidogyne* spp. and other root parasites can be avoided. Similarly, *Meloidogyne* spp, can be eliminated by cutting off existing roots or preferably by taking aerial division and rootings in nematode free soil. Conversely, nematode attacking aerial parts of the plants can sometimes be avoided by taking root cuttings. In this way *D.dipsaci* and *Aphelenchoides* sp. can be eliminated.

Nurseries are the main source of nematode spread. If the source is prevented then there will be little problems in the main fields after transplantation. It is, therefore, advisable to kill the nematode present in the planting material or select only healthy planting materials (seedlings, corms, bulbs, tubers) etc. In papaya citrus healthy and nematode free seedlings are to be planted. Similarly in banana the suckers are to be selected from nematode free mother plants.

17.8. Removal of Infected Plants and Weeds

Just after the harvest of the crop nematode infected / infested plants should always be destroyed to prevent further spread of nematodes. But alone destruction of crop residue will not give complete control for the nematodes in succeeding crop season. Along with that extensive destruction of weeds is also often recommended for phytonematodes. But for those nematodes which have very wide host range this method of nematode control may not be quite effective.

17.9. Biological Control

Biological control aims to manipulate the parasites, predators and pathogens of nematodes in the rhizosphere in order to control the plant parasitic nematodes. Addition of organic amendments such as farm yard manure, oil cakes, green manure, pressmud, etc. encourages the multiplication of nematode antagonistic microbes which in turn checks the plant parasitic nematodes.

17.9.1. Predacious nematodes

Predacious nematodes have specialized open stoma armed with teeth to catch and swallow the plant parasitic nematodes. Addition of organic amendments help to encourage the multiplication of predacious nematodes such as *Mononchus* spp. Other genera like *Diplogaster* spp. and *Tripyla* spp. also come under the group of predacious nematodes.

17.9.2. Predacious fungi

Most of the predacious fungi come under the order Moniliales and Phycomycetes. There are two types of predacious activities among these fungi. They are nematode trapping fungi and endozoic fungi.

Endozoic fungi

The endozoic fungi usually enter the nematode by a germ tube that penetrates the cuticle from a sticky spore. The fungal hyphae ramify throughout the nematode body, absorb the contents and multiply. The hyphae then emerge from dead nematode. *Caternaria vermicola* often attacks sugarcane nematodes.

Parasitic fungi

Paecilomyces lilacinus is an effective egg parasite on many nematodes. The parasitic fungus is particularly effective against *Meloidogyne, Heterodera, Rotylenchus* and *Tylenchulus*. The fungus attacks the eggs as they are deposited in groups as a mass. The parasitic fungus was found to be effective against potato cyst nematode, root - knot nematodes in tomato, brinjal, betelvine and banana and *T. semipenetrans* in citrus.

Antagonistic fungi

Trichoderma spp. are one among a group of beneficial fungi that are present in almost all soils and diverse habitats. They have proven commercially viable as a successful biological control agent against plant parasitic nematodes and pathogenic fungi. Many species of *Trichoderma viz., T. viride, T. harzianum, T. koningi, T. longibrachiatum* and *T. hamatum* were reported effective against root knot nematodes infesting tomato, brinjal and crossandra.

17.9.3. Bacteria

Spore forming bacteria - *Pasteuria penetrans*

Pasteuria penetrans was found to be very effective against the root-knot nematodes in many crops. The *P. penetrans* infested J_2 of root-knot nematodes can be seen attached with spores throughout the cuticle. 18-20 per cent increased in tomato yield when treated with *P. penetrans* in root - knot nematode infested field.

Plant growth promoting rhizobacteria for nematode management

Bacteria that colonize the rhizosphere soil under the chemical influence of the root are commonly referred to as rhizobacteria. Specific rhizobacteria that have the ability to improve plant growth are called as plant growth promoting rhizobacteraia (PGPR). The effectiveness of *P. fluorescens* in reducing the infestation of potato cyst nematode *Globodera rostochiensis*. The same study in banana indicated the efficacy of *Agrobacterium radiobacter, P. fluorescens* and *B. subtilis* in reducing the infestation of *M. incognita* race 1 population in banana and to enhance plant growth. The rhizobacteria showed significant change and reduction in root galls and pattern of gall formation and also induced root growth in the plants. *P. fluorescens* increased the growth and yield parameters of banana and caused significant reduction in *Helicotylenchus multicinctus* infestation under field conditions. Native isolates of *Pseudomonas fluorescens* (Pfbv 22) and *Bacillus subtilis* (Bbv 57) isolated from healthy betelvine plants were found to be very effective against the nematode and *Phytopthora* complex disease in betelvine.

Nematode problems are often overlooked since they are unseen enemies. The symptoms produced by nematodes are often confused with soil and micronutrient related problems. Hence proper investigation is required to identify the nematode problems by analyzing the crop root and soil samples. Integrated nematode management is a system approach to keep nematode population below economic threshold level. Hence judicious combination of possible non-chemical methods can be practices according to the availability and suitability in a given situation will be more appropriate for the management of nematodes.

Other ITK'S

Neem Cake

Neem cake 40-50 kg/acre powdered well. Then the powdered neem cake is applied manually in the tomato field as basal application during land preparation and also before flowering the same dose was applied manually to the individual crops.

Asafoetida

1 kilo of Asafoetida in a gunny bag and keep it in the entrance of irrigation channel. The effluent water when enter the field it takes care of protecting crops from nematode attack.

18

Organic Certification

In India, there are two accreditation systems for authorizing Certification and Inspection agencies for organic certification. National Programme on organic Production (NPOP) promoted by Ministry of Commerce is the core programme which governs and defines the standards and implementing procedures. National Accreditation Body (NAB) is the apex decision making body. Certification and Inspection agencies accredited by NAB are authorized to undertake certification process. The NPOP notified under FTDR act and controlled by Agricultural Processed Foods Export Development Authority (APEDA) looks after the requirement of export while NPOP notified under APGMC act and controlled by Agriculture Marketing Advisor, Directorate of Marketing and Inspection looks after domestic certification. Currently 20 certification agencies have been authorized to undertake certification process.

National Programme on Organic Production

National Program on Organic Production (NPOP) was launched during 2001 under the Foreign Trade & Development Act (FTDR Act). The document provides information on standards for organic production, systems criteria, and procedures for accreditation of Inspection and Certification bodies, the national organic logo and the regulations governing its use.

National Standards for Organic Production (NSOP)

National Standards for Organic Production are grouped under following six categories:

1. Conversion
2. Crop production
3. Animal husbandry
4. Food processing and handling
5. Labeling
6. Storage and transport

Standard requirements for crop production, food processing and handling are listed below:

18.1. Conversion Requirements

The time between the start of organic management and cultivation of crops or animal husbandry is known as the conversion period. All standard requirements should be met during conversion period. Full conversion period is not required where organic farming practices are already in use.

18.2. Crop production

18.2.1. Choice of crops and varieties

All seeds and planting materials should be certified organic. If certified organic seed or planting material is not available then chemically untreated conventional material can be used. Use of genetically engineered seeds, pollen, transgenic plants are not allowed.

18.2.2. Duration of conversion period

The minimum conversion period for plant products, produced annually is 12 months prior to the start of the production cycle. For perennial plants (excluding pastures and meadows) the conversion period is 18 months from the date of starting organic management. Depending upon the past use of the land and ecological situations, the certification agency can extend or reduce the minimum conversion period.

18.2.3. Fertilization policy

Biodegradable material of plant or animal origin produced on organic farms should form the basis of the fertilization policy. Fertilization management should minimize nutrient losses, avoid accumulation of heavy metals and maintain the soil pH. Emphasis should be given to generate and use own onfarm organic fertilizers. Brought in fertilizers of biological origin should be supplementary and not a replacement. Over manuring should be avoided. Manures containing human excreta should not be used on vegetation for human consumption.

18.2.4. Pest disease and weed management including growth regulators

Weeds, pests and diseases should be controlled preferably by preventive cultural techniques. Botanical pesticides prepared at farm from local plants, animals and microorganisms are allowed. Use of synthetic chemicals such as fungicides, insecticides, herbicides, synthetic growth regulators and dyes are prohibited. Use of genetically engineered organisms or products is prohibited. All equipments from conventional farming systems shall be properly cleaned and free from residues before being used on organically managed areas.

18.2.5. Soil and water conservation

Soil and water resources should be handled in a sustainable manner to avoid erosion, salinisation, excessive and improper use of water and the pollution of surface and ground water. Cleaning of land by burning (e.g. slash and burn and straw burning) should be restricted. Clearing of primary forest for agriculture (jhuming or shifting cultivation) is strictly prohibited.

18.3. Collection of Non-cultivated Material of Plant Origin and Honey

Wild harvested products shall only be certified organic, if derived from a stable and sustainable growth environment and the harvesting shall not exceed the sustainable yield of the ecosystem and should not threaten the existence of plant or animal species. The collection area should not be exposed to prohibited substances and should be at an appropriate distance from conventional farming, human habitation, and places of pollution and contamination.

18.4. Food Processing and Handling

General principles - Organic products shall be protected from co-mingling with nonorganic products, and shall be adequately identified through the whole process. Certification programme shall regulate the means and measures to be allowed or recommended for decontamination, clearing or disinfection of all facilities where organic products are kept, handled, processed or stored. Besides storage at ambient temperature the following special conditions of storage are permitted.

- 100% of the ingredients of agricultural origin shall be certified organic.
- Water and salt may be used as organic products.
- Preparations of micro organisms and enzymes commonly used in food processing may be used with the exceptions of genetically engineered micro organisms and their products.
- Co-mingling with inorganic products shall be prevented.
- Proper storage under controlled atmosphere, cooling, Freezing drying and humidity regulation is needed.
- Mechanical, physical and biological methods to control pests only recommended.
- Carcinogenic pesticides and disinfectants are not permitted.

- Use of minerals and vitamins as additives and processing aids shall be restricted.
- Processing by mechanical, smoking, extraction, precipitation, filtration only approved.
- Irradiation is not allowed.
- Eco-friendly, biodegradable packaging materials is to be used.
- Product integrity should be maintained.
- Proper labeling of products is must.

Controlled atmosphere, cooling, freezing, drying and humidity regulation

Pest and disease control

For pest management and control following measures shall be used in order of priority Preventive methods such as disruption, and elimination of habitat and access to facilities. Other methods of pest control are:

- Mechanical, physical and biological methods
- Permitted pesticidal substances as per the standards and
- Other substances used in traps.
- Irradiation is prohibited.
- Direct or indirect contact between organic products and prohibited substances (such as pesticides) should not be there.

18.5. Packaging

Material used for packaging shall be ecofriendly. Unnecessary packaging material should be avoided. Recycling and reusable systems should be used. Packaging material should be biodegradable. Material used for packaging shall not contaminate the food.

18.6. Labelling

When the full standard requirements are met, the product can be sold as "Organic". On proper certification by certification agency "India Organic" logo can also be used on the product.

18.7. Storage and Transport

Products integrity should be maintained during storage and transportation of organic products. Organic products must be protected from co-mingling with non-organic products and must be protected all times from contact with the materials and substances not permitted for use in organic farming.

18.8. General Requirement for Certification

1. A registered operator shall Comply with National Programme for Organic Production (NPOP) norms and shall adhere to the National Standards for Organic Production (NSOP) and TNOCD general standards for organic agricultural production, animal husbandry production, honey, wild collection, processing, packaging, storage, labelling and transport standards.
2. Prepare, implement, and update annually an organic production plan and submit to Tamil Nadu Organic Certification Department (TNOCD) every year.
3. Permit on-site inspections with complete access to the production and handling operation, including non certified production and handling operation, areas, structures, offices by the Organic Certification Inspectors and other higher officials of TNOCD and also officials of APEDA whenever required.
4. Maintain all records applicable to the organic operation for not less than 5 years after creation of such records and allow authorized representatives of TNOCD, State or Central Government officials of accrediting agency access to such records during normal working hours for review and copying to determine compliance with NPOP norms and TNOCD Standards.
5. Pay the prescribed fees charged by TNOCD within stipulated time.
6. Operator shall inform the TNOCD in case of any.
   a) Application, including drift, of a prohibited substances to any, production unit, site, facility, livestock, or product that is part of an operation and
   b) Changes in certified operations or any portion of a certified operation that may affect the organic integrity in compliance with standards of NPOP and TNOCD.

18.9. Application for Certification

A person seeking organic certification of production or handling operation shall submit application for registration in the prescribed format in triplicate. The application shall include the following information

1. An organic production or handling system plan.
2. All information requested in the application shall be completed in full i.e. name, addresses, details of contact person, telephone number of the authorized person etc.
3. The names of organic certification body to which application is previously made and out come, non-compliance noted if any, copy of such records and reason for applying shall be given.

4. Any other information necessary to determine the compliance with the standards specified.
5. The prescribed registration fee, one time inspection fee, one time travel cost shall be paid by the operator along with the application form. The other prescribed fees shall be paid by the operator as notified by TNOCD during the course of certification process.

18.10. Review of Application

1. Application shall be scrutinized.
2. Any information required shall be communicated to the operator and operator shall submit the requested information immediately.
3. Application without prescribed fee shall not be reviewed.
4. After review of application decision shall be made by TNOCD on acceptance/ rejection of the application.
5. The rejected application shall be returned to the applicant citing reasons for rejection along with the fees enclosed.
6. Fee paid for the applications accepted by TNOCD shall not be refunded at any circumstances.
7. An initial onsite inspection shall be fixed and communicated to the operator after registration or shall be noted in the registered copy of application itself.
8. An applicant can withdraw the application at any time but the fees paid shall not be refunded.

18.11. Scheduling of Inspection

1. Initial field inspection shall be fixed at a reasonable time so that the operator can demonstrate compliance or capacity to comply with the standards while conducting inspection of land, facilities and activities. Such initial onsite inspection shall be delayed up to six months from the date of registration so as to give time for the operator to comply with required standards including record keeping.
2. All onsite inspection shall be conducted only in the presence of operator or an authorized representative of the operator who is knowledgeable about the operation. However this requirement does not arise in the case of unannounced / surprise inspections.
3. There shall be one annual inspection and additional inspection shall be fixed based on the risk assessment carried out during initial inspection.

18.12. Verification During Inspection

1. During the field inspection, the OCI shall verify the compliance or the capacity to comply with the NPOP standards and TNOCD standards.
2. Verification of information on organic production plan submitted by the operator and practical implementation of the standards.
3. OCI shall ensure that the prohibited substances/ materials are not used and in case of suspicion the OCI, shall draw samples of soil, water, wastes, seeds, plant tissues, plant, animal and processed products.
4. The samples shall be tested in NABL accredited ISO 17025 laboratories. The operator shall bear the cost of samples sent for analysis.
5. During onsite inspection the OCI shall conduct interview with the person responsible for the organic production system to confirm accuracy of information gathered during inspection and completeness of inspection, observation gathered during the onsite inspection. The inspector shall also collect other required information as well as issues of concern.
6. After inspection the OCI shall prepare checklist and inspection report and obtain signature of the operator or his representative.
7. A copy of the check list and inspection report shall be sent to the concerned operator and Evaluator.
8. Inspection reports shall be evaluated by the evaluator within reasonable time and any additional information required shall be addressed to the operator.
9. In case of any non compliance to the prescribed standards an explanation shall be called from the operator and sanctions shall be imposed if required.

18.13. Continuation of Certification

1. To continue certification the operator shall renew registration by paying fees for renewal.
2. An updated annual report for production or handling operation shall be submitted by the operator.
3. An updated corrective action for minor non conformities previously identified shall be submitted by the operator.
4. TNOCD after receipt of renewal application for continuation of certification shall scrutinize the application and verify the facts.

Fair trade

All the operators shall perform their operation with social justice; they shall not employ child labour, and shall protect rights of women, smallholder, traditional agriculture and indigenous people's rights.

Appeal

1. Registered operator may appeal against the notice of denial of certification, proposed suspension or revocation to the appellate authority (Director, TNOCD).
2. An appeal shall be made within the time period mentioned in the notification or within 30 days from the date of receipt of the notification, whichever occurs later. The appeal shall be considered filed on the date of receipt in the office of Director, TNOCD. The decision of the appellate authority shall be final.

18.14. Intiating Organic Farming

Dry lands are the potential places where organic farming can be started because in dry lands

1. Less effect of high input agriculture thus least residue of pesticide and less time required for conversion.
2. Organic manures improve the fertility and water retention capacity of poor soils of dry lands.
3. Poor economic status of the dryland farmers limits them to purchase high cost inputs. But they can do the labour intensive operations to support organic farming.

18.15. Action Plan for Promotion of Organic Farming

The salient features of crop production management in organic farming include use of organic manures, recycling of organic wastes, proper crop rotation, intercropping, mixed cropping and poly-cropping, green manure cropping, use of biofertilizers, mulching of weeds, integrated pest management, judicious use of irrigation water. The action plan is follows.

18.15.1. Development of Organic Farming Practices

Model organic farms can be developed for the crops amenable for organic farming and package of practices can be developed for further dissemination and adoption.

18.15.2. Imparting Training on Organic Farming

The small and medium farmers, NGOs and private entrepreneurs have to be trained on the best organic farming practices, storage and processing of organic products. The benefits of organic farming can be elaborated to the participants of such training programmes. Demonstration in selected locations for specific crops can be conducted in the farmers' fields to enable the farmers to gain acquaintance with new technology packages.

18.15.3. Identification of areas and villages for organic farming

Organic farming can be promoted for a suitable region or a village or cluster of farms to derive maximum benefits of organic farming. This will also facilitate in transport and marketing of the crop produces. Organic farming societies / farmers' forums can be established at village levels. The forums can help the farmers in getting soft loans from banks and registration in the institute identified by the government for organic certification.

18.15.4. Facility creation for organic farmers

Facilities should be created for storage, processing, packaging, transport, quality control and sales mechanisms at taluk /district levels and to ensure premium price for the organic farm produces. To begin with agricultural marketing department can be entrusted for facility creation activities at district level and then to taluk and block levels. The market intelligence cells can be created at State Agricultural Universities which can co-ordinate with Government agencies, NGOs, farmers'forums in assessing the market rates, market facilities at national and international levels. Basic training on food processing, post harvest technology and market intelligence can be imparted to different stake holders at the State Agricultural Universities.

18.15.5. Organic certification

The growing awareness among the public and farmers on the ill effects of chemical agriculture and increasing demand for the safe and quality food necessitated the promotion of organic agriculture in India. Since, organic certification improves the image of organic agriculture and provides transparency in certification, Government agencies can be assigned with organic certification program. The certification facility also can be utilized by different stake holders including local bodies, NGOs, private entrepreneurs and self help groups (SHGs) who produce and market organic commodities in large quantities.

18.6. Participatory Guarantee System

Participatory Guarantee System (PGS) is a quality assurance initiative that is locally relevant, emphasize the participation of stakeholders, including producers and consumers and operate outside the frame of third party certification. As per IFOAM (2008) definition "Participatory Guarantee Systems are locally focused quality assurance systems. They certify producers based on active participation of stakeholders and are built on a foundation of trust, social networks and knowledge exchange". PGS is a process in which people in similar situations (in this case small holder producers) assess, inspect and verify the production practices of each other and collectively declare the entire holding of the group as organic.

PGS system has number of basic elements which embrace a participatory approach, a shared vision, transparency and trust. Participation is an essential and dynamic part of PGS. Key stakeholders (producers, consumers, retailers and traders and others such as NGOs) are engaged in the initial design, and then in the operation of the PGS. In the operation of a PGS, stakeholders (including producers) are involved in decision making and essential decisions about the operation of the PGS itself. In addition to being involved in the mechanics of the PGS, stakeholders, particularly the producers are engaged in a structured ongoing learning process, which helps them improve what they do. This process is facilitated by the PGS group itself or in some situations a supportive NGO. The learning process is usually 'hands-on' and involves field days or workshops. The idea of participation 4 embodies the principle of collective responsibility for ensuring the organic integrity of the PGS.

Guiding Principles for Organic Participatory Guarantee System In tune with the international trends and IFOAM's PGS Guidelines

PGS India system is also based on participatory approach, a shared vision, transparency and trust. In addition it gives PGS movement a National recognition and institutional structure without affecting the spirit of PGS.

1. Participation

Participation is an essential and dynamic part of PGS. Key stakeholders (producers, consumers, retailers, traders and others such as NGOs) are engaged in the initial design, and then in the operation of the PGS and decision making. The idea of participation embodies the principle of a collective responsibility for ensuring the organic integrity of the PGS. This collective responsibility is reflected through: (i) Shared ownership of the PGS (ii) Stakeholder engagement in the development process (iii) Understanding of how the system works and (iv) Direct communication between producers and consumers and other stakeholders Together these help shape the integrity based approach and a formula for trust. An important tool for promoting this trust is having operational processes that are transparent. This includes transparency in decision making, easy access to the data base and where possible farms are open to participation and visits of consumers. Participation of traders/ retailers or consumers in decision making may not be possible under all situations, but their participation in any form will increase the credibility and trustworthiness of the group.

2. Shared Vision

Collective responsibility for implementation and decision making is driven by common shared vision. All the key stakeholders (producers, facilitating agencies, NGOs, social organizations and even the State Governments) support the guiding principles and goals, PGS is striving to achieve. This can be achieved initially through their participation and support in the design and then by joining it. This may include commitment in writing through signing an application/ document that includes the vision. Each stakeholder organization (or PGS group) can adopt its own vision conforming to the overall vision and standards of PGS India.

3. Transparency

Transparency is created by having all stakeholders, including producers and consumers, aware of exactly how the guarantee system works to include the standards, the organic guarantee process (norms) with clearly defined and documented systems and how decisions are made. Public 5 access will be ensured to documentation and information about the PGS groups, such as lists of certified producers and details about their farms and non-compliance actions. These will be available through a dedicated National database websites. But still it does not mean that entire information on National PGS database will be available to everyone. At the grass roots level transparency is maintained through the active participation of the producers in the organic guarantee process which can include

- Information sharing at meetings and workshops
- Participation in internal inspections (peer reviews)
- Involvement in decision making.

4. Trust

The integrity base upon which PGS are built is rooted in the idea that producers can be trusted and that the organic guarantee system can be an expression and verification of this trust. The foundation of this trust is built from the idea that the key stakeholders collectively develop their shared vision and then collectively continue to shape and reinforce their vision through the PGS. The ways this trust is reflected may depend entirely on factors that are culturally/ socially specific to the PGS group. The idea of 'trust' assumes that the individual producer has a commitment to protecting nature and consumers' health through organic production. Mechanism for expressing trustworthiness includes:

- Declaration (a producer pledge) via a witnessed signing of a pledge document
- Written collective undertaking by the group to abide by the norms, principles and standards of PGS.

5. Horizontality

PGS India is intended to be non-hierarchical at group level. This will reflect in the overall democratic structure and through the collective responsibility of the PGS group with sharing and rotating responsibility, by engaging producers directly in the peer review of each other's farms; and by transparency in decision making process.

18.16.1. National networking

PGS India while keeping the spirit of PGS intact also aims to give the entire movement an institutional structure. This is proposed to be achieved by networking the groups under common umbrella through various facilitating agencies, Regional Councils and Zonal Councils. To make the system completely transparent and accessible to traders and consumers entire data will be hosted on a common platform in the form of a website. National Centre of Organic Farming shall be the custodian of data, define policies and guidelines and undertake surveillance through field monitoring and product testing for residues. Regional councils and facilitating 6 agencies will facilitate the groups in capacity building, training, knowledge/ technology dissemination and data uploading on the PGS website. But at every stage it will be ensured that these agencies including apex body do not interfere in the working and decision making of the group. Even if surveillance is done and reports are made, the same will also be put on website in public domain. What action is to be taken on adverse reports will be left to the group and Regional Council.

18.16.2. Advantages of PGS over third party certification system

In PGS organic farmers have full control over the certification process and are able to produce far more credible and effective system of quality assurance compared to third party certification. Important benefits of this system over third party certification system are as follows:

a) The procedures are simple; documents are basic and use the local language understandable to farmers.

b) All the members are local and known to each other. Being themselves practicing organic farmers have high degree of understanding on dayto-day knowledge or acquaintance of the farm.

c) Peer appraisers are among the group and live in the same village, therefore have better access to surveillance.

d) Peer appraisal instead of third party inspections reduces cost.

e) Mutual recognition and support between Regional PGS groups ensures better networking for processing and marketing.

f) Empowers farmers with increased capacity building g. Bring consumers to the farm without the need of middleman.

h) Unlike grower group certification system, PGS offer every farmer with individual certificate and each farmer is free to market its own produce independent of group.

i) Consumers and buyers are often involved in production and verification process.

j) Random residue testing at regular intervals ensures the integrity and increases the trust.

18.16.3. Limitations of PGS

PGS certification is only for farmers or communities that can organize and perform as a group within the village or in close-by villages with continuous territory and is applicable on, on-farm activities comprising of crop production, processing and livestock rearing (including bee keeping) and off-farm processing "by PGS farmers of their direct products".

Individual farmers or group of farmers having less than 5 members are not covered under PGS. They either have to opt for third party certification or join the existing PGS local group.

PGS is applicable on on-farm activities comprising of crop production,processing and livestock rearing and off-farm processing "by PGS farmers of their direct products". Off-farm processing activities such as, storage, transport and value addition activities by persons/agencies other then PGS farmers away from the group are not covered under PGS. Off-farm input approval granted by the group is applicable on the members of the same group and cannot be taken as a basis for universal approval for other groups. Off-farm inputs need to be approved by each group for their member's use on case to case basis.

PGS ensures traceability only up to end till it is in the custody of PGS group. Once the product leaves the custody of PGS group there is no control of PGS on its integrity, Therefore PGS is ideal for local direct sales/ direct trade between producer and consumer and direct trade of packed finished product with PGS logo between PGS group and traders/ retailers. But Local Groups and buyers in consultation with RC can devise some mechanism with full traceability records to allow use of PGS logo on products packed by traders/ retailers.

Operational Structure

Schematic operational structure of the PGS India is given below:

↓

Farmer/ Farm family

↓

Local groups

↓

Facilitating agencies

↓

Regional councils

↓

Zonal Councils (may be on Regional basis)

↓

National Centre of Organic Farming
(As Secretariat)

↓

National Advisory Committee at DAC

For further details contact

Tamil Nadu Organic Certification Department
1424A, Thadagam Road
Coimbatore-641003
Tamil Nadu, India
Phone: 0422-2435080

19

Success Stories

In food production remarkable progress has been made in the last fifty years in agricultural production and self – sufficiency (Food security) has been reached. Yet nutritional security seems to be a distant goal. With depleting natural resources and increasing threat to the environment, sustainable crop production is attracting the attention of the scientists, farmers and administrators. Due to over exploitation of natural resources and indiscriminate use of agro chemicals, green revolution also brought in quite a lot of miseries which focus the urgent need to have EVERGREEN REVOLUTION that is sustainable, leading to food and nutritional security to the rural poor. At this juncture it will be more appropriate to practicing organic agriculrure through the integration of traditional wisdom and modern scientific knowledge in synergy with nature.

19.1. Eco-friendly Technologies for Rice Cultivation

A sound package of eco-friendly technologies to grow rice is being successfully adopted by a few progressive farmers in Puliangudi village in Tirunelveli district of Tamil Nadu. "The technologies work well with indigenous rice varieties such as Kitchili Samba. The cost of cultivation is substantially reduced and the organic rice fetches a premium price in the market," says Mr. P. Gomathinayagam, a pioneer in organic farming in Puliangudi.

"I grew a medium-duration (140 days) Kitchili Samba rice in about 1.6 hectares. The seeds were treated with *Panchagavya,* and the nursery was treated with plenty of tank silt and a host of organic amendments. Liberal quantities of tank silt were applied and green leaf manure was incorporated a few days ahead of the final ploughing. Biogas slurry was applied through irrigation when the seedlings were just establishing in the main field," he explained.

One round of spray with 3 per cent solution of Panchagavya was given 20 days after transplanting. On the 30th day, a combination of coconut milk and butter milk, mixed in equal volume, in ten times their volume of water was sprayed on the crop to promote active plant growth and tillering. On the 40th day, another round of spray with *Panchagavya* (3 per cent solution in high volume spray) was given. A bio-insect repellent was sprayed on the 45th day of transplantation.

The crop was regularly irrigated, and a grain yield of about 6 tonnes was expected from the 1.6 hectare plot. He also was assured of high quality straw for his cattle. The cost of cultivation worked out to Rs. 14,000 for 1.6 hectares. "I sell the output as organic rice at a rate of Rs. 30 per kg, and it makes organic rice cultivation more rewarding economically as well environmentally," pointed out Mr. Gomathinayagam. He is championing the cause of organic farming in the southern districts of Tamil Nadu. Several farmers are following his advice.

I adopted the organic rice farming technologies and harvested about 9.25 tonnes of paddy a hectare from the bold grained Trichy 1 variety. I dumped liberal quantities of daincha in the field and allowed it to decompose well in the field ahead of planting. There was no need for any plant protection also. The cost of cultivation worked out to Rs. 12,500 per hectare. I also harvested plenty of healthy straw for our animals," said Mr. V. Antonysamy, a progressive farmer of Puliangudi village

19.2. Eco-friendly Method of Setting up Vermicompost Unit

Vermicompost is the basic ingredient for successful organic farming. More than 85 per cent of organic crop cultivation depends on it. Usually farmers across the country build a roof either with thatched straw or asbestos sheets as a cover for their vermicompost manufacturing unit.

The bottom of the unit will either have sand or plain cement or sometimes toughened red soil.

A progressive organic farmer, Mr. D. Bharani of Mayiladuthurai taluk in Tamil Nadu's Nagapattinam district, has used local tree trunks for the four poles supporting his rudimentary compost unit.

Climbing tendrils

"The tree trunks absorb the moisture from the compost unit and grow as individual trees," he says. For the roof, he has used the climbing tendrils of vegetable plants growing near the compost unit.

The plants grow well, absorbing the required moisture from the unit and their leaves provide shade to the manufacturing unit."In addition to making the compost which I sell at Rs 5-8 per kg, I am also able to sell vegetables such as bhendi, brinjal, snake gourd and bitter gourd grown on the roof of the compost shed," he says.

"Farmers, instead of spending money on constructing thatched sheds and asbestos, can follow simple methods like this. By doing so, they can get double income from the compost unit and the vegetables." He is also cultivating rasthali banana variety in about four acres. "Commercially, rasthali has a good demand in the market compared to other varieties and if consumers know that it is grown organically, then the farmer need not search for buyers. It will be vice-versa," he says.

Sturdy against winds

But why did he choose rasthali variety when there are so many other varieties? "Rasthali variety does not grow quite high and is often sturdy against strong winds." Strong winds often uproot banana trees and farmers have to tie each tree to a wooden pole to prevent the tree from falling or getting uprooted. "Secondly organic practices are found ideal for my banana orchard as banana is often found susceptible to wilt disease which is a major and fatal infestation.

"Chemical control methods have not been found successful in controlling this infestation, compared to organic methods," he explained."I had purchased the suckers from known sources and from healthy trees. The suckers, before planting were dipped in a solution of 10 per cent Panchagavya and 50 gm of pseudomonas for 3-5 minutes. For an acre, about 780-800 pits of 8x8 (row to row and plant to plant) were dug and the suckers were planted in them. About 3 kg of farm yard manure (FYM) was also applied in each pit.

Heat generation

The FYM was applied a little distance away from the pit, because if it were applied directly into the pit or near the suckers it would spoil the plant growth due to heat generation. Panchangavya spray was done once every month till the crop was about 5 months old. He was able to harvest his first yield in about 14 months after planting and this variety can be maintained for two years.

"One bunch was sold for Rs. 120-130 and I was able to get a net income of Rs. 80,000."The expenditure for maintaining one tree comes to about Rs. 35 and after deducting the expenses for all my plants I am still able to get a net profit of Rs. 50,000." he said.

For more information Mr. D. Bharani can be contacted at Kothangudi village, Komal post, Mayiladuthurai taluk, Nagapattinam district, Tamil Nadu: 609-805, phone: 04364-228711 and 04364-237415 and mobile: 9486278569.

19.3. Ginger Garlic Extract: A Bio-pesticide for Organic Cultivation

Organic practices avoid investment on costly chemicals, Ms. Rajareega at her farm in Sivaganga district, Tamil Nadu seen manufacturing the botanical pesticides.

There is a growing body of evidence to suggest that in the past 4-5 decades there has been an excessive dumping of chemical toxins on the soil. As a result the soil has become barren and ground water toxic, in many places. Contrast this with organic inputs that are safe, non toxic, and cost much less. For example, if using chemical pesticides and fertilizers for growing a crop in a hectare works out to about Rs.6,000-7,000 the cost of growing the same crop using organic inputs may come to only about Rs.500 - Rs. 1,000, according to Ms. Rajareega of Raasi organic farms at Muthupatti village in Sivaganga district, Tamil Nadu.

Lower cost

Even if some critics say that organic farming cannot provide the same high yields as chemical farming, the organic farmers argue that at least their land is safe; that they have not invested in buying the chemicals and increasing their cost of cultivation."If you look at the suicides by farmers, then you will understand that all those farmers who committed suicides have built up huge debts.

The debts kept growing because of borrowing at high interest rates for buying these chemicals which promised to increase the yield. In the end, it only increased their debts," she explains.

"If only farmers use safer and natural pest repellents and manures then where is the question of debt and suicides,?" she enquires. She has been using only organic manures and bio-repellents made from locally available resources.

Five leaf extract

For example she uses 5 different leaf extracts (*Eindhu ilai karaisal* in Tamil) derived from *Calotropis* (called *erukku* in Tamil), *Jatropha curcas* (*Kattu amanaku* in Tamil), Neem (*Vembu* in Tamil), Guduchhi/Amruth (*Seenthil kodi* in Tamil), Chaste tree (*Nochi* in Tamil), Malabar nut (*Adathoda* in Tamil), Kalmegh (*Siriyanangai* in Tamil), Clerodendron (*Peenarisanghu* in Tamil) and Usil (*Arappu* in Tamil). These plants are commonly found in all villages. About 1 kg of leaves from each plant is taken and powdered and then ground into a paste. It is then mixed with 5 litres of cow's urine.

The concoction is then diluted in 5 litres of water and left undisturbed for 5 days. When required for using about 500 ml of this concoction is diluted in 10 litres of water and sprayed over the plants, she explains.

Ginger garlic extract

Another tried and proven mixture she uses is ginger garlic extract (called *Inji poondu karaisal* in Tamil). About 1 gm of ginger and garlic each, 2gm of green chilli and 5 litres of cow's urine and water are taken. The garlic, ginger and green chilli are ground into a paste and mixed with cow's urine and water. After 10 days the mixture is filtered and used. The prescribed quantity is about 500 ml of this solution diluted in 10 litres of water which can be sprayed over the plants.

Ideal spraying time

The ideal time for spraying these karasals is during 6 am to 8.30 am and between 4 pm and 6.30 pm. Depending upon the soil, crop and other climatic factors the concentration can be raised or lowered. Farmers can contact their nearby organic farmers who are using these karaisals or can contact Ms. Rajareega for guidelines regarding the concentration.

Effective control

Both the above karaisals have been found effective in controlling leaf roller, thrips, mealy bugs, fruit, stem and bark borer, hairy caterpillar and aphids. Even if a farmer is not convinced about the benefits of organic inputs he can continue to grow his crops using chemicals, but at the same time he can set aside a small portion in his field to grow the same crop using organic inputs. By doing so he can find out for himself the cost benefit ratio. That itself can convince him of its efficacy.

Readers can contact Ms. Rajareega, Raasi organic farms, Muthupatti, via Kallal, A. Siruvayal (post), Sivaganga district, Tamil Nadu, email: rajareega@rediffmail.com, mobile: 9865-582142 and phone: 04565-284937.

19.4. Organic Farming in Banana a Profitable Venture

Organic farming is the in-thing now in Thanjavur district- the rice bowl of Tamil Nadu. Having used chemical fertilisers for some years, farmers are now reverting to organic farming.

Nearly one hundred farmers of Thanjavur district were given a first hand idea of how successful is organic farming by the Central Bank of India recently. They were taken to TARI high-tech horticulture farm at Marunkulam, where banana, maize and paddy were raised using organic farming.

According to Kulandaisamy, a progressive farmer and owner of TARI high-tech horticulture farm, organic farming helps to increase the yield and nutritive value. In his farm at Marunkulam he has raised banana crop using organic fertilisers. He found that the yield and quality of the crop was good.

He has raised Rasthali and robust variety of bananas on 1.5 acres and three acres respectively in his farm adopting organic farming methods. With respect to Rasthali, a bunch has five to six hands (hands means seeppu in Tamil), instead of three to four hands, which are normally seen in ordinary cultivation.

The bunch has adequate space between hands and hence the fruit is big, says Mr. Kulandaisamy pointing to the big bunch of bananas in his farm. A bunch weighs 20 kgs and fetches Rs. 200. In the case of Robust variety (green banana) he has resorted to a high density cultivation with 1600 plants in an acre. In the case of Rasthali he has gone for 1000 plants in an acre.

Robust variety of banana has 12 to 15 hands in a bunch and the bunch weighs 30 to 35 kgs. Fertiliser used by Kulandaisamy was composed organic matters, neem, pancha kavya prepared using cow's urine and cow dung, etc. As a farmer who is practising integrated farming, he has a dairy farm from where he produces panchagavya and uses the cowdung generated in the dairy for organic farming.

What about marketing of bananas produced under organic farming?

They fetch a premium price and already a shop in Chennai selling organic produce has booked for bananas. He is also selling the bananas to Tiruchi market.

He has also gone for maize in two acres of land under organic farming and paddy in two acres under organic farming. He says that 'Rasthali', which is normally prone to wilt attack is free from the disease in organic farming. "There is a tendency even now among farmers to go for chemcial fertilisers. This has to be discouraged and they should be enlightened about organic farming," says Mr. Kulandaisamy.

Another uniqueness of Mr. Kulandaisamy is that he produces liquid bio-fertilisers. He has put up a plant under the New Anna Marumalarchi Scheme for Rs.1 crore and is producing liquid bio-fertilisers. An innovative farmer, Kulandaisamy has raised Vanila on one acre of land.

He has spent Rs.9 lakhs for infrastructure alone to raise the sensitive crop. The crop is irrigated using micro-sprinklers under a shade net in a conditioned atmosphere. "Vanila can be a profit-oriented venture when raised as an inter-crop in coconut," says Kulandaisamy.

19.5. Organic Practices for Increasing Mango Yield

Mango trees respond well to organic manure applications. Organic manures such as vermicompost, *panchagavya* and vermiwash are used for promoting healthy tree growth and fruit formation.

Essential nutrients

From the initial planting stages to caring of full-grown trees, Panchagavya and vermicompost can be effectively used to supply essential nutrients to the trees and prevent pest infestations, according to Dr. S. Sundaravadivel,

Vermitechnologist and Environmentalist based in Chennai. Vermicompost is prepared by using earthworms. Vermiwash is the liquid collected after the passage of water through a column of activated earthworms. It is very useful as an organic spray for all crops.

Pest repellent

Panchagavya is an organic growth promoter, which is prepared by mixing cow dung, cow urine, cow's milk, curd and ghee in suitable proportions, and is sprayed on the plants. It contains several macro, micronutrients, beneficial bacteria and fungi, which aid in growth promotion and act as effective pest repellents.

It can be prepared by thoroughly mixing five kilos of fresh cow dung and one litre of cow's ghee in a plastic or cement tank or earthen pot. The mixture is stirred daily for three to four days."About three litres of cow's milk, two litres of cow's curd, three litres of sugarcane juice, three litres of tender coconut water and 10 to12 bananas are mixed well and added to the mixture. The entire concoction is allowed to ferment for fifteen days," said Dr. S. Sundaravadivel.

The container should be covered with a net (or) cotton cloth to allow aeration of the fermenting unit, according to him. The concoction is stirred two or three times a day for about fifteen days and then used. For mango trees of about 6-7 years age, vermicompost may be applied at the rate of 10 kilograms per tree and one litre of *panchagavya* diluted in 30 litres of water may be sprayed over the foliage (crown) and at the base of the tree. Spraying Panchagavya over the crown and at the base of the tree must be done four to five times, according to Dr. Sundaravadivel. The first spraying must be done before the flowering season (January-March) to increase flower formation.

A second spraying must be done after 15-20 days. The process must be repeated till the flowers turn into small sized buds. Once the buds start forming then the application can be done once a month, according to him. Use of Panchagavya and vermicompost has been found to increase the size, number and enhance the colour of the fruits.

Recommended practice

The recommended practice for one hectare of mango trees is about 25 litres of panchagavya (mixed in 750-800 litres of water) and four to five tonnes of vermicompost. Spraying panchagavya has been found effective in the control of fruit fly menace, a common infestation in all fruit bearing trees, according to Dr. Sundaravadivel. According to him, trees treated with organic manures bore large sized leaves and formed a dense canopy with profuse rooting systems. The taste and shelf life of the fruits were also found to be more satisfactory.

Nitrogen-fixing bacteria

"The interaction of the root hairs of these trees with the organic manures also increased the activity of the nitrogen-fixing bacteria in the soil.

"The organic manures also act as a carrier medium for the development of several beneficial microorganisms such as *azospirillum, azotobacter, rhizobium and phosphobacteria*," he said. Dr. S. Sundaravadivel can be reached by mobile at 98843-90104 and email: sundaravadivel66@hotmail.com.

19.6. Alphonso Mango Ideal for Organic Farming

Alphonso, the king of mango varieties, does well under organic farming conditions in Kanchipuram and Thiruvallur districts of Tamil Nadu, and its full potential should be exploited by the small and marginal farmers in the districts, says Mr. S. S. Nagarajan, Vice President (Agricultural Research), TAFE (Tractors and Farm Equipment Limited), Chennai.

An authority on mango cultivation, and a pioneer in establishing the largest 'Rumani' orchard in Kanchipuram district at 'J-Farm', Kelambakkam, Mr. Nagarajan has successfully demonstrated the promises held out by the superior mango variety 'Alphonso'. "After working with 'Rumani' variety in the last thirty five years, I shifted my focus to 'Alphonso' variety in 1998, and it has done exceedingly well in the farm. It responded favourably to drip irrigation and organic inputs, and yielded high quality sweet fruits of attractive aroma," explains Mr. Nagarajan. In about 0.8 hectares in 'J-Farm' he planted 140 Alphonso grafts got from a nursery in Dharmapuri district. This variety is suited for high density planting, and it will make up for the low yields in the initial few years of bearing.

The regular bearing commences from the ninth or tenth year of planting," he points out. Every young plant was regularly manured in August with liberal quantities of ripe farmyard manure along with 400 g each of Azospirillum and Phosphobacterium. Drip irrigation to provide 25 litres of water per tree per day was established. All other regular orchard practices such as clipping the sprouts below the graft union, weeding and hoeing in the basins, and ploughing the interspaces and plant protection with eco-friendly botanical insecticides were adopted.

"During the first five years Daincha was raised as rainfed crop, and ploughed *in situ* as an organic nutrient supplement. Groundnut was grown as an irrigated crop in the interspaces, and haulms were incorporated in to the soil as green leaf manure. Of the 140 trees, 112 trees started yielding in the fifth year after planting, and on an average each tree yielded about 70 fruits each weighing about 250 g. In all about 2000 kg fruits were harvested and it fetched

Rs. 12 per kg at the farm gate. The gross income in the first year of bearing is Rs. 24,000, and the cost of cultivation in the first five years worked out to Rs. 24,000'', he said.

"The quality of the fruits was of superior quality, and they were sweet and free of spongy tissues. The results were quite encouraging," says Mr. Nagarajan. The trees would yield as high as 3000 kg from the 0.8 hectares from the sixth year of planting, and it would fetch a handsome profit to the growers. As the tree grows, the yield will go up, and a thirty year-old tree would produce as much as 2500 quality fruits, according to him. `J-Farm' has perfected the organic farming practices for raising Alphonso mango, and the farmers should benefit from it, according to him. Farmers, however, should avoid using chemical fertilizers and fruiting hormones to get quick returns from the variety, as they may prove harmful in the long run, he says "Alphonso is an excellent variety for export and if grown organically, its value in the export market will go up significantly," explains Mr. Nagarajan.

20

Steps to A Successful Organic Transition

The transition from conventional to organic farming requires numerous changes. One of the biggest changes is in the mindset of the farmer. Conventional approaches often involve the use of quick-fix remedies that, unfortunately, rarely address the cause of the problem. Transitioning farmers generally spend too much time worrying about replacing synthetic input with allowable organic product instead of considering management practices based on preventative strategies. Here are a few steps new entrants should follow when making the transition to organic farming:

20.1. Understand the Basics of Organic Agriculture and the Organic Farming Standards

Since organic production systems are knowledge based, new entrants and transitional producers must become familiar with sound and sustainable agricultural practices. Transitional producers should be prepared to read appropriate information, conduct their own trials and participate in formal and informal training events. As mentioned, switching from conventional to organic farming is more than substituting synthetic materials to organic allowed materials. Organic farming is a holistic system that relies on sound practices focused on preventative strategies. Since there are often few organic remedies available to organic producers for certain problems, prevention is the key element in organic production.

20.2. Identify Resources That Will Help You

Existing organic farmers are generally very helpful in sharing valuable technical information. A good mentor should be able to provide transitional producers with knowledge, practical experience and suggest appropriate reading materials. Mentors are able to identify some of the most important challenges transitional farmers will be confronted with. Mentors may also help source production materials that are otherwise difficult to find. Producers should also contact agrologists, veterinarians and other agricultural and financial consultants, in order to learn ways to improve their current farming practices.

The Internet is a valuable source of information, especially to new organic farmers. A broad range of reading materials are available from many organic/ ecological organizations such as the Organic Agriculture Centre of Canada (OACC), the Atlantic Canadian Organic Regional Network (ACORN), the Canadian Organic Growers (COG), the Certified Organic Associations of British Columbia (COABC), the National Sustainable Agriculture Information Services/ Appropriate Technology Transfer for Rural Areas (ATTRA), the Sustainable Agriculture Research and Education (SARE), and the Agri-réseau/agriculture biologique- Quebec. Consider joining an organic organization or network to access these valuable resources and establish good working contacts.

20.3. Plan Your Transition Carefully

Develop a transitional plan with clear and realistic goals. The plan should clearly identify various steps to be taken in making the transition to organic and be sure to include realistic timeframes. Identify your strengths and weaknesses. Consider ways to address any weaknesses, while building on strengths. The business side of the transitional plan should contain a multiple year budget and an effective/realistic marketing strategy. Make sure your list of expenses is comprehensive. Include all prerequisites to begin the transition; such as, mechanical weeding equipment, specialized composting equipment and applicators, additional handling equipment dedicated to the organic products, and processing equipment. Although the demand for organic products is continually growing, growers need to make sure they have a reliable market for the organic products they plan to produce.

Careful planning is very important. During the early part of the transitional period, yields are often depressed and premium prices for certified organic products are generally not yet obtainable. Use realistic yields and prices when evaluating the feasibility of your project.

In some instances, it is preferable to continue using conventional measures early on in the transitional process in order to avoid dramatic yield reduction which could jeopardize the financial well-being of the operation. Farmers who are planning to convert their livestock operation should consider certifying their fields first. This allows time to learn more about organic livestock management requirements while, at the same time, starting to produce organic feeds.

Although organic certifiers generally want to see the entire farm become organic, certifiers generally allow new entrants several years of transition time before the whole farm is fully certified.

Parallel production is the simultaneous production, processing or handling of organic and nonorganic crops, livestock and other products of a similar nature. Although this type of activity is highly discouraged by certifiers, some allow it, especially during the transition period. If permitted to practice parallel production, producers must be prepared to deal with significant record keeping in order to ensure traceability and organic integrity.

20.4. Understand Your Soils and Ways to Improve them

Since soil is the heart of the organic farming system, it is crucial that new entrants understand the various characteristics and limitations of the soils found on their farm. Soil suitability may vary significantly from one field to the next. Fields with good drainage, good level of fertility and organic matter, adequate pH, biological health, high legume content, and with less weed and pest pressure, are excellent assets. Often these fields are the first ones ready for transition and certification.

Many tools exist to assess soils. Soil chemical, physical and biological analyses, soil survey and legume composition field assessments, and field yield histories are very important and should be considered early in the transition. Unhealthy soils require particular attention.

If farmers plan to grow crops without raising any livestock, it may be necessary for them to source allowable soil amendments such as composted manure, limestone, rock dust, and supplementary sources of nitrogen, phosphorus, potassium and micro-nutrients. Even with the best of crop rotations that include green manure crops like legumes (nitrogen fixing crops), transitional growers will be challenged if they want to obtain optimal yields without additional livestock manure, compost and/or other off-farm soil inputs. When these inputs are scarce or expensive, producers may benefit from integrating livestock on their farm.

Let's not forget, under organic production, farmers must be able to recycle nutrients through proper nutrient management practices: recycling through good manure and compost utilization, crop rotations, cover crops (green manure, catch, and nitrogen fixing crops), and by reducing nutrient losses due to leaching, over-fertilization, as well as poor manure and compost management (storage, handling, and spreading).

20.5. Identify the Crops or Livestock Suited for Your Situation

Before growing a crop or raising any livestock, consider the following: degree of difficulty to grow or raise the product organically, land and soil suitability, climate suitability, level of demand for the product, marketing challenges, capital required, current prices for conventional, transitional and organic products, and profitability over additional workload.

20.6. Design Good Crop Rotations

Once the crops are chosen, carefully plan the crop rotation(s) and select the most suitable cover crops (green manure, winter cover crops, catch crops, smother crops, etc.). Crop rotations are extremely important management tools in organic farming. They can interrupt pest life cycles, suppress weeds, provide and recycle fertility, and improve soil structure and tilth. Some rotational crops may also be cash crops, generating supplemental income.

On some farms, land base availability may be a limiting factor when planning your crop rotations. The transitional plan should, therefore, include crop rotation strategies. Responding to external forces such as new market opportunities may also have a significant impact on crop rotations, so farmers need to consider the effect that growing new crops has on their crop rotations and land base availability.

20.7. Identify Pest Challenges and Methods of Control

It is important to know the crop's most common pests, their life cycles and adequate control measures. For instance, Colorado potato beetle may be a pest of significant importance when growing potatoes; cucumber beetles in cucurbitaceous crops (cucumber, squash, and melons); flea beetle in many seedlings crops; clipper weevil and Tarnish Plant Bug in strawberry crops.

There are several measures available to reduce pest pressure: crop rotation, variety selection, sanitation, floating row covers, catch crops, flamers, introduction of beneficial insects, bio pesticides, and inorganic pesticides. Transitional growers should be prepared to use and experiment with some of these options. When considering a new type of production, discuss pest issues with your agrologists, IPM specialists and/or other existing organic producers to optimize your chances of success.

Availability of organic supplies has improved significantly over the past few years. New pest control products containing B.t., spinosad, kaolin clay are effective and currently available to organic growers. It is often reported that the types of weeds found on the farm evolve with time as growers change the way they grow their crops and control their weeds. By keeping track of the weed population, growers will be able to refine their crop rotations and improve their control measures.

Under organic livestock management, cattlemen must provide attentive care that promotes health and meets the behavioral needs of various types of livestock. With good herd health practices, farmers rarely need to rely on conventional medicine. Organic cattlemen should, however, try to familiarize themselves with alternative remedies such as herbal/aroma therapies, homeopathy, and immune system promoters.

20.8. Be Ready to Conduct Your Own On-farm Trials

Successful organic farmers continuously try new and/or innovative management practices. Practices such as cover cropping, inter-planting, and use of various soil and pest control materials need to be evaluated regularly by organic farmers. Be prepared to try new approaches.

20.9. Be Ready to Keep Good Records

Record keeping is one of the most important requirements to maintain organic integrity. Farmers are expected to keep detailed production, processing and marketing information. This information includes everything that enters and exits the farm. Third party, independent inspectors require farmers to present the above mentioned documentation when inspecting the farm operation. Once the record-keeping requirements are understood and the reporting procedure established, paperwork becomes routine.

20.10. Avoid these Common Mistakes

- Underestimating the need for good transitional and marketing plans.
- Underestimating the need to fully understand the Organic Standard. Organic producers must understand the standard in order to know what is permitted and prohibited.
- Failing to think prevention. Transitional farmers should consider improving their crop rotation, soil and crop management skills, livestock management practices (feeding program, heard health program, grazing system, housing facilities, and husbandry).

Annexures

Annexure I

Common Queries in Organic Farming

1. How are crop diseases managed on organic farms?

- Soil-borne diseases are managed by improving organic matter and biological activity.
- Cultural, biological, and physical methods such as rotation, sanitation, pruning, and selection of disease resistant varieties are all part of organic disease management.
- Some natural substances, such as clays, and a few synthetic fungicides such as copper sulfate are permitted by National Organic Program Standards when used in conjunction with the farm plan and used according to the restrictions found on the National List.

2. What are the requirements for converting to organic dairy production?

2 Ways

- Animals from conventional sources must be maintained under organic management for 12 months prior to sale of any products as organic. Replacement animals may be added to the herd after a similar 12-month conversion period.
- If an entire, distinct herd is converted, a one-time allowance is granted to permit feeding of up to 20% non-organic feed for the first 9 months, followed by 100% organic feed for three months. If this type of conversion is made, all replacement animals must be managed organically from the last third of gestation.

3. How do organic animals meet their nutritional requirements?

- All agricultural products provided in the feed ration must be organic, with a limited amount of supplementation and additives.
- Ruminant livestock must have a significant portion of their feed needs met by pasture

4. How do producers maintain the health of organic animals?

- Livestock health care must be based on preventive practices, such as balanced nutrition from organic feed, stress reduction, and preventative practices.
- Medications on the National List may be used only when necessary, and may not be administered in the absence of illness.
- Antibiotics are not permitted and products from animals treated with any prohibited medication must be diverted from organic marketing channels

5. What methods are available to manage parasites in organic livestock?

- Organic livestock producers rely on cultural practices to minimize parasite infestations.
- Synthetic parasiticides may be used only if they are on the National List and are prohibited for use in slaughter stock.

6. What are the living conditions for organic livestock?

- All organic animals are required to have **access to the outdoors** and exercise areas, and must be provided with healthy living conditions.
- Ruminants are also required to have access to pasture.

7. What is the National Organic Program?

- The National Organic Program (NOP) consists of the regulations and regulatory agents to establish and protect the standards for agricultural products labeled as 'organic.'
- These standards are known as the National Organic Standards.
- All organic food label claims must now be backed by valid certification according to the NOP Rule.

8. Who has to be certified?

- All producers and handlers that make an organic claim for their products must be certified by accredited certification agency.

- There are few exceptions for small farmers or handlers; handlers that buy and sell without repackaging or changing form, and retailers that do not process food.
- Exempt operations must maintain records and follow the exact same production practices as certified farmers in order to label their products as organic.

9. Who does the certifying?

Accredited state, private and international agencies will certify agricultural products and food as organic under the NOP.

10. How long does it take to transition land farmed conventionally to organic status?

In order to be eligible for organic certification, land must have had no prohibited materials applied to it for three years immediately preceding harvest.

11. Must an entire farm be converted, or can a farm make the transition field by field?

- A farm can be converted field by field.
- However, to be certified, a field must have distinct, defined boundaries and buffer zones to protect it from runoff and unintended contamination from adjoining land.
- The farm also needs to have facilities and recordkeeping in place to ensure and document that organic and non-organic crops are not commingled.

12. What are acceptable sources of animals used for organic meat production?

Slaughter animals can come from any breeding stock that has been organically managed from the last third of gestation.

13. What sources of poultry are acceptable for organic poultry products?

Poultry must be managed organically from the second day of life.

14. Can animals be converted to organic production at the same time as the land?

Yes. Livestock operations may convert animals with the land on which they are pastured.

15. What does it mean to be 'certified organic'?

- Certified organic means that the food has been grown and handled according to the National Organic Program Standards and inspected by independent state or private organizations.
- Periodic unannounced inspections are also conducted.
- Certification includes annual inspection of all farm fields and facilities, farm activity records, plus periodic testing of soil, water and produce to ensure that growers and processors meet National Organic Standards.

16. How much does certification cost?

- Each certification body is required to establish and publish fee schedules that are applied fairly to all applicants.
- Fees vary considerably from agency to agency, and depend on the size and type of organic operation to be certified.

17. Are farms outside the U.S. subject to the National Organic Standards?

- In order to market agricultural products as organic in the United States, they must have been produced and handled in accordance with the Organic Foods Production Act and the National Organic Standards.
- The same system of inspection and certification to the standards is applied to foreign operators who export their products into the United States.

18. Is organic certification automatically recognized in other countries?

- All states and accredited certifiers accept certifications issued by accredited or recognized certification programs.
- Foreign governments and international certification bodies have similar organic standards but may insist on additional certification to confirm operators who export organic products meet their standards.

19. How does a farm get certified?

- The operator obtains and reads the National Organic Standards, and conducts a self-assessment to see if the operation meets these requirements in terms of land history, production practices, materials used and recordkeeping procedures.
- In many cases, some practices and systems need to be modified to comply.

- Once an operation complies, the operator then selects an accredited certifier, submits an application, gets inspected, meets any conditions identified by the certifier, and obtains a certificate.

20. What are the penalties for misuse of the term "organic?"

Any operation that knowingly sells or labels an agricultural product as "organic," not in accordance with the Organic Foods Production Act and the National Organic Standards may be subject to a civil penalty and criminal sanctions based on violation of laws governing fraud and false statements.

21. Must organic farmers use organic seeds?

- The NOP Rule requires that organically produced and handled seeds be planted when such seed is commercially available for the variety.
- Annual transplants must always be organically grown unless the Secretary of Agriculture in response to a natural disaster or other major interruption issues a temporary variance.
- All seed used in organic production must be untreated, or treated only with substances (such as microbial products) that are on the *National List*.

22. What is an Organic Systems Management Plan or Organic Farm Plan or Organic Handling Plan?

- These documents identify who is responsible for the organic operation and describe the management and recordkeeping practices to monitor implementation of that plan.
- The plan serves as a contract between the operator and the certifier.
- Most certifiers assist operators in developing their plan by providing forms and guidance documents.
- Organic Systems Management Plans must be updated at least annually in order to maintain certified organic status.

23. How does an Organic Systems Management Plan relate to soil management?

- Farmers are required to demonstrate that they use appropriate tillage and cultivation practices without negative impacts on soil structure, and manage crop nutrients and fertility using crop rotations, cover corps, and application of organic materials.
- There is also a requirement that soil organic matter be maintained or improved in a manner that does not contaminate crops, soil, or water by

plant nutrients, pathogenic organisms, heavy metals, or prohibited materials.

- The farmer must have a system for monitoring all practices and procedures, as well as records for all farm inputs, harvest products, and storage facilities.

24. What farm inputs are allowed and what materials are prohibited in organic production?

In general, the NOP allows natural (non-synthetic) substances and prohibits synthetic substances, unless they appear on the National List.

25. What is the National List?

It is not a comprehensive list of all approved materials, rather it can be described as an "open" list since it contains only

1. Synthetic materials allowed for use in crop and livestock production and
2. Nonsynthetic (natural) materials prohibited for use in crop and livestock production.

26. Who determines if a specific product is acceptable for use on an organic farm?

In most cases, the **certification agencies** determine whether or not the use of a given input on a farm complies with organic standards.

27. Should an organic farmer plan on what inputs to use in the coming year?

- Organic farmers should anticipate production needs and determine the practices and inputs needed to achieve that production.
- All fertilizers and pesticides that a farmer intends to use over a season must be included in the farm plan.
- In all cases, a certified farmer should have any products used on the farm are approved by his/her certification agent *before* the input is used.

28. Are there any further restrictions on the use of fertilizers and pesticides in organic farming?

- Farmers need to be aware of the limitations of the *National List* when it applies to farm inputs.
- Producers may only use substances listed for crop use on crops.
- For example, phosphoric acid is permitted in livestock sanitation, but not as a fertilizer.

- A material listed for a specific use is restricted to that use, i.e. soap is listed for insect control but not disease control.
- Some materials have specific restrictions, for instance - copper must be used in a manner that minimizes accumulation in the soil.

29. Are there any further restrictions on the use of fertilizers and pesticides in organic farming?

- Farmers need to be aware of the limitations of the *National List* when it applies to farm inputs.
- Producers may only use substances listed for crop use on crops.
- For example, phosphoric acid is permitted in livestock sanitation, but not as a fertilizer.
- A material listed for a specific use is restricted to that use, i.e. soap is listed for insect control but not disease control.
- Some materials have specific restrictions, for instance - copper must be used in a manner that minimizes accumulation in the soil.

30. Can pesticides be used on an organic farm?

- Most pesticides are prohibited for use in organic production, but a number are allowed with restrictions.
- In crop production, **pesticides** must have active ingredients that are either non-synthetic or on the *National List*, and all inert ingredients must be non-synthetic or inerts of minimal concern.

31. What feed additives and supplements are permitted?

- Natural (nonsynthetic) feed additives and supplements are permitted, as are synthetic substances that are on the *National List.*
- These include synthetic vitamins and minerals, which are limited to the amount necessary for adequate nutrition.
- Slaughter by-products are prohibited for feeding to mammals and poultry, and urea and manure re-feeding is prohibited for all livestock.
- Synthetic amino acids are not included on the *National List*, with the exception of a temporary allowance granted for methionine for use in poultry until October 2005.

32. What are we supposed to do when we need to treat a sick animal?

- Animals that are sick must be treated.

- A producer who withholds treatment from a sick animal to maintain its organic status can be decertified.
- If a synthetic animal drug used to treat an animal does not appear on the *National List*, then the animal must be diverted to conventional channels.

33. Are any parasiticides allowed?

- Only one parasicitide, Ivermectin is on the National List.
- It is restricted for use in dairy and breeding stock only.
- Parasiticides are categorically prohibited on slaughter stock.

34. Will it be possible to tell if a product meets organic standards just by reading the label?

- The EPA has implemented a new voluntary labeling program to help identify products that meet NOP requirements.
- Approved registered pesticide in this voluntary program can include the phrase "for organic production" on their labels.
- Not all products that are compliant with organic rules will be so identified.

Annexure II

Certification Agencies

IMO Control Pvt. Ltd.
Mr. Umesh Chandrasekhar DirectorNo. 1314, Double RoadIndiranagar 2nd Stage Bangalore-560 038.(Karnataka)
Phone No.: 080-25285883, 25201546
Fax: 080-25272185 Email:imoind@vsnl.com

Bureau Veritas Certification India Pvt. Ltd. (Formerly known as BVQI (India) Pvt. Ltd.) Mr. R. K. Sharma Director Marwah Centre, 6th Floor Opp. Ansa Industrial Estate
Krishanlal Marwah Marg Off Saki-Vihar Road Andheri (East) Mumbai-400 072 (Maharashtra) Phone No.: 022-56956300, 56956311Fax No. 022-56956302 / 10Email: scsinfo@in.bureauveritas.com

Indian Organic Certification Agency (INDOCERT) Mr. Mathew Sebastian Executive Director Thottumugham P.O. Aluva-683 105 Cochin, (Kerala)
Telefax:0484-2630908-09/2620943
Email:Mathew. Sebastian@indocert.org

ECOCERT India Pvt. Ltd
Dr. Selvam Daniel(C.R.) Sector-3, S-6/3 & 4, Gut No. 102 Hindustan Awas Ltd. Walmi-Waluj Road Nakshatrawadi, Aurangabad – 431 002 (Maharashtra) Phone No.: 0240-2377120, 2376949
Fax No.: 0240-2376866
Email: ecocert@sancharnet.in

Lacon Quality Certification
Pvt. Ltd Mr. Bobby Issac Director Chenathra, Theepany, Thiruvalla - 689101. (Kerala) Telefax: 0469 2606447
Email: laconindia@sancharnet.in

Natural Organic Certification Agency
Mr. Sanjay Deshmukh, CEOChhatrapati House Ground FloorNear P. N. Gadgil Showroom Pune-411 038 (Maharashtra)
Phone No.: 020-25457869, 56218063
Fax:020-2539-0096
Email: contact@nocaindia. com

SGS India Pvt. Ltd.
Dr Manish Pande Divisional Manager – Food, Retail & CSRS250 Udyog Vihar Phase – IV Gurgaon – 122 015 (Haryana)
Phone No.: 0124-2399990-98
Fax No.: 0124-2399764
Email: namit_mutreja@sgs.com

One Cert Asia Agri Certification
Pvt. Ltd. Mr. Sandeep Bhargava
Chief Executive Officer Agrasen Farm, Vatika Road,Vatika P.O., Off Tonk,Jaipur-303 905, (Rajasthan)
Phone No. : - 0141-2770342
Telefax No: - 0141-2771101
Email: info@onecertasia.in

Control Union Certifications (Formerly known as Skal International (India))
Mr. Dirk Teichert Managing Director"Summer Ville"8th Floor 33rd – 14th Road Junction Off Linking Road, Khar (West)
Mumbai –400052 (Maharasthra)
Phone 022-67255396/97/98/99
Fax 022-67255394/95
Email: cuc@controlunion.incucindia@ controlunion.comcontrolunion@vsnl.com

Uttranchal State Organic Certification Agency (USOCA)
Director 12/II Vasant Vihar Dehradun-248 006 (Uttaranchal)
Phone No.: 0135-2760861
Fax: 0135-2760734
Email: uss_opca@rediffmail.com

Contd.

APOF Organic Certification Agency
(AOCA) Mr. K. Dorairaj Chief Operating Officer#3, 1st floor, 9th cross, 5th main, Jayamahal Extn, Bangalore – 560046
Phone No: 080-55369888 Fax: 080-23430155
Email: aocabangalore@yahoo.co.in

Rajasthan Organic Certification Agency (ROCA) 3rd Floor, Pant Krishi Bhawan, Janpath, Jaipur 302 005(Rajasthan)
Phone No: 0141-2227104
Tele Fax: 0141-2227456
Email: dir_rssopca@rediffmail.com

Vedic Organic Certification Agency
Plot No. 55, Ushodaya Enclave, Mythrinagar, Miyanagar, Hyderabad – 500 050
Mobile No.: 09290450666
Tel. No.: 040-65276784
Fax: 040-23045338Email: voca_org@yahoo.com;usha_preetham@yahoo.co.in

ISCOP (Indian Society for Certification of Organic Products) Rasi building,162/163 PonnaiyarajapuramCoimbatore – 641 001
Tamil NaduMob. No.: 094432 43119

Food Cert India Pvt. Ltd
Quality House, H. No. 8- 2- 601/P/6, Road No. 10, Banjara Hills, Panchavati Colony, Hyderabad – 500 034
Tel. No.: +91- 40-23301618, 23301554, 23301582
Fax: +91-40-23301583
Email: foodcert@foodcert.in

Aditi Organic Certifications Pvt. Ltd
No. 531/A, Priya Chambers
Dr. Rajkumar Road, Rajajinagar,1st Block, Bangalore - 560010
Tel.: +91-80-32537879
Fax: +91-80-23373083
Mobile: +91-9845064286
Email: aditiorganic@gmail.com
Website: www.aditicert,net

Chhattisgarh Certification Society,India (CGCERT), RaipurA-25, VIP Estate
Khamhardih, Shankar Nagar
Raipur- 492007,Chhattisgarh (India)
Telefax: +91-771-2283249
Email: cgcert@gmail.com

Tamil Nadu Organic Certification Department (TNOCD), Coimbatore
Thadagam Road, Coimbatore-641013,Tamil Nadu (India)
Tel.: +91-422-2405080
Fax: +91-422-2457554
Email: tnocd@yahoo.co.in

Annexure III

Contact Farmers and Agencies

S.No	Name of the Organic Farmer	Address of the farmer
1.	Mr.Sivaprakasam	Aranarai, Perambalur, Thiruvalluar District, Tamil Nadu
2.	Mr.G. Balakrishnan	Putharam Farm, Nemam, Thirukkathipalli (Via), Thanjavur District, Tamil Nadu
3.	Mr. UmeshChandrasekar and Mr.Meenakshi -Puvidham	Nagarkoodal Village (Po), Indur (Via), Dharmapuri - 636 803, Tamil Nadu. Ph.: 04342-311641, e - Mail: puvidhamtrust@yahoo.com.
4.	Mr.Ganapathy	Sakthi Farm, Veerapathy, Puliyur Post, KulathurTaluka, Pudukottai District - 622 504, Tamil Nadu.
5.	Mr. N.S.A. VeluMudaliar	4, Chidampara VinayagarKovil Street, Puliangudi, NellaikattabommanDistrict - 627 855 TamilNadu.
6.	Mr. P. Thangasamy	KarpagaSolai, Sendhangudi, Nagaram Post, Alangudi Taluk, Pudukottai District -614 624, Tamil Nadu.
7.	Gloria Land	Sri Aurobindo Ashram, Pondicherry - 605 002, Tamil Nadu. Ph: 0413 2666337, 2339017, Cell: 094432 87531, 094432 72780, e-Mail:glorialand@sancharnet.in
8.	Bhagyadhan Estate	P.O. Box 63, Kodaikanal - 624 101, TamilNadu. Contact: Arthur Steele
9.	Mr. N. Chokkalingam	No. 82, Virattipattu, Madurai - 625 010, Tamil Nadu.
10.	Mr. Gomathinayagam	No. 18, Uchi Magaliamman, Koil Street, Puliangudi, Nellaikattabomman District - 627 855, Tamil Nadu. Vivasaya Seva Sangam, C.B.Complex, Gandhi Bazaar, Puliankudi-627 855,

Contd.

S.No	Name of the Organic Farmer	Address of the farmer
		Thirunallvelli District, Tamil Nadu. Ph: 0463 6233235, Cell: 9629952636
11.	Khoram Estates	Fleurette,' Sivanadi Road, Kodaikanal - 624 101, Tamil Nadu. Contact person: MinooAvari
12.	Annapurna Farm	Bharat Nivas P.O., Auroville - 605 101, Tamil Nadu. Ph.: 0413 3155660, e-Mail: brooks@auroville.org.in
13.	Kolunji Farm	Kudumbam, Ezhil Nagar, Keeranur, Pudukottai District - 622 502, Tamil Nadu
14.	Santosh Farm	OoruppannadiNivas, Kottur, Malayandipattanam, Pollachi, Coimbatore District - 642 114, Tamil Nadu. Ph.: 04259 - 286499 to 286504 Cell: 09442416543 e-Mail: santoshfarms@gmail.com Contact: Mr. MadhuRamakrishnan
15.	Mr. N. Nagaraj	N. N. Farms, 244, Bahuttampalayam, Ekkaraithatthapalli P.O. Bhavanisagar Via - 638451, Erode District, TamilNadu. Ph.: 04295 221895, Cell: 09443071495
16.	Mrs. S.Poongodi /	PudhuNilavu Organic Farm / Manonmani Vermi Farm, Thalavumalai,
17.	Mr. R.Selvam	Arachaloor, Erode District - 638 101, Tamil Nadu. Ph.: 0424 2357537 e-Mail : manpulu@rediffmail.com
18.	Mr. V.S.Arunachalam	Elunkathir Organic Farm, P.Vellalapalayam Post, Gobichettipalayam, Erode District -638476, Tamil Nadu. Ph.: 04285 246301, Cell: 094433 46323
19.	Mr. K.Mohanasundaram & Mrs.Pushparani	AmudhaSurbhi Organic Farming Training Centre, 12, Thingalur Road, Nasiyanur Post, Erode District - 638107, TamilNadu. Ph.: 0424 2555227

Contd.

S.No	Name of the Organic Farmer	Address of the farmer
20.	Mr.R. Nallusamy& Mrs.Shanti	Koppampatty Post, ThuraiyurTaluka, Trichy - 621 012, Tamil Nadu. Ph.: 04327-253366
21.	Mr.M. Krishnamurthy	KullampalayatharThottam, Perumapalayam, Nagalur Post, Athani Via - 638 502, Bhavani Taluk, Erode District, Tamil Nadu. Ph.: 042567-261463
22.	Mr.V.C. Kannan	220, Thirumalirunsolai, Pudukaraipudur Post, Gobichettipalayam - 638 313, Erode District, Tamil Nadu. Ph.: 04285 - 266303, Cell: 098427 61232
23.	Mrs.ReetaGanapathy	Illuppakkorai - 614202, Ganapathiagraharam Via, Papanasam Taluka, TanjoreDistrict, TamilNadu
24.	Mr.P.C.Subramaniam	Nasiyanur, Pallivalayam, Erode District - 638 107, Tamil Nadu. Ph.: 04326-240555
25.	Mr.K.S.Raghavan	No. 2 Uppukinar Street, Kottur, M. Patnam PO, Pollachi, Coimbatore - 642 114, Tamil Nadu. Ph.: 04259-2522271
26.	Mr.S. Thangaraju and Mrs.Banumathi	236/4, AkkaraiKodiveri Post, Kasipalayam Via, GobichettiPalayam, Erode-District-638 454 Ph.: 04285 264150
27.	Mr.V. Ravi	RajchettyarThottam, Uppupallam, Kenjanur PO, Sathyamangalam via, Erode District - 638 401, Tamil Nadu. Ph.: 04295 24779
28.	Mr.N.Gopalakrishnan	No. 4/19, Akila Nagar First Cross,Ganapathy Nagar, South Extension, Mambazhasalai, Thiruvanaikoil, Trichy- 620 005, Tamil Nadu. ? Farm: Panickampatty village, Kuliathali Taluk, Karur District, Tamil Nadu.

Contd.

S.No	Name of the Organic Farmer	Address of the farmer
		Cell: 094431 48224, 099421 67789, e-Mail: dngopal2003@gmail.com /dngopal2003 @yahoo.co.in
29.	Mr.P.B. Mukundan	H. No. 92, Rajaji Street, Chingalpet 603001, Kanchipuram Dist., Tamil Nadu. Ph.: 044 27423902, Cell: 09382337818
30.	Mr.Ilangovan& Beauty Trust	39, Thirumanjana Street, Tal: Lalgudi, Dist.-Tiruchirapalli, Tamil Nadu 621601. Ph.: 0431-2543755, Cell: 09842411953
31.	Mr.V. Antony Samy	#53, Westcar Street, Sinthamani, Pulliyangudi, Taluk Sivagiri, District Thirunallvelli, Tamilnadu. Ph.:04636 233343, Cell: 9443582076
32.	Mr.R. Srinivasan	Vil: Kurumbarai, Post: Polambakkam, Taluka Cheyyur - 603 309, KancheepuramDistrict, Tamil Nadu. Cell: 09884756090
33.	Mr.A. Raja Pandian	No. 2, U.V. Saminathan St., Maruthi Nagar, Raja Keelpakam - 600 073, Chennai, Tamil Nadu. Ph.: 044-22270687, Cell: 09840620660
34.	Mr.P. Adaikalaraju	82/36, S.M.E.S.C. Colony, II Cross, K.K. Nagar (P.O.), Tiruchirapalli - 620021, TamilNadu. Cell: 09443581704
35.	Mr.A. DhanrajPatil& Arivagam Trust	Arivagam Trust, 36 Thirumangalam Road, Santhaipettai, Lalgudi - 621 601, Tamil Nadu. Ph.: 0431-6541986, Cell: 09943018554, e-Mail:humenreach@gmail.com, Web:www.tamilwriters.com Contact Person: Mr.A.Dhanaraj, Project Director
36.	Mr.R. Ramakrishnan	Vaipoor Village & Post Tiruvanamalai Tal. & Dist.- 606774, Tamil Nadu. Ph.: 04175-244791, Cell: 09787179096
37.	Mr.S.A.Dharmalingam	250 Suneisandai , TalukaSenumpatti Bhavani, Erode Dist.-638 504, Tamil Nadu. Ph.: 0425-8258527.

Contd.

S.No	Name of the Organic Farmer	Address of the farmer
38.	Mr.K.V. Palaniswamy	*VenkataswamyIllam, Kethanur - 641 671, Tal. Tiruppur, Dist. TamilNadu. Ph: 04225 279220/279241, Cell: 09843059241, 09943979791 * PrakashPaper Mills, Pattam, Pottapalayam - 630611, SivagangaiDistrict, TamilNadu. Ph. 0452 2465744, 3092767.
39.	Mr.K.Nallasamy	ErrikariThottam, Perumalkovilpudur, Nakalur post, Anthiyur - 638502, Erode District, Tamil Nadu. Cell: 9842729596, 9842829596
40.	Mrs.Parvathi Venkateshwaran	W/o Venkateshwaran, PallatuThottam, Alukuli post, Karatadipalayam, GobichettipalayamTaluka, Erode District,TamilNadu. Ph: 04285 264103, 264646
41.	Senthil Farm	Bangalapudur, 28, New Hospital Road, Gobichettipalayam-638452, Erode District, Tamil Nadu. Cell: 9842089273, e-Mail: mailto:visualv@gmail.com visualv@gmail.com Contact: Mr.V.Senthilkumar
42.	Mr.Ramalingam	AnnurGounderThottam, Sathyamangalam-638402, Tamil Nadu. Cell: 9965699288.
43.	Mr.K.R.Sundaram	1/1A Kovilthottam, Annur, Coimbatore -641 653, Tamil Nadu. Ph.:04254 264278, Cell: 9600916166
44.	Mrs. IndraRamanathan	VadakkuThottam, KanakkamPalayam Post, Kallipatti, Erode- 638505, Tamil Nadu. Ph: 04285 263333, Cell: 9345981081
45.	Mr.S. A. Myilsamy	Kannakanthottam,Chettipalayam, Aalathur post, Kavanthapaddi via, Bhavani Taluka-638455, Erode District, Tamil Nadu. Cell: 9965154640, 9965154642

Contd.

S.No	Name of the Organic Farmer	Address of the farmer
46.	Mr.P. Balasubramanium	S/o Palanisamygounder, Thoppampatti,Narasimmanayakanpalayam, Mettupalayam, Coimbatore,Tamil Nadu. Cell: 9944475099
47.	Mr.Arun Lakshminarayanan	AIG Farms, Vedapatti, Thondamuthur road, Coimbatore, Tamil Nadu. Cell: 9843015951, 9442222908
48.	Mr.KasthuriRangian	Malankuli, Thalavadi, Sathyamangalam- 638461, Tamil Nadu. Cell: 9443388145, 9442222908, 9842255908.
49.	Mr.Nandhakumar	Sri venkateswara Blue Metal, MangalakaraiPudur, Karamadai Via, Mettupalayam, Coimbatore - 641301, Tamil Nadu. Cell: 98452 72439
50.	Mr.R.Ramamurthy	52-Gt, Annur Road, Mettupalayam - 641301, Coimbatore, Tamil Nadu. Ph.: 04254 223952, Cell: 98422 23952
51.	Mr.K. Sureshkumar	B-4 Lakshmi Narayana Apartments, Dr.Suburayan Street, Tatabet, Coimbatore-12, TamilNadu. Ph: 0422 2490612
52.	Mr.S.Chinappan	Vattakkalvalasu, Malayampalayam, Karumandampalayam, Erode District, Tamil Nadu. Ph: 0424 2351136, Cell: 94430 19781
53.	Mr.C.V.Subramanium	3' R Spiritual farm, Polumvampatti road, Thanamanallur post, Thondamuthur, Coimbatore-641109. Ph.: 0422 2906200, Cell: 9994368998
54.	Mr.T.S.Raman	Green Kovai farm, AIMS for SEVA, Annaiketti post, Coimbatore- 641108, Tamil Nadu. Ph: 0422 265 7001, Cell: 94426 46713
55.	Mr.Selvaraj	M S organic Farm, Uvasimangalam, Alanthurai, Siruvani, Coimbatore-641101, Tamil Nadu. Cell: 9344451474, 9965975534.

Contd.

S.No	Name of the Organic Farmer	Address of the farmer
56.	Mr.C.Ganeshan	21,Senguntapuram, IInd. cross, Ramanutar Nagar, Karur-639 002, Tamil Nadu. Cell: 9865209217
57.	Mr.Venkat Rasa	S/o K.R.Radhakrishnan, No. 2, Erode Main Road, ValayuthamPalayam Post, Namakkal, Karur Taluk & District- 639 117, Tamil Nadu. Ph.: 04324-270398, Cell: 9842770398, e-Mail: rasaokr@gmail.com
58.	Mr.K.Ramakrishnan	Bangalore Thottam, Poyankuttai, Bhavani Taluka ,Erode District, Tamil Nadu.
59.	Mr.A.R.Selvaraj	S/o. RamasamyReddiyar, Aathirediyur, Kettisamuthuram post, Anthiyur via, Bhavani Taluka, Erode-638501, Tamil Nadu. Ph.: 0456 263331, Cell: 9750050605
60.	Mr.R.Raju	S/o. Ramappa Gounder, D.No:467, Karuppana MothiliyarThottam, Anna Nagar, Bommanpattti, BhavaniTaluka, ErodeDistrict, Tamil Nadu. Ph.: 04256 253152, Cell: 9976723940
61.	PerumalKovilThottam	Bommanpatti, Vellithiruppur, Bhavani, Erode-638 314, Tamil Nadu
62.	Mr.Karupanaraj	Bejalatti, Thalamalai, Thalavadi, Sathyamangalam, Erode District, Tamil Nadu
63.	Mr.P.B. Murali	77, Luz Church, Allvarpattai, Myllapur, Chennai, Tamil Nadu. Ph.: 044 24991203, Ph.: 9380691203, e-Mail: pbm50@rediffmail.com
64.	Mr.M. Sethuraman	Kidathalimeadu, Kalai post, Mailladuthurai - 609811, Tamil Nadu. Ph: 04364236467, Cell: 9952844467
65.	Mrs. Kalaivani	AnthiyurBommanpatti, Annanagar, Vellithiruppur, Bhavani TK, Erodedistrict, TamilNadu. Cell: 9865485221

Contd.

S.No	Name of the Organic Farmer	Address of the farmer
66.	Mr.G. Sither	Agriculturist, Layco's Nature Food Shop, 71/1516, Vanakkara Street, M. Chavady, Thanjavur 613 001, Tamil Nadu. Ph: 04362- 239788/272417, Cell: 09443139788, e-Mail: sither_1960@yahoo.co.in
67.	Auro Annam	Grace, Auroville 605 101, Tamil Nadu. Ph.: 0413-2622044, Res: 0413-2623391, e-Mail: auroannam@auroville.org.in, margarita@auroville.org.in
68.	Mr.R.Raju	S/o. RamappaGounder, D.No:467, KaruppanaMothiliyar Thottam, Anna Nagar, Bommanpattti, BhavaniTaluka, Erode District, Tamil Nadu. Ph.: 04256 253152, Cell: 9976723940
69.	Mrs.Vanya Orr	Project Director, Earth Trust, Ketty Post, Nilgiris, Tamilnadu. . Mob.: +919787749943, office: +914232517036 www.earthtrustnilgiris.org
70.	Mr.T. Navaneetha Krishnan	Poonthottam, 21/25, Vellipalayam Road, Mettupalayam, Coimbatore - 641301
71.	Mr.V. Natarajan	66D, Rajapuram Road, Banglamedu Mettupalayam Coimbatore District- 641301 Ph.: 04254 224888, Cell: 9442019955
72.	Mr.A.Yuvaraj	KurichanValasu, Nasiyanur, Erode district. Cell: 9865873751
73.	Mr.R. Jayachandran	Ariyanoor Village Periyavenmani Post Maduranthakam Kancheepuram- 603311 Ph.: 044 - 27539608

Annexure IV

Conditions for Products Used in Fertilization and Soil Conditioning in Organic Farming

Items	Conditions for use
Material from plant and animal origin	
Matter produced on an organic farm unit	
• Farmyard and poultry manure, slurry, urine	Permitted
• Crop residues and green manure	Permitted
• Straw and other mulches	Permitted
• Composts and Vermicompost	Permitted
Matter produced outside the organic farm unit	
• Blood meal, meat meal, bone meal, feather meal without preservatives	Restricted
• Compost made from plant residues and animal excrement	Restricted
• Farmyard manure, slurry, urine• Fish and fish products without preservatives	Restricted
• Guano	Restricted
• Human excrement	Restricted
• Wood, bark, sawdust, wood shavings, wood ash, wood charcoal	Restricted
• Straw, animal charcoal, compost and spent mushroom and vermiculate substances	Restricted
• Compost from organic household	Restricted
• Compost from plant residues	Restricted
• Sea weed and sea weed products	Restricted
By products from the industries	
• By-products from food and textile industries of biodegradable material of microbial, plant or animal origin without any synthetic additives	Restricted
• By products from oil palm, coconut and cocoa (including fruit bunch, palm oil mill effluent, cocoa peat and empty cocoa pods.	Restricted
• By-products of industries processing ingredients from organic agriculture	Restricted
• Extracts from mushroom, Chlorella, Fermented product from *Aspergillus*, natural acids (vinegar)	Restricted
Mineral origin	
• Basic slag	Restricted
• Calcareous and magnesium rock	Restricted
• Lime, limestone, gypsum	Permitted
• Calcified sea weed	Permitted
• Calcium chloride	Permitted
• Mineral potassium with low chlorine content (e.g. sulphate of potash, kainite, sylvinite, patenkali)	Restricted
• Natural phosphates (rock phosphate)	Restricted
• Trace elements	Permitted
• Sulphur	Permitted
• Clay (bentonite, perlite, zeolite)	Permitted

Annexure V

Conditions for Products used in Plant Pest and Disease Control

Items	Conditions for use
Material from plant and animal origin	
• Plant based repellents (Neem preparations from *Azadirachta indica*)	Permitted
• Algal preparations (gelatin)	Permitted
• Casein	Permitted
• Extracts from mushroom, chlorella,fermented products from *Aspergillus*	Permitted
• Propolis	Restricted
• Beeswax, Natural acids (vinegar), plant oils, Quassia	Permitted
• Rotenone from *Derris elliptica*, *Lonchocarpus*, *Tephrosia* spp.	Restricted
• Tobacco tea (pure nicotine is prohibited)	Restricted
• Preparation from *Ryania* species	
Mineral origin	
• Chlorides of lime/soda	Restricted
• Burgundy mixture	Restricted
• Clay (bentonite, perlite, vermiculite, zeolite)	Permitted
• Copper salts/ inorganic salts (Bordeaux mix, copper hydroxide, copper oxychloride)	Not allowed
• Quick lime	Restricted
Mineral origin	
• Diatomaceous earth	Permitted
• Light mineral oils	Restricted
• Permangnate of potash	Restricted
Insects origin	
• Release of parasites, predators of insect pests	Restricted
• Sterilized insects	Restricted
• Sterilized insect males	Not allowed
Microorganisms used for biological pest control	
• Viral, fungal and bacterial preparations (biopesticides)	Restricted
Others	
• Carbon dioxide and nitrogen gas	Permitted
• Soft soap, soda, sulphur dioxide	Permitted
• Homeopathic and ayurvedic preparations	Permitted
• Herbal and biodynamic preparations	Permitted
• Sea salt and salty water	Permitted
• Ethyl alcohol	Not allowed
Traps, barriers and repellants	
• Physical methods (e.g. chromatic traps, mechanical traps)	Permitted
• Mulches, nets	Permitted
• Pheromones – in traps and dispensers only	Permitted

Annexure VI

Awareness Creation

a) Individual contacts

i. Farm and home visits
ii. Office calls
iii. Telephone calls
iv. Personal letters
v. Result demonstrations.

b) Group contacts

i. Method demonstration meetings
ii. Leader training meetings
iii. Lecture meetings
iv. Conferences and discussion meetings
v. Meetings at result demonstrations
vi. Tours
vii. Schools
viii. Miscellaneous meetings.

c) Mass contacts

i. Bulletins
ii. Leaflets
iii. News stories
iv. Circualrletter's
v. Radio
vi. Television
vii. Exhibits
viii. Posters.

Who has to do?

Besides media, research community, research organizations and scientific insisutions concerned with organic crops must also take up the responsibility of promoting awareness.

Media

- FM station with SAUs
- Gyanvani
- UGC scientific programmes
- Scientific journals
- Science supplements in newspapers
- Roping in advertisement agencies.

Research community

- Exclusive public awareness units on organic crops in all R& D Departments
- Public – private awareness campaigns
- Sensitization of scientists' about consumer preferences
- Utilization of school and college students in conducting organic food safety exhibitions.
- Vision approach.

Annexure VII

Organic Agriculture Scenario

Organic Revolution

Organic agriculture started in 1920's in Europe

The development of the international organic movement over the last 70 years can be broadly classified into 3 Years.

- **1924 to 1970** – The period of struggle – during this period establishment of organic agriculture became tought due to financial crisis. But a lot of core works was written on organic farming.
- **1970 – 1980** – period of organic symbols – this era made awareness at the consumer level. Hence the demand for organic produce was created that led to multiplication of more retail outlets
- **From 1980 onwards** – acceptance of organic agriculture, national and international standards were set. Later in 1980 organic farming was framed on legal terms.

The international development of organic cultivation started with five renowned personalities *viz*., Rodolf Steiner from Australia, Han's Muller from German – Swiss, Lady Eve Balfour from Britain, J.I.Rodale in US and Masnobu Fukuoka from Japan.

The Global World of Organic Agriculture – 2010

As per the details released at BioFach 2010 at Nuremberg, the organic agriculture is developing rapidly, and statistical information is now available from 160 countries of the world. Its share of agricultural land and farms continues to grow in many countries. The main results of the latest global survey on certified organic farming are summarized below:

Growing area under certified organic agriculture

- 35 million hectares of agricultural land are managed organically by almost 1.4 million producers.
- The regions with the largest areas of organically managed agricultural land are Oceania (12.1 million hectares), Europe (8.2 million hectares) and Latin America (8.1 million hectares). The countries with the most organic agricultural land are Australia, Argentina and China.
- The highest shares of organically managed agricultural land are in the Falkland Islands (36.9 percent), Liechtenstein (29.8 percent) and Austria (15.9 percent).

- The countries with the highest numbers of producers are India (340'000 producers), Uganda (180'000) and Mexico (130'000). More than one third of organic producers are in Africa.
- On a global level, the organic agricultural land area increased in all regions, in total by almost three million hectares, or nine percent, compared to the data from 2007.
- Twenty-six percent (or 1.65 million hectares) more land under organic management was reported for Latin America, mainly due to strong growth in Argentina. In Europe the organic land increased by more than half a million hectares, in Asia by 0.4 million.
- About one-third of the world's organically managed agricultural land – 12 million hectares is located in developing countries. Most of this land is in Latin America, with Asia and Africa in second and third place. The countries with the largest area under organic management are Argentina, China and Brazil.
- 31 million hectares are organic wild collection areas and land for bee keeping. The majority of this land is in developing countries – in stark contrast to agricultural land, of which two-thirds is in developed countries. Further organic areas include aquaculture areas (0.43 million hectares), forest (0.01 million hectares) and grazed non-agricultural land (0.32 million hectares).
- Almost two-thirds of the agricultural land under organic management is grassland (22 million hectares). The cropped area (arable land and permanent crops) constitutes 8.2 million hectares, (up 10.4 percent from 2007), which represents a quarter of the organic agricultural land.

Total area under organic agriculture continent wise growth

Country	Area (million ha)	% organic Area to world
Australia / Oceania	12.1	35
Latin America	8.1	23
Europe	8.2	23
North America	2.5	7
Asia	3.3	9
Africa	0.5	2.5

(NPOF, 2008)

Land area of major countries under organic agriculture during 2008-2009

Country	Area under organic agriculture (ha)	Percentage of total agricultural land	Number of organic farms
Australia	12,294,290	2.8	1550
China	2,300,000	0.4	1600
Argentina	2,220,489	1.7	1486
USA (2005)	1,620,351	0.5	8493
Italy	1,148,162	9.0	45,115
Uruguay	930,965	6.1	630
Spain	926,390	3.7	17,214
Brazil	880,000	0.3	15,000
Germany	825,539	4.8	17,557
UK	604,404	3.8	4485
Canada	604,404	0.9	3571
France	552,824	2.0	11,640
India	528,171	0.3	44,926
World total	30,418,261	0.65	718,744

(Willer, 2009)

World wide scenario

S.No	Indicator	Global Totals	Leading Countries
1.	Organic agricultural land	37.2 mil. ha	Australia (12 mil. ha),Argentina (3.8), US (1.9)
2.	Countries with > 5% organic agricultural land	25	Falklands (35.9%), Liechtenstein (29.3%), Austria (19.7%)
3.	Producers	1.8 million	India (574,591), Uganda (188,625 as of 2010), Mexico (169,570)
4.	Organic market size	$US 62.9 bn.	US ($US 29 bn.), Germany(9.2), France (5.2)
5.	Organic per capita consumption per year	$US 9	Switzerland ($US 250), Denmark (226), Luxemburg (187

Source: The World of Organic Agriculture 2013, IFOAM and FiBL

Annexure VIII

State Wise Organic Farming Area

State-wise farm area under organic certification (excluding forest area) during 2013-14

Name of States	Area (ha)
Andaman and Nicobar Islands	321.28
Andhra Pradesh	12325.03
Arunachal Pradesh	71.49
Assam	2828.26
Bihar	180.60
Chhattisgarh	4113.25
Delhi	0.83
Goa	12853.94
Gujarat	46863.89
Haryana	3835.78
Himachal Pradesh	4686.05
Jammu & Kashmir	10035.38
Jharkhand	762.30
Karnataka	30716.21
Kerala	15020.23
Lakshadweep	898.91
Madhya Pradesh	232887.36
Maharashtra	85536.66
Manipur	0.00
Meghalaya	373.13
Mizoram	0.00
Nagaland	5168.16
Odisha	49813.51
Pondicherry	2.84
Punjab	1534.39
Rajasthan	66020.35
Sikkim	60843.51
Tamil Nadu	3640.07
Tripura	203.56
Uttar Pradesh	44670.10
Uttarakhand	24739.46
West Bengal	2095.51
Total	723039.04

National programme on organic production (NPOP) in India (2012-2013 to 2014-2015) (In MT)

States/UTs	2012-13	2013-14	2014-15
Andaman and Nicobar Islands	0.00	0.00	0.00
Andhra Pradesh	5935.63	8361.60	9556.78
Arunachal Pradesh	60.62	17.34	0.00
Assam	4832.72	7923.87	7515.37
Bihar	0.00	0.00	0.00
Chhattisgarh	273.36	2276.93	4579.94
Delhi	0.00	0.00	0.00
Goa	2547.62	4088.66	3427.63
Gujarat	37941.99	56217.21	60621.45
Haryana	3353.17	3823.24	2833.38
Himachal Pradesh	36.02	123.58	227.26
Jammu and Kashmir	3724.39	6084.57	11234.83
Jharkhand	0.00	0.00	0.00
Karnataka	322455.22	258335.82	254664.37
Kerala	5784.10	6926.40	9122.56
Lakshadweep	0.00	0.00	0.00
Madhya Pradesh	425848.82	397015.12	321586.83
Maharashtra	270743.69	296846.63	217112.20
Manipur	28.00	17.70	0.00
Meghalaya	1446.03	200.31	1738.29
Mizoram	0.00	0.00	0.00
Nagaland	0.00	9.68	391.67
Odisha	48725.49	33668.27	28678.65
Puducherry	0.00	1.00	3.00
Punjab	880.12	2177.54	506.77
Rajasthan	90021.46	54118.55	59449.77
Sikkim	0.00	90.62	67.63
Tamil Nadu	17016.77	8854.55	17218.40
Telangana	0.00	0.00	164.00
Tripura	36.92	0.00	262.82
Uttar Pradesh	47732.99	55575.44	50165.63
Uttarakhand	28835.61	15573.11	23695.63
West Bengal	10738.47	10774.53	10929.11
India	1328999.22	1229102.28	1095754.12

State-wise consumption of organic manures and chemical fertilizers in India

State/UTs	Organic Fertilisers (in lakh MT)	Urea	DAP	MOP	Complex
Andhra Pradesh	72.03	1806.27	371.41	223.18	1122.33
Telangana		1680.72	243.54	103.16	771.26
Andaman and Nicobar Islands	0.00	0.59	0.59	0.31	0.26
Arunachal Pradesh	1.00	0.97	0.15	0.25	0.00
Assam	9167.50	281.51	34.79	111.33	0.00
Bihar	12.17	1870.64	360.15	139.99	157.94
Chandigarh	0.00	0.00	0.00	0.00	0.00
Chhattisgarh	104.25	651.07	236.87	66.56	84.41
Dadra and Nagar Haveli	0.00	0.85	0.72	0.01	0.00
Daman and Diu	0.00	0.15	0.03	0.00	0.00
Delhi	0.80	7.38	0.00	0.00	0.00
Goa	0.82	4.73	1.83	0.71	2.53
Gujarat	365.81	2144.87	438.59	107.01	398.20
Haryana	0.09	1907.18	396.76	23.59	11.40
Himachal Pradesh	17.65	63.69	0.00	7.93	23.27
Jammu and Kashmir	6.57	124.38	54.14	17.73	1.54
Jharkhand	525.02	162.56	28.77	3.27	15.83
Karnataka	11.65	1479.19	458.28	251.48	1009.41
Kerala	3.85	174.93	31.36	136.37	204.96
Lakshadweep	0.00	0.00	0.00	0.00	0.00
Madhya Pradesh	5.66	2223.53	897.49	51.34	182.91
Maharashtra	9.96	2655.43	639.91	329.96	1573.33
Manipur	0.61	17.14	2.14	1.94	0.00
Meghalaya	16.35	7.20	1.08	0.45	0.00
Mizoram	0.10	9.47	4.57	4.31	0.00
Nagaland	0.88	1.61	1.20	0.51	0.55
Odisha	23.63	532.18	150.00	87.81	190.74
Puducherry	3.23	21.87	1.42	2.00	7.08
Punjab	3.06	2690.54	685.28	35.55	17.15
Rajasthan	0.02	1828.92	462.84	1.07	29.26
Sikkim	0.35	0.00	0.00	0.00	0.00
Tamil Nadu	14.39	911.95	219.49	246.90	414.90
Tripura	0.00	20.17	3.83	7.61	0.00
Uttar Pradesh	0.22	5805.42	1364.55	109.52	330.94
Uttarakhand	23.57	274.63	21.80	1.43	34.72
West Bengal	151.41	1238.74	243.84	207.13	678.60
India	10542.65	30600.48	7357.42	2280.41	7263.52

Area covered under organic and green manures in india

(In Lakh Hectare)

Items	2008-09	2010-11
I. Organic Manure		
Rural Compost	40.72	120.37
Urban Compost	24.37	17.04
FYM	126.74	292.44
Vermicompost	24.58	48.87
Other Manures	29.43	47.36
Total Organic Manures	**245.83**	**526.08**
II. Green Manure	-	**20.62**

State-wise Number of Fruits and Vegetable Compost Units under Credit Linked

Back-ended Subsidy under National Project on Organic Farming

States/UT	Number of unit			Total subsidy released (Rs. in Lakh)
	2013	2014	2015	
Assam	2	2	2	12.75
Bihar	0	1	1	4.13
Delhi	1	1	1	20.00
Goa	1	1	1	11.60
Gujarat	1	1	1	4.77
Karnataka	4	5	5	184.53
Kerala	2	2	2	21.29
Maharashtra	0	1	1	35.71
Tamil Nadu	2	2	2	57.00
Tripura	1	1	1	20.00
Uttar Pradesh	1	1	1	19.19
West Bengal	0	1	1	15.13
India	16	19	19	406.09

Selected country-wise export of organic food products from India

Countries	2011-2012		2012-2013		2013-2014	
	Volume(MT) Export Qty.	Value (Rs.in Crore) Export Value	Volume(MT) Export Qty.	Value (Rs.in Crore) Export Value	Volume(MT) Export Qty.	Value (Rs.in Crore) Export Value
Australia	349.14	5.15	468.26	6.60	749.95	14.58
Bangladesh	330.75	4.60	805.48	23.56	0.00	0.00
Brazil	0.02	0.00	0.50	0.02	0.00	0.00
Canada	19848.91	66.66	33645.80	146.05	38545.57	182.41
Chile	0.26	0.01	5.00	0.16	6.75	0.14
China	327.59	5.13	4.42	0.14	76.35	1.57
Croatia	0.00	0.00	0.00	0.00	46.50	0.31
Dominican Republic	144.00	0.59	0.00	0.00	0.00	0.00
Egypt	1.50	0.07	7.00	0.17	12.00	0.31
European Union	51138.86	505.29	82835.37	678.51	56946.72	553.85
Hong Kong	0.36	0.02	50.14	0.50	19.57	0.38
Indonesia	299.73	2.85	40.00	0.10	0.00	0.00
Iran	0.00	0.00	0.00	0.00	38.00	1.21
Israel	871.48	2.57	610.07	2.34	312.93	3.72
Jamaica	0.00	0.00	0.00	0.00	10.00	0.42
Japan	232.77	8.79	199.22	11.11	309.07	16.12
Kenya	10.06	0.17	4.40	0.11	1.02	0.02
Korea Dem. Rep.	50.70	0.28	14.62	0.10	14.40	0.08

Contd.

Countries	2011-2012		2012-2013		2013-2014	
	Volume(MT) Export Qty.	Value (Rs.in Crore) Export Value	Volume(MT) Export Qty.	Value (Rs.in Crore) Export Value	Volume(MT) Export Qty.	Value (Rs.in Crore) Export Value
Korea Republic	8.05	0.36	332.11	1.90	143.48	2.33
Kuwait	12.00	0.09	66.00	0.48	18.00	0.19
Lao Pd. Dem. Rep.	14.40	0.08	0.00	0.00	0.00	0.00
Lebanon	40.00	0.13	0.00	0.00	0.00	0.00
Malaysia	42.87	0.74	560.06	6.62	43.44	0.91
Mauritius	1.14	0.06	20.00	0.09	40.00	0.21
Nepal	0.00	0.00	0.00	0.00	12.00	0.07
New Zealand	499.00	2.97	409.68	2.57	599.79	4.23
Nicaragua	0.00	0.00	0.00	0.00	0.03	0.01
Nigeria	0.10	0.00	0.00	0.00	0.00	0.00
Norway	0.03	0.03	0.00	0.00	4.11	0.28
Philippines	0.00	0.00	0.00	0.00	110.11	1.88

Selected state-wise total organic manures produced in India

State/UTs	Total Organic Manure Produced* (In Lakh MT)			
	2009-10	2010-11	2011-12	2012-13
Andhra Pradesh	93.550	118.450	106.000	97.650
Arunachal Pradesh	0.120	0.120	0.010	0.430
Assam	33.910	5.850	2.850	880.908
Bihar	5.500	66.250	66.250	11.910
Chhattisgarh	128.730	144.480	129.150	103.290
Delhi	-	-	-	0.666
Goa	1.354	3.900	4.300	4.710
Gujarat	21.000	40.000	363.500	366.70
Haryana	10.050	18.400	18.400	18.400
Himachal Pradesh	40.550	40.550	40.550	40.550
Jammu and Kashmir	459.950	22.200	22.200	22.2071
Jharkhand	23.000	23.000	234.450	23.000
Karnataka	2001.270	1442.090	1108.620	1560.640
Kerala	131.870	131.870	84.990	11.945
Madhya Pradesh	97.500	136.000	136.000	136.000
Maharashtra	91.320	95.470	0.820	95.740
Manipur	0.500	0.500	0.500	0.500
Meghalaya	N.A.	0.950	10.570	14.900
Mizoram	0.210	0.210	0.080	0.081
Nagaland	0.090	0.160	0.160	0.729
Odisha	85.450	131.820	11.490	19.857
Puducherry	-	-	-	-
Punjab	92.190	379.620	341.290	342.080
Rajasthan	5.070	294.520	294.520	380.820
Sikkim	22.500	27.600	0.006	0.150
Tamil Nadu	9.060	56.390	8.370	56.390
Tripura	N.A.	N.A.	N.A.	0.000
Uttar Pradesh	38.760	327.780	327.780	0.086
Uttarakhand	0.380	0.380	10.640	0.385
West Bengal	92.190	162.840	162.840	162.840
India	3486.070	3671.400	3486.330	4353.294

Selected State-wise Production of Organic Manure and Availability in India (2014-2015)

State/UTs	Rural Compost	FYM	City compost	Organic manure	Vermicompost	Other manure	Total manure	Area cover (Lakh ha)	Green manure (ha)
Andhra Pradesh	49.50	0.21	21.30	0.00	1.02	0.00	72.03	7.20	28.54
Arunachal Pradesh	0.11	0.25	0.00	0.01	0.65	0.06	1.08	0.11	0.07
Assam	0.92	940.00	0.03	0.00	1.75	0.00	942.70	94.27	2.90
Bihar	3.83	0.00	0.29	4.91	3.14	0.00	12.17	1.22	362000.00
Chhattisgarh	49.00	45.50	3.25	0.00	3.50	3.00	104.25	10.43	420000.00
Delhi	0.00	0.00	0.80	0.00	0.00	0.00	0.80	0.08	0.00
Goa	0.04	0.00	0.50	0.04	0.03	0.22	0.83	0.08	3.00
Gujarat	0.00	361.00	0.00	0.00	0.51	4.20	365.71	36.57	3.20
Haryana	0.00	0.00	0.00	0.08	0.06	0.00	0.15	0.01	0.00
Himachal Pradesh	2.00	0.00	0.23	0.01	26.40	0.00	28.64	2.86	0.50
Jammu and Kashmir	1.76	0.00	0.19	0.00	0.00	0.00	1.95	0.20	1080.00
Jharkhand	0.00	0.00	0.00	0.00	511.00	0.00	511.00	51.10	0.00
Karnataka	0.00	0.00	10.32	0.00	6.10	0.00	16.42	1.64	0.00
Kerala	0.00	1.11	1.86	0.00	0.88	0.00	3.85	0.39	0.00
Madhya Pradesh	5.85	0.01	0.06	0.00	0.00	0.14	6.05	0.61	60000.00
Maharashtra	0.00	0.00	5.50	0.00	1.42	3.13	10.05	1.00	0.65
Manipur	0.00	0.65	0.00	0.00	0.05	0.00	0.70	0.07	0.00
Meghalaya	0.00	17.00	0.00	0.00	0.00	0.00	17.00	1.70	0.00
Mizoram	0.00	0.06	0.00	0.00	0.06	0.00	0.12	0.01	0.00

Contd.

Contd.

Nagaland	0.04	0.84	0.00	0.00	0.05	0.00	0.93	0.09	0.90
Odisha	23.27	0.00	0.07	0.00	0.00	0.00	23.34	2.33	0.00
Puducherry	0.00	0.00	3.23	0.00	0.00	0.00	3.23	0.32	0.00
Punjab	0.00	2.96	0.00	0.01	0.33	0.03	3.33	0.33	199000.00
Rajasthan	0.00	0.00	0.00	0.02	0.00	0.01	0.02	0.00	0.00
Sikkim	0.32	0.00	0.00	0.00	0.03	0.00	0.35	0.04	0.00
Tamil Nadu	0.00	0.00	6.38	5.04	0.00	0.00	11.42	1.14	0.00
Tripura	0.00	0.00	0.00	0.00	0.00	0.00	0.00	0.00	0.00
Uttar Pradesh	0.00	0.00	0.00	0.00	0.11	0.00	0.11	0.01	238805.00
Uttarakhand	9.40	13.60	0.00	0.00	0.30	0.00	23.30	2.33	0.00
West Bengal	79.84	17.20	10.21	0.07	25.81	3.99	137.12	13.71	16.10
India	225.88	1400.39	64.22	10.19	583.20	14.76	2298.62	229.86	1280941.00

State-wise number of organic units (Compost, Biofertilizer, and Vermiculture)* set up and capacity generated in India (As on March, 2010)

States/UTs	FVMWC		Biofertilizer		Vermiculture	
	Number	Capacity	Number	Capacity	Number	Capacity
Andhra Pradesh	0	0	8	1068	5	600
Arunachal Pradesh	0	0	0	0	40	1500
Assam	1	32	0	0	67	2450
Bihar	0	0	0	0	44	2230
Chhattisgarh	0	0	1	37.5	108	4635
Delhi	1	100	0	0	0	0
Goa	1	26	1	150	0	0
Gujarat	1	44	3	405	86	3870
Himachal Pradesh	0	0	2	300	37	1470
Jammu and Kashmir	2	200	1	37.5	25	937
Jharkhand	0	0	2	75	23	975
Karnataka	2	110	1	150	54	5650
Kerala	2	50	2	300	1	10
Manipur	0	0	1	37.5	20	750
Maharashtra	0	0	10	1035	35	3625
Madhya Pradesh	1	100	2	100	83	5512
Mizoram	1	100	1	37.5	62	2325
Meghalaya	0	0	1	84	0	0
Nagaland	0	0	1	37.5	103	3862
Orissa	0	0	1	37.5	147	5512
Punjab & Haryana	1	25	2	280	187	20437
Rajasthan	0	0	1	81	144	12600
Sikkim	0	0	0	0	8	300
Tripura	0	0	0	0	72	2700
Tamil Nadu	2	110	9	490	45	2128
Uttar Pradesh	2	125	1	37.5	138	18562
Uttarakhand	0	0	2	270	78	3037
West Bengal	0	0	2	210	7	753
India	17	1022	55	5260.5	1619	106430

Production of organic products in India

Products	2010-11	2011-12	2014-15
Cereals and Millets (Excluding Rice)	171684.66	40785.61	159500.16
Dry Fruits	52369.09	521.46	7348.39
Fibre Crops	-	-	208931.40
Fodder Crops	-	-	44.06
Fruits and Vegetables	335863.10	8227.74	31042.77*
Medicinal and Herbal & Aromatic Plants (Crops)	1792014.86	189.27	32663.25

Contd.

Contd.

Oil Seeds Excluding Soyabean	360837.17	2849.80	228414.22
Ornamental Plants and Flowers	-	-	2358.41
Plantation Crops	-	-	33929.77
Pulses	42721.61	12956.69	34717.45
Spices and Condiments	-	-	18176.21
Sugar Crops	-	-	338192.74
Tuber Crops	-	-	165.92
Miscellaneous	221191.96	27.36	269.40
Tea	27684.26	5273.34	-
Coffee	13122.03	1376.54	-
Cotton	552388.47	111382.54	-
Total	3746560.38	206264.05	1095754.14

Major products produced in India by organic farming

Type	Products
Commodity	Tea, Coffee, Rice, Wheat
Spices	Cardamom, Black pepper, White pepper, Ginger, Turmeric, Vanilla, Mustard, Tamarind, Clove, Cinnamon, Nutmeg, Mace, Chilli
Pulses	Red Gram, Black Gram
Fruits	Mango, Banana, Pineapple, Passion fruit, Organge, Cashew nut, Walnut
Vegetables	Okra, Brinjal, Garlic, Onion, Tomato, Potato
Oil seeds	Sesame, Castor, Sunflower
Others	Cotton, Sugarcane, Herbal extracts

Organic products produced and exported as per ranking

Rank	Product
1	Tea
2	Spices
3	Fruits
4	Vegetables
5	Rice
6	Coffee
7	Cashew
8	Oil seeds
9	Wheat
10	Pulses

Annexure IX ITK's District Wise

Dindigul

Sl.No	Name of the village and block	ITK identified	Rationale	Benefit
1	Manjanaickenpatty village of Palani block	About 250 gm of sand mixed with 100 g of neem seed powder placed at the base of leaf sheath after removing the old spathe in coconut to control Rhinoceros bettle.	Sand with neem powder will enter into the neck of the bettle and due to coarse nature; the neck will be cut away from the body as beetle moves the head.	Low cost, easy to adopt and prevents the Rhinoceros bettle havoc effectively and efficiently.
2	Nallampillai village of Nilakottai block	Dried leaf powder of tulsi, neem and adathoda mixed with pulses seeds for preventing damage of storage pests.	The chemicals present in the leaf powder deter the pest from damaging seeds of pulses.	Low cost, easy to adopt & replace the use inorganic pesticide during storage. Does not impair the germination potential of seeds.
3	Pudhupatty of Reddiarchatram block	Tying plastic carry bags to sticks of 2' height and stacked in the boundary with an escapement of 2 m x 2 m This technique is carried out to scare away the squirrel from damaging the sorghum seeds sown.	The sound of the whirling wind and bags caused will scare away squirrel and birds.	Low cost, germination of sorghum seeds will be maintained up to 98 %.
4	Pallapatti village of Nilakottai block	For preventing indigestion in dairy animals, beetal leaf and nuts of arecanut are crushed well, mixed and given to the animals.	Beetal leaf along with arecanut nut enhances the digestion.	Improves digestion.
5	Kattamanaickenpatty village of Nilakottai block	Spraying bio- gas slurry mixed with water 1:10 ratio in jasmine prevents flower drop and corrects nutrient (Fe) deficiency.	Nutrients are supplied to correct irondeficiency and prevents flower drop.	Easy to adopt andlow cost.
6	Mallayapuram of Athoor blcok	Vanaspathi is smeared in udder to cure mastistis.	Secondary infection is devoured.	Easy to adopt.

Kanchipuram

S. No.	Crop/ Enterprise	ITK Practiced	Purpose of ITK
1	Animal Science	Oral Administation of powered gingelly with palm jaggery	Retention of placenta of cattle
		Aloe vera – ½ kg, Aanai neringe – ½ kgMixed in equal proposion	Treating in fantile gentilia
		Aloe vera, Aanai neringe, Amukkra root, Mixed in equal proposion	Urine treatment of cattle
		Peraindai – 300 grams,Salt – 100 g,Dry chillies – 50 g,Pepper – 50 g, Garlic – 50 g,Kandatippili – 50 g,Elakai – 10 g	Indigestion in cattle
		Neem bark sivanar venpu, karundal kailngu, jaggery,Malaivempu each 50 grams – twice for 3 days	Deworming
		Perandai leaves boiled in water cooled, filtered	Bloat
2	Fisheries	Covering the rostrum of brood stock of freshwater prawn with plastic tube	While transport to avoid damage of packing material
		Use of banana leaves	For deposition of eggs of egg laying ornamental fishes
		Use of teak leaves	To maintain acidic ph in the ornamental fish tank during breeding of the ornamental fishes
		Placing the carrot and potato slices and banana peals in the tanks where the ornamental fish larvae are reared	To enhance the formation of live feed, Infusoria.
3	Agriculture	Kerosene @ 1 lt mixed with soap & water is sprayed	To control leaf folder & stem borer in paddy
		Use of trap crop (marigold)	To control pests in chilli & tobacco crops
		Mulching the field with trashes	To control shoot borer in sugarcane, leaf miner in groundnut
		NSKE (Neem Seed Kernal Extract)25 kg neem seed kernel, 500 lts of water, soaking for 8 hours	To control rice sheath rot(5%), Blackgram powdery mildew, green leaf hopper (vector of rtv) at 2 sprays at 15 days interval;
4	Homescience	Use of cow dung in floors	To control mites & ants.
		Use of tamarind in pickles	As preservative to improve keeping quality

Coimbatore

S.No	Crop / Enterprise	ITK Practiced	Purpose of ITK
1.	Animal Science	Equal quantity of Napthalene balls and camphor were mixed with water into paste and apply on the body of cattles for 2 hours.	To control ecto parasite in animals.
2.	Home Science	Storage containers of pulses in the houses were filled with red chillies, neem leaves at the bottom part.	To control spoilage of pulses by insects in household level.

Erode

Sl. No.	Crop / Enterprise	ITK Practiced	Purpose of ITK
1	Goat (Attu oottam)	Goat dung- 5kg, Goat urine- 5 litres, Goat milk- 2 litres, Goat curd- 2 litres, (Cow)Ghee -1 litre, Sugarcane juice- 2litres, Banana -10 nos, Tender coconut water-2 litres, Toddy – 2 litres/ yeast -800g, Jaggery – 800g, water- 5 litres.Preparation:Goat dung and urine to be mixed together and soaked over night. The next day morning mix cows ghee and keepit for 4 days. On the fourth day mix all the other ingredients and stir well(morning and evening). Keep the mixture for 2 weeks. The preparation will be ready for use on the 18th day.	Crop growth promoter
2	Vermiculture	A gunny bag dipped and coated with cow (Country breed animals) dung and placed on a shaded land for a day.	Country Earthworm collection
3	Cotton,cowpea	1% Turmeric+ ash powder solution (each one kg in 100 liters of water) and ground nut.	Management of sucking pest aphids
4	Paddy, fruit trees	5% fermented coconut milk solution(1litre in 20 liters of water)	Growth promoter
5	Cattle	Leaves of jack-200g and neem bark 300g are boiled in 5 liters of water for 1 hr until it becomes 3 litres. Half a liter of this decoction is given orally for 2 times.	Management of FMD
6	Large and small ruminants	Banana inflorescence -3 for large animals 1 for small ruminants	Management of diarrhoea
7	Groundnut, sugarcane	Fermented castor solution trap:5kg of castor seeds to be pulverized and mixed with 5 liters of water. Keep it for 7-10 days for fermentation Add 2 liters of fermented castor solution in each buried mud pot and fill with water up to neck portion.	Management of white grubs

Kanyakumari

S. No.	Crop / Enterprise	ITK Practiced	Purpose of ITK
1.	Paddy	Spraying of bird's eye chilli leaf extract is used at five per cent concentration. Spraying was usually done after the appearance of the incidence.	Control of ear head bug
2.	Paddy	*Cycus* inflorescence is placed in the field at flowering stage of the cop	Control of rat damage
3.	Brinjal	Before planting the seedlings are dipped in water, which contains one gram of aseophoteda and 10 grams of turmeric powder per litre of water.	Control of fusarium wilt

Karur

S. No.	Crop / Enterprise	ITK Practiced	Purpose of ITK
1	Paddy	Application of tamarind seed @ 100kg / ac at last ploughing	To prevent the *Cyperus rotandus* infestation
2	Paddy	Spraying of 3% panchagavya (300ml / 10litre of water) at tillering stage and subsequent spraying at 15 days	To increase number of tillers per hill To control pest and disease
3	Paddy	Spraying of lemon juice solution @ 100ml / 10litres of water (10 egg + required lemon juice to submerge all eggs + 250gm of jaggery – allowed 10 days for fermentation then mixed well and add powdered jaggery equal to juice solution)	To increase crop yield
4	Water salinity	Keeping Amla trunks in well water	To reduce salt content in irrigation water
5	Paddy	Spraying of Jatropha leaf extract (20Kg of Jatropha leaf + + 200 litres of water)	To control leaf folder
6	Paddy	Chilli powder solution spray (2kg of chilli powder + 200 litres of water)	To control aphids and hairy caterpillars
7	All crops	Soaking seeds in mint leaf extract for 1 hour (Extract from 100 gm of mint leaves + 1 litre of water)	To prevent seed borne diseases
8	Pulses	Spray turmeric powder and ash solution (2Kg of turmeric powder + 8 Kg of ash + 200 litre of water per acre)	To control sucking pests like aphids, hoppers etc.,
9	Vegetables	Spraying of garlic extract and kerosene solution (2 kg of garlic + 200 litres of water + 400ml of kerosene)	To control fruit borers
10	Paddy	Spraying of *Prosopis* leaf extract (20Kg of leaves + 200 litres of water)	To control blast
11	Paddy	Spraying of salt and ash solution (2Kg of salt + 8 kg of ash + 200 litres of water per acre)	To control leaf folder
12	All crops	Spraying of papaya leaf extract (10Kg of papaya leaves + 200 litres of water)	To control bacterial and viral diseases
13	Vermicompost production	Gunny bags were dipped in *Calotrophis* leaf extract solution before used it to cover the vermicompost bed (Leaves of the *Calotrophis* are soaked with water with small amount of cow dung for 3 days) then used it for dipping gunny bags	To control termites
14	Paddy	Spraying of the extract of tamarind and vadhanaryana leaf	To over come Zn deficiency
15	All crops	Spraying of the extract of calotropis.	To overcome the Boron deficiency

Nagapattinam

S. No.	Crop / Enterprise	ITK Practiced	Purpose of ITK
1.	Paddy	Nochi leafs along with stored paddy grain. News paper clippings and herbal leaf mixture.	To repell stored product pests.
2.	Pulses	Use of neem oil / red earth	To repell stored product pests.
3.	Vegetables	Neem extract/ Pungam Oil/ Panchaghavya	To control sucking pests and borers.
4.	Vegetable seeds	Preserving vegetable seeds by treating with cow dung.	To maintain the viability and ** of the seeds.

Nilgiris

S. No.	Crop / Enterprise	ITK Practiced	Purpose of ITK
1.	Vegetables	Spraying plant extracts of all sap oozing weeds (eg.) Perandai, Erukam, Vilvam etc.,	To control pests in vegetables
2.	Composting	Addition of salt – ½ kg and ayesenduram- 10 gms.	For quick composting and enrichment of nutrients
3	Vermicomposting	Placing egg shells in the compost beds	To avoid lizard damage to worms
4	Floriculture	Spraying diluted solution of lime (CaCo3 soaked in cow urine for one day @ 1kg/20lt.	To control mites in carnation
5	Vegetable seeds	Seeds stored in bags are placed in lofts above the fire place	Seed storage
6	Vegetables	Wide spacing while transplanting seedlings	To tide over plant decay due to water logging
7	Tea	Spraying soda with turmeric (5:2) in 15 lts.of water once in fifteen days.	For controlling blister blight in tea
8.	Tea	Mixing of diesel with weedicides.	Weed control
9.	Silver oak	Spraying of cow dung over Silver oak saplings.	To prevent cow damage to silver oak plants in tea field.

Puducherry

S.No.	ITK	Scientific rationale	Validation by KVK
1.	Application of Custard Apple extract for the control of rice green leaf hopper.	Alkaloid annonaine found in seed is responsible for suppressing leaf hopper.	-do-
2.	Spraying of Garlic extract to suppress Spodoptera litura	Diallyl disulphide, diallyl tri-sulphide active molecule is responsible to control Spodoptera litura.	-do-
3.	Water Soaked Bengal gram for infantile genitalia	Protein rich diet improves the growth and early maturity. of the genitalia.	-do-
4.	Spraying of plain water for thrips management in rice.	Delivery of water at high speed washes off soft bodied thrips.	Validated in farmers field of IPM Field School
5.	Spraying of green Chilli extract control sucking pest	Capsaicin active ingredient in green chillies is responsible to Farmers' Field School.	Success observed in IPM
6.	Use of 'Panchagavya'	Growth promotion is favoured by the beneficial micro organisms multiplied in large quantities after application	Yet to be validated by this Kendra

Pudukottai

S.No.	Crop / Enterprise	ITK Practiced	Purpose of ITK
1.	Groundnut and pulses	Used polyethylene covers in sticks in the field	Scare off birds in Groundnut and pulses field
2.	Culinary preparation	Addition of raw papaya pieces in mutton kurumas	Tenderize meat by papain enzymatic action
3.	Banana	Covering of banana bunches with its tree leaves	Retain its original colour and increase size
4.	cow	Oral application of Neem oil as dewormer	Removing the worms in cows
5.	Eucalyptus	Application of dried Poultry manure while planting Eucalyptus	Augments the growth and establishment of newly planted Eucalyptus seedlings
6.	cow	Oral administration Aloevera & Aanai nerunji leaves	induce heat in cows
7.	goat	Oral administration of Betelvines, omam and ani seeds to goats	Correct indigestion
8.	Banana	Smoking banana bunches with dried banana leaves	Uniform & quick ripening of banana bunches
9.	Amla	Beating of branches of Amla tree before flowering season	Boosting yield
10.	Jasmine	Pinching of current season shoot in Jasmine	Obtaining higher flower yield at the appropriate time

Salem

S. No.	Crop / Enterprise	ITK Practiced	Purpose of ITK
1.	Sugar cane	Sugar cane Harvesting knife	A flat shaped iron blade is attached with wooden pole, tools like axeThe length of the blade can be 15 cm, width is 12 cm. It is traditionally used for harvesting the sugarcane.
2.	Cereals & Oil seeds	Measuring equipment-Vallam	This equipment is made up of iron and it is round in shape. The height of the equipment is 23 cm, the diameter is 20 cm.This equipment is traditionally used for measuring the agricultural grains (Paddy, cumbu, sorghum, ragi, groundnut and millets).The capacity of equipment is 4kg and the worth of this implement ranges from Rs.80 to Rs.100.

Sivaganga

Sl.No.	Crop / Enterprise	ITK Practiced	Purpose of ITK
1	Black gram	NSK extract + Neem leaf + Cow dung slurry + Urine Spray	Pest and Disease management
2	Paddy	Seedling dipping in Panchagavya+Pseudomonas	Better establishment
3	Paddy	Fertilizer application after subjected to soil incubation	To arrest wastage of nutrients
4	Groundnut	Trap cropping with Bhendi+Cowpea+Chrysanthemum+Maize	To control Pests

Thanjavur

S. No.	Crop / Enterprise	ITK Practiced	Purpose of ITK
1	Dairy	For the prevention of prolapse of uterus in milching animal by used the leaves of Karekaki soppu (*solanum nigrum*)	Applying and feeding of Karekaki soppu prevent the secretion of prolapse causing hormone in animals after calving
2	Dairy	**Simple anorexia:** One hand full of leaves of Tylophora indica, datura metel, aegle marmelos, thespesia papulnea, allium sativum and fruits of piper nigrum mustard the above ingredients are crushed and made in to bolus and given orally twice a day for 4 days.	It will increase the secretion of digestive juices and rumen motility of animal thereby increased the intake of feed

Thiruvallur

S.No.	Crop / Enterprise	ITK Practiced	Purpose of ITK
1	Rice	Soaking of vites and neem leaves in cows urine in pot and kept in soil burried for 3 moths filtered and used	The solution can be used for management of leaf folder in rice
2	Chillies	The paste prepared with the combination of green chillies and asafoetida is dissolved in water in the ratio of 1:1:1	The solution is used for control of muranai caused by thrips and white aphid
3	Sugarcane	The trash mulching along the furrows is discouraged in soil.	The practice encourage termatorium in red soils

Thiruvarur

S.No.	Crop / Enterprise	ITK Practiced	Purpose of ITK
1.	Agro Forestry	Teak seeds soaked in boiled water	To increase the germination
2.	Agro Forestry	Storing of tree seeds in earthen pots	Increases the viability of seeds for six months
3.	Pulses	Pelleting the pulse seeds with materials like arappu (Albizzia amara) leaf powder	Against the attack from ants and birds
4.	Rice	Spraying of leaf extract with detergent	Effective against rice leaf folder
5.	Rice	Using Thanjavur kitty trap	Control of field rats
6.	Black gram and green gram	Spraying of leaf extract of different indigenous plants species viz., notchi, aduthoda and datura .	Plant protection measures in many areas.
7.	Pulses	Drying of blackgram seeds during amavasya (dark moon)	Against bruchid (pulse beetle) infestation in pulses
8.	Pulses	Storage of seeds along with neem leaves or dried chilli in earthen pot	Protection measure against storage pest and disease

Trichy

S.No.	Crop / Enterprise	ITK Practiced	Purpose of ITK
1.	Rice, Banana & Vegetables		Spraying Panchagavya@ 3% To get high yield without external chemical inputs
2.	Rice	Soaking dry cowdung in kerosene and placing in rice fields	For rat control
3.	Poor quality water	Putting amla tree wood logs in alkali water wells.	To neutralize alkalinity and improve taste of drinking water.
4	Pulses	Soaking fresh ***Eichornia*** tender leaves for 5 hours and spraying @ 10% concentration	To control pulse beetle

Tuticorin

S.No.	Crop / Enterprise	ITK Practiced	Purpose of ITK
1.	Banana	Application of waste cotton from ginning factory	Mulching Reduce the weed growthWithstand water logged condition

Virudhunagar

S.No.	Crop / Enterprise	ITK Practiced	Purpose of ITK
1.	Onion	Seed onion spread over the stone bed over which bamboo stakes covered with straw	Seed storage
2.	Redgram	Seeds coated with til oil and then sun drying	Seeds coated with sesame oil facilitates in splitting of pulses
3.	Pulses	Red earth treatment	Red earth treatment helps in preventing bruchid oviposition
4.	Sunflower and millets	Creating sound with steel box	Bird scaring
5.	Rice	Pungam and neem leaves mixed with grain	All grain moths are repelled

Annexure X

ITK from Worldwide

1. IMO – Indigenous Microbial Organism (For composting inoculant)

a) Mix 1 kilo cooked rice with 1 kilo muscovado sugar.

b) Place in earthen jar or plastic pail.

c) Cover with clean Manila paper and fasten with rubber strip.

d) Allow to ferment for 7 to 14 days.

e) Separate the juice in clean container and seal, ready for use.

f) **Dosage and usage:** Mix 4 tbsp I 1 liter of water or 1 litter IMO to 100 liters water and spray on plants and soil root zone. Spray on hog feeds and animal manure to eliminate malodor. Use IMO as inoculants in composting degradable organic matter.

2. OC – Organic Compost Formulation and Making (For composting inoculant)

a) Materials to be used:

1. 100 kilos or 2 bags of rice or corn brand.
2. 100 kilos or 2 bags of top soil.
3. 0.5 kilo IMO (Indigenous Microbial Organism).
4. 0.5 kilo FFAA (Fermented Fish Amino Acid).

b) Mix thoroughly the above materials and cover with plastic sheet.

c) Ferment the materials for 7 to 14 days.

d) IMO and FFAA can also be used as inoculants in making compost with the use of sawdust or hammer milled corncobs with chicken dung or other animal manure.

3. FFAA – Fermented Fish Amino Acid Formulation (For foliar fertilizer and growth activator)

a) Mix 1 kilo unwashed fresh trash fish with 1 kilo muscovado sugar or molasses.

b) Place in earthen jar or plastic pail.

c) Cover with clean Manila paper and fasten with rubber strip.

d) Allow the materials to ferment for 7 to 14 days

e) Squeeze out the juice and place in a clean container and seal.

f) Collect the solid fishbone to be used for making calcium nutrient spray formula for plants.

g) Juice is used as foliar fertilizer to induce vegetative growth.

h) **Dosage:** 1 liter FFAA to 1 drum (200 liters) of water or 1 ml FFAA to 1 liter of water.

4. CPN – Calcium for Plant Nutrient Formulation (For Foliar Fertilizer)

a) Crush 1 kilo egg shell and burn.

b) Mix with 10 liters of pure coconut vinegar.

c) Place in a jar and cover with clean Manila paper. Fasten with rubber strip.

d) Let it stay in the jar for 3 weeks, and ad 2 kilos fishbone. (Fishbone from making FFAA can be used.)

e) After 4 weeks, the liquid can be used as Calcium Nutrient spray on plants.

f) **Dosage:** 1 ml to 1 litre of water or 1 litre to 1 drum (200 litres) water

5. FFJ – Fermented Fruit Juice Formulation (For foliar Fertilizer and drench fertilizer for seedlings)

a) Mix 1 kilo chopped banana or other fruits (except citrus), and mix with 1 kilo muscovado or molasses.

b) Place in an earthen jar or plastic pail.

c) Cover with clean Manila paper and tie with rubber strip.

d) Allow to ferment for 7 to 14 days and separate the juice in clean container and seal.

e) **Usage:** Animal drink nutrient enhancement.

f) **Dosage:** Mix 1 liter FFJ to 1 drum (200 liters) of water or 1 ml FFJ to 1 liter of water

6. LABS – Lactic Acid Bacterial Serum Formulation (For Foliar Fertilizer or seedling drench)

a) Mix 1 kilo uncooked brown rice and or fresh milk with 1.5 liters water inside a jar.

b) Cover the jar with clean Manila paper and tie with rubber strip.

c) Allow to ferment for 7 to 14 days.

d) **Usage:** The juice can be used as soil conditioner or fertilizer.

e) **Dosage:** Mix 2 ml juice with 1 liter of water. (1 tbs. Per gallon water).

7. OHN – Oriental Herbal Nutrient Formulation using Garlic (For Foliar insect repellant and fungicide)

a) Mix 1 kilo clean ginger, crushed by stone of wood (no metal implement should be used), with 1 kilo muscovado sugar or molasses and place in a jar.

b) Pour in a bottle of gin, Ginebra San Miguel 40% proof.

c) Cover the mouth of the jar with a clean Manila paper and tie it with a rubber strip.

d) Allow to ferment for 7 to 14 days.

e) **Usage:** OHN is used as spray against insects and fungi.

f) **Dosage**: 3 ml OHN (garlic) mix with 1 liter of water. (1.5 tbs per gallon)

8. OHN – Oriental Herbal Nutrient Formulation using Ginger (For foliar insect repellant and fungicide)

a) Mix 1 kilo clean ginger, crushed by stone of wood (no metal implement should be used), with 1 kilo muscovado sugar or molasses and place in a jar.

b) Pour in a bottle of gin, Ginebra San Miguel 40% proof.

c) Cover the mouth of the jar with a clean Manila paper and tie it with a rubber strip.

d) Allow to ferment for 7 to 14 days.

e) **Usage:** OHN is used as spray against insects and fungi.

f) **Dosage:** 3 ml OHN (ginger) mix with 1 liter of water.(1.5 tbs. per gallon

9. ST – Seed Treatment for Germination

Dosage and treatment of liquid formulations

0.2 % FPJ Fermented Plant Juice

0.2 % BRV (Brown Rice Vinegar) or Coconut vinegar.

0.2 % OHN (0.1 % OHN Garlic + 0.1 % OHN Ginger)

Mix the above formulations together with water.

How to use: Soak the seeds to be germinated for 4 to 8 hours. For slow germinating seeds, soak the seeds for a longer time.

10. SW – Sea Water Usage as Spray for Plants against Diseases

Get sea water from the blue colored area or deep portion where water is clear and uncontaminated with land pollution. Mix 1 liter of seawater with 30 ml fresh water in a plastic container and let it stay for a duration of 2 days. The mixture can then be used as spray on disease infected plants.

Storage Guidelines for the Fermented Preparations

Some of the fermented preparations are living cultures of microorganisms and special storage measures must be taken to keep them healthy. The most important measure is to keep the cap loose on the bottle to allow some air circulation. Leaving the cap loose also prevents the liquid from exploding out of the bottle when removing the cap since there may be a buildup of gases formed as a by-product of microbial metabolic activities. Shaking the bottles provides oxygen to the microorganisms. The preparations can be stored in glass bottles and should be protected from excessive heat and direct sunlight.

Preparation	Bottle cap	Shaking	Sugar	Shelf life
IMO	Keep the cap loose	Shake once a week.	Add sugar everymonth (about 20% of the volume of the solution)	6 months, or until it starts to smell bad.
FPJ	Keep the cap loose and loosen it even more after 2 weeks.	Shake once a week.	Add sugar every month (about 20% of the volume of the solution).	6 months, or until it starts to smell bad.
FFJ	Keep the cap loose and loosen it even more after 2 weeks	Shake once a week.	Add sugar every	6 months, or until it starts to smell bad
FAA	Keep the cap loose	Shake once a week.	month (about 20%	6 months, or until
KAA	Keep the cap loose	Shake once a week.	of the volume of	it starts to smell
$CaCO_3$	Leave the cap loose for the 1st 2 weeks. At this time tighten the cap for 20 minutes and then reopen. If gas escapes the bottle upon reopening, leave itloose, if not, tighten the cap.	Do not shake	Do not add sugar tothe solution.	6-12 months
OHN	Leave the cap loose forthe 1st 2 weeks. At this time tighten the cap for 20 minutes and then reopen. If gas escapes the bottle upon reopening, leave it loose, if not, tighten the cap.	Do not shake	Do not add sugar tothe solution.	6-12 months

11. Water Soluble Potassium

Materials needed

Mud pot, porous paper, tobacco stems, water & rubber band/thread

Procedure

- Dry tobacco stems and cut them into pieces
- Put 1 kg of tobacco stem in cloth bag and dip it in 5 lit of water
- Cover the jar with paper
- It takes 7 days to get crude liquid of natural potassium
- Dilute 0.7 of this solution with 20 lit of water

12.Water Soluble Phosphoric Acid

Materials needed

Charcoal from sesame stems, porous paper, glass jar, water & rubber band/ thread

Procedure

- Take sesame stems
- Burn the sesame stems
- Put 1 kg of sesame stem charcoal in 10 litres of water
- Stir the jar for 7 days to allow the air into water
- In 20 lit of water, 700 ml of this liquid is added and mixed well for spraying.

13. Water Soluble Phosphoric Acid

Materials needed

Egg shells, brown rice vinegar, porous paper, glass jar, water & rubber band/ thread

Procedure

- Crush the egg shells/sea shells in to small pieces
- Lightly roast the shells to remove any organic substances that may not and deteriorate during process
- Put the roasted shells and add brown rice vinegar
- With in 2-3 days the water soluble calcium is ready

14 Herbal Tea Preparations for Plant Protection

Materials Needed:

200 liters capacity plastic drum.

Grinder / chopper and mortar & pestle (lusong pambayo)

Strainer/screen/cloth (salaan)

Dipper (tabo).

Wooden ladle / paddle (Kahoy na panghalo)

Fresh clean water (tubig na malinis)

Herbal materials (Halamang panghalo)

10 kilos Ginger (Luya)

5 kilos Garlic (Bawang)

5 kilos Aloe vera (Sabila)

10 kilos Hot pepper (Siling labuyo)

30 kilos Neem tree leaves (Dahon ng Neem Tree)

30 kilos Madre de Cacao leaves (Dahon ng Kakawati)

5 kilos Derris (Tubli)

5 kilos Bitter vine (Panyawan//Makabuhay)

Other herbs with insecticide, fungicide and pest repellant properties.

Procedure

1. Prepare the above materials.
2. Grind or pond the herbs separately.
3. Place all ground and pounded herbs in the plastic drum.
4. Fill the drum with fresh clean water.
5. Mix the materials with a wooden ladle
6. Stay overnight or one day to allow the herb juice to mix with water. Herbal tea..
7. Get herbal tea from drum pass through screen strainer
8. Add equal amount of fresh clean water to the herbal tea.
9. Place in sprayer or sprinkler.
10. Spray on plants, drench from base, trunk, branches and leaves.
11. Repeat spraying 3 or 7 days interval as the need arises.

15. Botanical Pest Control

1. **Goat Weed (*Aegaratum conisoides*) Leaves-** Extract juice and spray against diamond black moth and cotton Steiner.
2. **Damong Maria (*Artemesia vulgaris*) Leaves** – Pound, extract juice and spray at the rate of 2 to 4 tbs. per 16 litters of water wit detergent or AZ41 and spray against cotton borer and mango tip borer.
3. **Lantana (*Lantana camara*) Flowers** – Pound and store around the grains to serve as repellant against weevils.
4. **DITA (*Derris philippinensis*) Roots** – Pound and extract juice. Spray at the rate of 1 cup per gallon of water or powder, mix with detergent or AZ41 and spray at the rate of 120 grams powder + 250 to 300 grams detergent per 4 gallons of water against diamond black moth and other insect pests.
5. **Wild Sunflower (*Tethornia diversifolia*) Leaves** – Pound and extract juice and use as spray at the rate of 1 to 2 kg. Fruit per litter of water against cotton Steiner, black armyworm and diamond black moth.
6. **Marigold (*Targetes erecta*) Roots** – A mixture from the pounded leaves, flowers and roots soaked in water at a proportion of 500 grams/liter of water has been found to be effective against lipidopterous pests, leafhoppers, beetles and house flies. The remaining cake can be used as a mulch or mixed with the soil to control nematodes and other soil pests. Marigold inter-cropped with vegetables like eggplants are said to repel insects from the plantation. Extract juice and spray at the rate of 2 to 4 teaspoon juice per litter of water mix detergent or FAA (Fish Amino Acid) against green leafhopper, brown plant hopper, diamond black moth and aphids.
7. **French Marigold (*Targetes patula*) Roots** – Pound and extract juice at one-kilogram roots mix with one litter of water and detergent or AZ41 then spray directly into the soil against green aphids and grain borer.
8. **Black Pepper (*Piper nigrum*) Fruits** – Pulverize seeds and mix with water and spray. Spread powder around stored grains against cotton Steiner, diamond black moth, common cutworms and corn weevil.
9. **Makabuhay (*Tinospora rumpii*) Vines** – Pound leaves and stem to extract juice. Mix the juice with water and stir thoroughly. The mixture can be used as spray for black bugs, steam borer, diamond back moth and leafhoppers. Extract juice and spray at the rate of 15 to 20 tbsp. juice per 5 gallons water against diamond black moth and green leafhoppers.

10. **Hot Pepper (*Capscium frutesens*)** Fruit – This can be effective for the controlling of lepidopterous persts, other chewing insects and pest for stored products. Mash mature fruits, add water, strain and use the mixture as spray. For stored product pests, pulverize the fruits and spread in storage area. Pound and extract juice and spray at the rate of 2 to 3 cups fruit per litter of water against rice moth.

    Hot Pepper - Researches from the University of the Philippines at Los Banos, Laguna have found that Siling Labuyo (Hot Pepper) fruit, skin and seeds are all effective against ants, aphids, caterpillars, Colorado beetle, cabbage worms, warehouse and storage pests, cucumber mosaic, ring spot virus, tobacco virus and other crop diseases. Briefly, siling labuyo serves as an insecticide, repellant, antifeedant, fumigant and anti-viroid.

11. **Custard (*Annona aquamosa*) Seeds** – Powder and disperse in water, then strain and use as spray against rice pest.

12. **Neem (*Azadiracta indica*)** Seeds – Remove husk of two to three handful of mature seeds, winnow or put in water to float away the husk, Grind seeds into fine particles. Soak ground seeds in 3 to 5 litters of water for at least 12 hours.Filter the solution, add detergent or AZ41, then use the spray against rice pest, diamond black moth and mango leafhoppers.

13. **Tobacco** – Chop or grind tobacco leaves, stalk and root. Soak in water for 13 to 36 hours. Strain tea solution; mix detergent, AZ41 or Aloe vera extract and spray against a wide species of insects including hoppers and worms.

14. **Madre de Cocas (*Gliricidia spium*).** Pound the roots, leaves and bark and soak in water at a proportion of 500g/liter of water. Let it stand overnight. Use the concoction as spray for lepidopterous pests and fleas. Example of lepidopterous pests are the larvae of moths, and butterflies that are usually seen as worms eating the leaves and fruits of many vegetables.

Annexure XI

Current Government Schemes on Organic Agriculture

Scheme names and Particulars	Eligibility	Contact
Production and distribution of Green manure seeds — 25 % subsidy	All farmers	Assistant Agriculture Officer / Assistant Director of Agriculture at the Block level
Production and Distribution of *Rhizobium, Azospirillum* and Phosphobacteria- Rs.6/200 gm packet	All farmers	Assistant Agriculture Officer / Assistant Director of Agriculture at the Block level
Production and distribution of Blue green algae Rs.2.75/kg	All farmers	Assistant Agriculture Officer / Assistant Director of Agriculture at the Block level
Production and distribution of parasites to control Black headed caterpillar. Subsidy charge Rs.35/ha	All farmers	Assistant Agriculture Officer / Assistant Director of Agriculture at the Block level
Release of parasite in sugarcane to control internode borer. Rs.35.75/ha	All farmers	Assistant Agriculture Officer / Assistant Director of Agriculture at the Block level
Production of NPV for the control of prodenia in cotton. Rs.53/ha	All farmers	Assistant Agriculture Officer / Assistant Director of Agriculture at the Block level
Composting of farm waste through pleurotus. Distribution of kits at free cost (1 kg of pluerotus, 5 kgs of urea and a leaflet containing technical information at a cost of Rs.140 per kit)	All farmers	Assistant Agriculture Officer / Assistant Director of Agriculture at the Block level
Vermicompost production scheme implemented to conduct demonstration. Supply input at the cost of Rs.1200	All farmers	Assistant Agriculture Officer / Assistant Director of Agriculture at the Block level
Vermicompost production scheme implemented to conduct training. Rs. 50 for each person participates in training	All farmers	Assistant Agriculture Officer / Assistant Director of Agriculture at the Block level

Vermicompost unit

Estimated value: Rs.60, 000/unit, 50% subsidy (ie) Rs.30, 000/- unit

1 unit means 1000 sq metre. The yearly production should be 25 tonnes.

Scheme guidelines

- In order to construct a shed of 40 x 25 ft, cement pillars should be erected at 10ft intervals. The pillars at the centre of the shed should be 16ft high while these at the edges should be 10ft. The pillars should be supported with wooden reapers and it should be laid with coconut fronds. The cost of making this 40 x 25ft shed is Rs.30, 000/-.

- Constructed with the following dimensions: 2ft length, 4ft breadth & 2ft depth. The tub should be separated at 10ft intervals with a 10t high wall. Hollow bricks can be used to construct the walls, they should be plastered with cement. The base should be plastered with cement with a light slope. The cost to construct a two vermicompost tubs of size 20' x 4'x2' = 160 cubic feet. is Rs.25,000/-
- Cow dung @ 5 tonnes (Rs.400/tan) is Rs.2,000/-
- Cost of earthworms & accessories is Rs.2,700/- (10 kg @ Rs.270/kg)
- Cost of Azospirillum to enrich vermicompost (10 kg 2 Rs.30/kg) is Rs.300 Total Rs.60, 000/-.

References

Anandkumar, S. 1998. Motivating farmers to convert to organic farming and strategies for organic extension. In: Organic Agriculture Ecology and Farming No. 18: 21.

Asian Agri-History (Vol. 16(1) January – March 2012), 2012. Asian Agri-History Foundation, Secunderabad – 500 009. pp. 108.

Asian Agri-History (Vol. 17(3) July – September 2013), 2013. Asian Agri-History Foundation, Secunderabad – 500 009. pp. 296.

Balasubramanian, A.V and T. D. Nirmala Devi. 2006. Traditional Knowledge Systems of India and Sri Lanka. Centre for Indian Knowledge Systems held at Chennai during September 2006.

Balasubramanian, V. 2006. Organic rice: myths, facts and challenges. In: Abstracts, 26th International Rice Research Conference, Oct 9-13th 2006 held at New Delhi, India: p.119 -120.

Barooah, M. and A. Pathak. 2009. Indigenous knowledge and practices of Thengal Kachari women in sustainable management of bari system of farming. Indian Journal of Traditional Knowledge 8(1): 35-40.

Bond, W. and Grundy, A.C. 2001. Non-chemical weed management in organic farming systems. Weed Research, 41: 383-405.

Centre for Indian Knowledge Systems (CIKS), 2006. Package of Organic Practices fromTamil Nadu, Tharamani, Chennai, India.

Chandy, K.T. 2013. Biological control of pests. Plant Pest Control: PPCS-7 Booklet No. 344.

Cherr, C.M., Scholberg, J.M.S. and McSorley, R. 2006. Green manure approaches to crop production. Agronomy Journal, 98: 302-319.

Chhonkar, P.K. 2003. Organic farming: science and belief. Journal of Indian Society of Soil Science,51: 365-377.

Chittapur, B.M. 1998. Composting of organic wastes. In Organics in Sustaining Soil Fertility and Productivity, Univ. Agril, Sci., Dharwad, India : 164 – 172.

CIKS Vrkshayurveda Experiments on Seed Treatment Techniques, 1993 - 2011. Centre for Indian Knowledge Systems, Chennai. pp.324.

Deka, M.K., M. Bhuvyan and L.K. Hazarika. 2006. Traditional pest management practices of Assam. Indian Journal of Traditional Knowledge, 5(1): 75-78.

Farm innovators. 2010. Division of Agricultural Extension. Indian Council of Agricultural Research.

Farming (NCOF), Government of India and Food and Agriculture Organisation (FAO), United Nations. pp. 174.

Giraddi, R.S. 1998. Vermitechnology and its scope in Agriculture. In Organics in Sustaining Soil Fertility and Proudctivity, Univ., Agrl., Sci., Dharward, India : 173–181.

Golakia, B.A. 1992. Proverbs for Predicting the Moods of Monsoon. Honey Bee. 3(1):12.

http://agritech.tnau.ac.in

http://www.rkmp.co.in

Inventory of Indigenous Technical Knowledge in Agriculture Document 1/2/3 by Mission Unit, Division of Agricultural Extension, Indian Council for Agricultural Research, New Delhi 110 012.

Inventory of Indigenous Technical Knowledge in Agriculture, (Document 1, 2 and Supplementary 2). Indian Council of Agricultural Research, New Delhi.

Joshi, C.P. and B.B. Singh. 2006. Indigenous Agricultural Knowledge in Kumaon hills of Uttaranchal. Indian Journal of Traditional Knowledge, 5(1): 19-24.

Kanani, P.R., Munshi, M.A., Makwana, D.K. and V.J. Savaliya. 1995. Bhadli nu Bhantar Ketlu Sacchu? (vernacular). Krishi Jivan. (Special issue on Varshad Agahi. pp.26.

Kapoor, K.K., P.K.Sharma, S.S.Dudya and B.S.Kundan. 2005. Management of organic wastes for crop production. Dept. of Microbiology, CCS, Haryana Agric.Univ., Hissar –125 004.

Krishna Chnadra, S.Greep and R.S.H. Srivathsa. 2009. Biocontrol agents and Bio-Pesticides (Liquid formulations). Published by Regional Director, Regional center of Organic Farming, Bangalore-24.

Li, Z. 1999. Organic certification for small farmers in China. Paper presented at the 4th Organic Agriculture-Asia Conference, 18-20 November, 1999, Tagaytay, Philippines.

Linderman, R.G. 1994. Role of AM fungi in biocontrol. In: Pfleger FL, Linderman RG (eds). Mycorrhizae and plant health. The American Society of Phytopathology, St. Paul, Minn. USA.

Mathu Ramakrishnan. 2013. First aid box for organic farming. Own products of farmer. Ravi Publications, Udumalpet, Tamil Nadu.

Mishra, P.K. 2002. Indigeneous technical knowledge on Soil and Water Conservation in Semiarid India (Eds: P.K. Mishra, G. Sastry, M. Osman, G.R. Maruthi Sankar and N. Babjee Rao). NATP, CRIDA, Hyderabd. 151p.

Muthukrishnan, P., K.Siddeswaran and P.M. Shanmugam. 2012. Organic farming. TNAU, CBE-3.

Muthuraman, P and S.N.Meera. Indigenous Technical Knowledge in Rice Cultivation. Directorate of Rice Research, Rajendranagar, Hyderabad 500030.

Namvazhi Velanmai. Tamil Nadu Iyarkai Velanmai Arakattalai (Tamil Nadu Organic Farming Trust), Virattipattu, Madurai – 625 016.

Narayanansamy, P. 2006. Traditional Knowledge of Tribals Crop Protection. Indian Journal of Traditional Knowledge, 5(1): 64-70.

Natarajan, S., P. Devasenapathy, R. Kalpana and C. Sudhalakshmi. Organic Farming: an overview. Published by Centre for Soil and Crop Management Studies and Water Technology Centre (IAMWARM), TNAU, Coimbatore, Tamil Nadu.

NCOF, 2009. Biofertilisers and organic production statistics 2010-11 to 2011-12.

Organic Cotton Cultivation, March 2005. Vijayalakshmi, K., Subhashini Sridhar, Sridevi, R. And Arumugasamy, S. (eds.). Centre for Indian Knowledge Systems, Chennai. pp. 39.

Organic Cultivation Techniques – Bhendi, December 2006. Nirmala Devi, T. D., Sridevi, R. And Vijayalakshmi, K. (eds.). Centre for Indian Knowledge Systems, Chennai. pp.17.

Organic Cultivation Techniques – Groundnut, December 2006. Nirmala Devi, T. D., Sridevi, R. and Vijayalakshmi, K. (eds.). Centre for Indian Knowledge Systems, Chennai. pp. 21.

Organic Cultivation Techniques – Tomato, Brinjal, Lady's finger and Chilli, February 2008. Subhashini Sridhar, Kiruthika, P. and Sridevi, R. (eds.). pp. 29.

Organic Cultivation Techniques – Tomato, December 2006. Nirmala Devi, T. D., Sridevi, R. And Vijayalakshmi, K. (eds.). Centre for Indian Knowledge Systems, Chennai. pp. 25.

Organic Cultivation Techniques for Chillies (Iyarkaivazhi Sagupadi Thozhilnutpangal-Milagai pamphlet), August 2010. Subramanian, K., Amirtha Nishanth, Abarna Thooyavathy, R. And Vijayalakshmi, K. (eds.). Centre for Indian Knowledge Systems, Chennai. pp. 8.

Organic Farmers Association of India Survey, 2009.

Organic Paddy Cultivation, December 2004. Vijayalakshmi, K., Nirmala Devi, T. D., Subhashini Sridhar and Arumugasamy, S. (eds.). Centre for Indian Knowledge Systems, Chennai. pp.102.

Organic Vegetable Gardening, 2003. Sridhar, S. Arumugasamy, S. and Saraswathy, H. And Vijayalakshmi, K. (eds.). Centre for Indian Knowledge Systems, Chennai, pp. 46.

Package of Organic Practices from Tamil Nadu for Rice, Groundnut, Tomato and Okra., 2006.

Pisharoty, P.R. 1993. Plant that Predicts Monsoon. Honey Bee. 4(4): 12.

Prasada Rao, G.S.L.H.V. 2003. Agricultural Meteorology. Kerala Agricultural University, Thrissur, Kerala.

Prasant Kumar Mishra. 2002. Indigenous technical knowledge on soil and water conservation in semi-arid India. CRIDA. Hyderabad-9.

Quioz, C., 1996. Local knowledge systems contribute to sustainable development indigenous knowledge and development monitor, 4 (1): 3 – 5.

Rajasekaran, B. 1993. Indigenous technical practices in a rice based farming system. Ames, IA: Center for indigenous knowledge for agriculture and rural development draft.

Rathakrishnan, T., et al. 2009.Traditional Agricultural Practices: Applications and Technical Implementations. New India Pub. Agency, New Delhi.

Rathnam, R. 1966. Agricultural development in Madras state prior to 1900. New Century Book House, Madras.

Sabarathinam, V.E., 1997. Rationalization of indigenous technical knowledge in Indian Agriculture, In : 3rd IFOAM – ASIA Science conference and general assembly "Food security in harmony with nature" (Shivashankar, K. ed.) held at UAS, Bangalore, during 1st – 4th Dec. p.97.

Savaliya, V.J., Kher, A.O., Kanani, P.R. and M.A. Munshi. 1991. Pashu Paxi ni Chestha ne Adhare Megh ne Endhan. Narmada Kisan Parivar Patra (Guj.). Pp.8.

Sharma, P., P. Khandelwal and P. Khandelwal. 2010. Krishi sutra Profiles of Agricultural Innovations in India. Small Farmers' Agribusiness Consortium (SFAC), New Delhi.

Sharma, S.K. 2002. A synoptic view of linkages of organic farming with productivity and sustainability of India. In : CAS training on organic agriculture – a paragon for sustainability held at JNKVV, Jabalpur, March 26th – April 15th, 2002, p.29.

Somasundaram E., M. Mohamed Amanullah and K. Vaiyapuri. Indigenous Technical Knowledge (ITK) in Organic Agriculture. CASA Training manaual, Deparmtent of Agroonomy, TNAU, Coimbatore.

Somasundaram, E. 2003. Evaluation of organic sources of nutrients and panchagavya spray on the growth and sustainable productivity of maize – sunflower – greengram system : Ph.D Thesis submitted to Department of Agronomy, Tamil Nadu Agricultural University, Coimbatore.

Sridhar, S., S. Arumugasamy, H. Saraswathy and K. Vijayalakshmi. 2006. Organic vegetable gardening. Published Centre for Indian Knowledge Systems (CIKS), Chennai, Tamil Nadu.

Suman Gupta and A. K. Dikshit. 2010. Biopesticides: An eco-friendly approach for pest control Journal of Biopesticides, 3(1): 186 – 188.

Sundaramari, M and Viswanathan. 2005. Indigenous Agricultural Practices for Sustainable Farming. Agrobios Publishers. Jodhpur.

Validation of Indigenous Technical Knowledge in Agriculture, (Document 3 and 4) 2003. Indian Council of Agricultural Research, New Delhi. pp. 505

Villegas-Panga, G. 2010. Kakawate (Gliricidia sepium, Leguminosae) as a soil amendment and biological control of soil-borne pathogens: The Philippines Experience. ActaHort. (ISHS)883:309-315. http://www.actahort.org/books/883/883_38.htm.

Vrkshayurveda – Ayurveda for Plants, August 2001. Subhashini Sridhar, Arumugasamy, S., K. Vijayalakshmi and Balasubramanian, A.V. (eds.). Centre for Indian Knowledge Systems, Chennai. pp.46

Wieringa, J. and J.Lomas. 2001. Lecture notes for Training Agricultural Meteorological Personnel. WMO No. 551

Willer, H. 2009. Organic agriculture worldwide: current statistics. In The World of Organic Agriculture. Statistics and Emerging Trends. IFOAM, Bonn and FiBL, Frick, pp. 23–46.

www.agritec.com

www.apeda.gov.in

www.ciks.org

www.cpreec.org

www.ncipm.org.in

www.sciencedirect.com

www.sristi.org

www.tnau.ac.in

www.tnocd.org

Yadav, A.K and V.K. Verma. Production Technology of Organic Inputs. National Centre of Organic Farming, Dept of Agriculture and Cooperation, Ghaziabad, Uttar Pradesh.

Yadav, A.K. Organic Agriculture (Concept, Scenario, Principals and Practices) National Centre of Organic Farming, Dept of Agriculture and Cooperation, Ghaziabad, Uttar Pradesh.

Other Publications on Agronomy

S.No.	Title	Author	ISBN	Year
1	A Handbook of Minerals,Crystals,Rocks and Ores	Alexander, P.O.	9788190723787	2009
2	A Handbook of Soil-Plant-Water-Fertilizer and Manure Analysis	Durai, M.V.	9789381450185	2014
3	A Handbook on Irrigation and Drainage	Panigrahi, Balram	9789381450888	2013
4	Acid Soils: Their Chemistry and Management	Sarkar, A.K.	9789381450383	2013
5	Advances in Nutrient Dynamics in Soil-Plant Systems for Improving Nutrient use Efficiency	Elanchezhian, R.	9789385516962	2017
6	Advances in Soil Borne Plant Diseases	Naik, Manjunath	9788189422813	2008
7	Agricultural Microbiology	Balachandar & Vendan	9789386546142	2018
8	Agriculture and Waste Management for Sustainable Future	Sannigrahi, Asoke Kumar	9789380235530	2011
9	Agrometeorology: A Simplified Textbook	Kathiresan, G.	9789383305568	2015
10	Agronomy: Principles and Practices	E. Somasundaram M. Mohd. Amanullah	9789385516740	2017
11	Bioinoculants: A Step Towards Sustainable Agriculture	Gupta, R.P. & A.Kalia: et.al.	9788189422219	2007
12	Climate Change and Agricultural Food Production	Kibria, Golam et.al.	9789381450512	2013
13	Climate Change and Agroforestry: Adaptation, Mitigation and Livelihood Security	C.B.Pandey	9789386546067	2018
14	Climate Change and Chemicals: Environmental & Biological Aspects	Kibria, Golam et.al.	9789380235301	2010
15	Climate Change and Environment: Concepts and Strategies to Mitigate Impacts	Devesh Sharma & K C Saha	9789385516375	2016
16	Climate Change and Food Security	Datta, M.& N.P.Singh	9788189422387	2008
17	Climate Change and Natural Resources Management	Lenka, S.& N.K.Lenka	9789381450673	2013
18	Climate Change and Plantations in the Humid Tropics	GSLHV Prasada Rao	9789385516368	2016
19	Climate Change and Sustainable Agriculture	P Suresh Kumar	9789385516726	2017
20	Climate Change and Water Security: Impacts, Future Scenarios, Adaptations and Mitigations	Golam Kibria	9789385516269	2016
21	Climate Mitigation and Carbon Finance: Global Initiatives & Challenges	Sahoo, A.K.	9789381450024	2012
22	Climate Resilient Animal Agriculture	GSLHV Prasada Rao	9789386546180	2018
23	Climate Resilient Crops for the Future	Peter, K.V.	9789383305599	2015
24	Climatic Variability: Impacts on Agriculture and Allied Sectors	Datta, M.:Ed.	9789381450949	2014
25	Clinical Veterinary Medicine: Practical Manual Series Vol 03	Sharma, Neelash	9788190851251	2009
26	Conducting An Effective and Successful Training Programme	Sontakki, Bharat S.	9789383305223	2015
27	Conservation Agriculture for Carbon Sequestration and Sustaining Soil Health	Somasundaram	9789383305322	2014
28	Droughts in Agricultural Production: Monitoring & Management	Rao, G.G.S.N.	9789385516009	2015

70	Pulses: Problems and Solutions	A. K. Singh	9789385516887	2018
71	Objective Agronomy	Vishwakarma,A.	9789380235127	2010
72	Organic Farming	Singh, A.K.	9789385516139	2015
73	Organic Farming: Scope and Uses of Biofertilizers	Panwar, JDS	9789385516184	2016
74	Organic Spices	Parthasarathy, V.A	9788189422844	2008
75	Pest Management and Residual Analysis in Horticultural Crops	Gulati, Rachna	9789381450710	2013
76	Pesticides: Methods of Their Residues Estimation	Kumari, Beena	9789380235394	2010
77	Physical Climatology *(Recommended by ICAR)*	Sellers, Williams	9789383305575	2014
78	Plant Nutrient Disorders: Diagnosis and Management	Sarkar, A.K.	9789385516023	2015
79	Practical Agricultural Meteorology	Srivastava, A.K.	9789380235776	2011
80	Practical Isotope Hydrology	Rao, S.M.	9788189422332	2006
81	Precision Farming in Horticulture	Singh, Jitendar	9789381450475	2013
82	Principles and Applications of Agricultural Meteorology	Patra, Alok Kumar	9789385516245	2016
83	Question and Answers in Agricultural Sciences	Rupinder Singh	9789385516993	2018
84	Rainfed Agriculture	R.K.Nanwal	9789386546043	2018
85	Recycling of Industrial Effluents	Manivanan, R.	9788189422127	2006
86	Remember Your Humanity: Pathway to Sustainable Food Security	Swaminathan,M.S.	9789381450178	2012
87	Remote Sensing Applications in Dryland Natural Resource Management	Gaur, Mahesh	9789381450321	2013
88	Remote Sensing in Geomorphology	Ramasamy, SM	9788189422059	2005
89	Renewable Energy Sources for Sustainable Development	Rathore, N.S.	9788189422721	2007
90	Research Methodology in Social Sciences	Patil, Shridhar	9789385516405	2016
91	Rural Livelihood and Food Security	Wani, M.H.	9789380235936	2012
92	Samekit Krishi Pranali	Kumar, Sanjeev	9789381450154	2012
93	Social Ecological Diversity and Traditional Food Systems	Singh, Ranjay	9789383305360	2014
94	Soil Conservation: Fully Revised and Updated: 3rd ed.	Hudson, Norman	9789383305971	2015
95	Soil Fertility and Nutrient Management in Horticulture	Chitranjan Sarangi	9789386546098	2018
96	Soil Health: Methods to Sustain and Improve Quality	Rajeev Padbhushan	9789386546562	2018
97	Soil Microbiology and Biochemistry	Hassan, G.Dar	9789380235134	2010
98	Soil Resources and Its Mapping Through Geostatistics Using R and QGIS	Priyabrata Santra	9789386546265	2018
99	Soil Sampling and Methods of Analysis	Pal, Sushant	9789381450574	2013
100	Soil Science: An Elementary Textbook	Puri, A.N.	9789383305063	2015
101	Soil Testing and Analysis: Plant, Water and Pesticide Residues	Brajendra, Patiram	9788189422707	2007
102	Soil, Plant and Water Analysis of Horticulture Crops	Chittranjan Sarangi	9789386546081	2018
103	Solving the Pulses Crisis	Singh, Anil	9789381450482	2013
104	Sustainability of Small Farms in Coastal Ecosystems	Satapathy, C	9789385516276	2016
105	Sustainable Agriculture: A Vision for Future	Desai, B.K.	9788189422639	2007
106	Sustainable Environmental Science	Sahu, D.D.	9789381450208	2012
107	System Based Integrated Nutrient Management	Gangwar, B.	9789381450055	2012
108	Systematics of Fruit Crops	Sharma, Girish	9789380235066	2009
109	The Chemistry of Soil Constituents	Greenland, D.J.	9789385516214	2016
110	The Chemistry of Soil Processes	Greenland, D.J.	9789383305926	2015
111	Water Resources Management for Enhancing Water Productivity	N.K.Gontia	9789385505728	2018
112	Water Quality Modeling: Rivers,Streams and Estuaries	Manivanan, R.	9788189422936	2008
113	Weed Science	Das, P.C.	9789383305261	2015
114	Women in Sustainable Agriculture	C.Satapathy	9789383305056	2014